Contemporary Perspectives on Shopping, Retail and Tourism

ASPECTS OF TOURISM

Series Editors: **Chris Cooper** *(Leeds Beckett University, UK)*, **C. Michael Hall** *(University of Canterbury, New Zealand)* and **Dallen J. Timothy** *(Arizona State University, USA)*

Aspects of Tourism is an innovative, multifaceted series, which comprises authoritative reference handbooks on global tourism regions, research volumes, texts and monographs. It is designed to provide readers with the latest thinking on tourism worldwide and in so doing will push back the frontiers of tourism knowledge. The series also introduces a new generation of international tourism authors writing on leading edge topics.

The volumes are authoritative, readable and user-friendly, providing accessible sources for further research. Books in the series are commissioned to probe the relationship between tourism and cognate subject areas such as strategy, development, retailing, sport and environmental studies. The publisher and series editors welcome proposals from writers with projects on the above topics.

All books in this series are externally peer reviewed.

Full details of all the books in this series and of all our other publications can be found on http://www.channelviewpublications.com, or by writing to Channel View Publications, St Nicholas House, 31–34 High Street, Bristol, BS1 2AW, UK.

ASPECTS OF TOURISM: 94

Contemporary Perspectives on Shopping, Retail and Tourism

Edited by
Dallen J. Timothy

CHANNEL VIEW PUBLICATIONS
Bristol • Jackson

DOI https://doi.org/10.21832/TIMOTH8830
Library of Congress Cataloging in Publication Data
A catalog record for this book is available from the Library of Congress.
Names: Timothy, Dallen J., editor.
Title: Contemporary Perspectives on Shopping, Retail and Tourism/Edited by Dallen J. Timothy.
Description: Bristol; Jackson: Channel View Publications, [2025] | Series: Aspects of Tourism: 94 | Includes bibliographical references and index. | Summary: "This book offers traditional perspectives on shopping and tourism and updates current thinking in relation to experiences, and internal and external forces that affect retail change and shopping behaviour. It provides empirical examples on current issues, opportunities, challenges and paradigms in the relationship between shopping and tourism"— Provided by publisher.
Identifiers: LCCN 2024050633 (print) | LCCN 2024050634 (ebook) | ISBN 9781845418830 (hardback) | ISBN 9781845418823 (paperback) | ISBN 9781845418854 (epub) | ISBN 9781845418847 (pdf)
Subjects: LCSH: Tourism and shopping. | Souvenirs (Keepsakes)
Classification: LCC G156.5.S64 C66 2025 (print) | LCC G156.5.S64 (ebook) | DDC 381/.14—dc23/eng/20241220
LC record available at https://lccn.loc.gov/2024050633
LC ebook record available at https://lccn.loc.gov/2024050634

British Library Cataloguing in Publication Data
A catalogue entry for this book is available from the British Library.

ISBN-13: 978-1-84541-883-0 (hbk)
ISBN-13: 978-1-84541-882-3 (pbk)

Channel View Publications
UK: St Nicholas House, 31–34 High Street, Bristol, BS1 2AW, UK.
USA: Ingram, Jackson, TN, USA.
Authorised Representative: Easy Access System Europe - Mustamäe tee 50, 10621 Tallinn, Estonia gpsr.requests@easproject.com.

Website: https://www.channelviewpublications.com
Bluesky: https://bsky.app/profile/channel-view.bsky.social
X: Channel_View
Facebook: https://www.facebook.com/channelviewpublications
Blog: https://www.channelviewpublications.wordpress.com

Copyright © 2025 Dallen J. Timothy and the authors of individual chapters.

All rights reserved. No part of this work may be reproduced in any form or by any means without permission in writing from the publisher.

The policy of Multilingual Matters/Channel View Publications is to use papers that are natural, renewable and recyclable products, made from wood grown in sustainable forests. In the manufacturing process of our books, and to further support our policy, preference is given to printers that have FSC and PEFC Chain of Custody certification. The FSC and/or PEFC logos will appear on those books where full certification has been granted to the printer concerned.

Typeset by Techset Composition India(P) Ltd, Bangalore and Chennai, India.

Contents

	Tables and Figures	vii
	Contributors	ix
1	Shopping, Tourism and Consumer Societies *Dallen J. Timothy*	1

Part 1: Consumer Behavior and Cultural Aspects of Shopping Tourism

2	Shopping Tourism and Consumer Emotional Loyalty: Destination and Programmatic Perspectives *Antónia Correia and Paulo Martins*	19
3	Intercultural Perspectives on Shopping and Tourism *Erdogan Koc and Ahu Yazici Ayyildiz*	37
4	The Evolving Tourism Marketplace: Changing Tourist Shopping Markets *Gianna Moscardo, Haipeng Jin and Laurie Murphy*	57
5	Luxury Retail, Place Branding and Destination Identity through Shopping Tourism *Cemile Ece and Efnan Ezenel*	78

Part 2: Economic, Urban and Spatial Perspectives

6	Tax-Free and Duty-Free Shopping: Benefits to Communities and Tourists *Azila Azmi, Azrul Abdullah and Mohammad Fadhili Yahaya*	93
7	Border Crossings, Retail and Shopping Tourism *Dallen J. Timothy and Gülsel Çiftci*	107
8	Shopping Tourism and Retail-led Urban Regeneration in Historic Cities *Azadeh Lak and Pantea Hakimian*	124
9	Economic Success in Unexpected Places: Spatial Anomalies in Swedish Retailing *Roger Marjavaara, Elin Nilsson and Magnus Strömgren*	143

10 Rethinking the Retail and Tourism Nexus as a
    Heritage-Making Performance: A European Perspective   159
    Chiara Rabbiosi

Part 3: Experiential and Niche Aspects of Shopping Tourism

11 Leisure Shopping, Retail Experiences and Destination
    Satisfaction   177
    Tim Coles

12 The Ancillary Role of Shopping in Other Types of Tourism 195
    Jennifer Frost and Warwick Frost

13 Shopping Festivals: High Impact Events for Shopping Tourism 213
    Sangeeta Peter and Victor Anandkumar

14 Food Retail and Food Souvenirs in Tourism: Bringing
    Home a Taste of Place   228
    Matthew J. Stone

Part 4: Conclusions: Past, Present and Future Perspectives

15 Shopping Tourism and Tourist Shopping: Looking Backward,
    Looking Forward   249
    Dallen J. Timothy

Index   266

Tables and Figures

Tables

Table 3.1	Characteristics and dimensions of national cultures	39
Table 4.1	Key themes in reviews of research into leisure shopping and tourist consumption patterns	59
Table 8.1	Participants' demographic data	131
Table 8.2	Composite reliability measures of constructs, bootstrapping with 2000 samples	133
Table 8.3	Average values of constructs, bootstrapping with 2000 samples	133
Table 8.4	Heterotrait and Monotrait (HTMT) ratios of constructs	134
Table 8.5	Variance inflation factor (VIF) values of the constructs	134
Table 8.6	Path coefficients, t-statistics and p-values of the structural model	136
Table 13.1	A selection of international shopping festivals, general information	216

Figures

Figure 4.1	Social practice framework for tourist shopping markets	62
Figure 4.2	Social practice framework for Chinese outbound tourist shoppers	65
Figure 4.3	Social practice framework for Indian outbound tourist shoppers	69
Figure 4.4	Social practice framework for Gen Y and Gen Z tourist shoppers	70
Figure 7.1	Distance–decay model of cross-border shopping (after Timothy & Butler, 1995)	111
Figure 8.1	The conceptual framework of the study	129
Figure 8.2	Examples of traditional Iranian bazaars	130
Figure 8.3	Path coefficients and outer loadings for the direct relationship between the successful regeneration and shopping tourism development	134
Figure 8.4	t-statistics for the inner and outer models	135

Figure 9.1	Population density in Sweden and the 25 largest cities in 2020, alongside the location of the described cases. Source: Statistics Sweden, 2023	149
Figure 9.2	Customers queuing in line to enter Gekås in Ullared during peak season in July 2021. Photo: Roger Marjavaara, 2021	151
Figure 10.1	Souvenirs created by a designer company from obsolete wooden pilings from the canals of Venice. Museums' store of the Doge's Palace and Museo Correr, Venice. Photo: Chiara Rabbiosi, 6 December 2022	166
Figure 13.1	Framework for understanding and planning shopping festivals	222

Contributors

Azrul Abdullah, PhD, is an Associate Professor at the Faculty of Accountancy, Universiti Teknologi MARA, Perlis Branch. As a certified accountant, he has accumulated diverse experience in sectors such as accounting, manufacturing, entrepreneurship and tourism. He was a visiting research fellow at Universitas Airlangga and Universitas Sumatera Utara. He is also an active researcher, writer and consultant in the areas of financial reporting, corporate governance, information disclosure, accounting education, tourism accounting, capital structure, financial leverage, debt structure and entrepreneurship.

Victor Anandkumar is a professor of marketing in the Department of Management Studies at Pondicherry University, India. He specializes in tourism marketing, inclusive marketing and marketing technology. He has published widely in many areas of tourism management, and has been a visiting professor at universities in Mauritius, Thailand and Germany. He had a brief stint in the industry and recently completed 27 years of teaching in higher education. Victor spearheads international initiatives in the Business School at Pondicherry University.

Ahu Yazici Ayyildiz is Associate Professor at Aydin Adnan Menderes University's Kusadasi Faculty of Tourism in Turkey. She has extensively published journal articles, books and book chapters on various aspects of tourism and hospitality marketing and management. She is particularly interested in investigating the influence of technology and culture on both tourists/guests and the service staff.

Azila Azmi, PhD, is an Associate Professor at the Department of Tourism, Faculty of Hotel and Tourism Management, Universiti Teknologi MARA, Pulau Pinang Branch. Over the years, she has worked on research and consultancy projects locally and internationally, funded by major grant-awarding bodies in the areas of tourism destination management, community-based tourism, tourist behaviour, cross-border, shopping tourism, entrepreneurship and ubiquitous learning. Much of her research and contributions to the field have been published in several academic journals.

Gülsel Çiftci is an Associate Professor at Trakya University's Faculty of Applied Sciences in Edirne, Turkey. She is also a visiting research scholar at the School of Community Resources and Development, Arizona State University. She received her PhD in business management and holds a master's degree in tourism management. With extensive experience in cross-border tourism, crisis management in the hospitality industry, and various research projects, she has been honoured with multiple academic awards. In addition, she is a licensed tour guide in Turkey.

Tim Coles is Professor in the Department of Management at the University of Exeter Business School. His research interests are in the sustainable development of tourism, currently focusing on how weather and climate impact the visitor business at heritage attractions. He has a longstanding interest in retailing and shopping having completed his PhD on the historical geography of evolving retail systems in Germany (1848–1914) and his early work examined the tourism–leisure–shopping–culture interface in east German towns and cities after unification.

Antónia Correia is a professor and researcher at the University of Algarve, Portugal. Her work focuses on behavioural economics, marketing, tourism and consumer behaviour, with over 5550 citations. She has published extensively in top-tier journals, special issues, books, and book chapters and projects. She has also contributed by supervising nearly 40 master's dissertations and 13 doctoral theses. She is part of several international research projects, and she sits on organization and scientific committees of conferences and on the editorial boards of peer-review journals. She is also the President of the unique collaborative laboratory in tourism in Portugal.

Cemile Ece is a PhD candidate in the Department of Tourism and Hotel Management at Eskişehir Osmangazi University, Turkey, and project expert. She currently researches ecotourism, sustainability, diversity, equity, inclusion, destination management, destination branding, consumer behaviour. Her recent research has been supported by the Scientific Research Centre of ESOGÜ. She has scientific articles and projects on tourism. She has practical experiences in tourism field for more than 10 years as a professional. She worked in the front office and guest relations departments at multi-national hotels, and she has been working as a project expert in NGOs and takes part in NGOs activities and takes an active role especially on environmental, sustainable, ecological issues, inclusion issues. She has experiences in EU+ projects and National Funded Projects.

Efnan Ezenel is a PhD candidate in the Department of Tourism and Hotel Management at Eskişehir Osmangazi University, Turkey, and a licensed tourist guide licensed by the Turkish Ministry of Culture and Tourism.

She currently researches inclusive tourism, accessibility, social inclusion, diversity, equity, inclusion, destination management, destination branding and consumer behaviour. She also studies rural tourism. She has practical experiences in tourism field for more than 12 years as a professional. She worked in guest relations manager at multi-national hotels in the hospitality sector. She takes an active part in disability, inclusion, accessibility issues and has been working for NGOs on disability and disability rights. In addition, she has EU+ projects and National Funded Projects

Jennifer Frost (née Laing) is an Adjunct Associate Professor at La Trobe University, Australia, and Visiting Professor at Sheffield Hallam University, UK. Her research interests include rural and regional regeneration through tourism and events; travel narratives; heritage tourism; royal events and health and wellness tourism. Jennifer is co-editor-in-chief of the *Journal of Heritage Tourism* and a co-editor of the Routledge Advances in Events Research book series. She has co-written six research books with Warwick Frost, including *Royal Events: Rituals, Innovations, Meanings* (Routledge, 2018) and *Medieval Imaginaries in Tourism, Heritage and the Media* (Routledge, 2023).

Warwick Frost is an Adjunct Professor of Heritage, Tourism and the Media at La Trobe University, Australia and Visiting Professor at Sheffield Hallam University, UK. Originally a historian, he has worked in tourism studies since 1996. His research interests include nature-based tourism, national parks, travel-writing, cultural heritage tourism and environmental history. He was the inaugural co-convenor of the International Tourism and Media conference series, co-editor-in-chief of the *Journal of Heritage Tourism* and co-editor of the Routledge Advances in Events Research book series. His latest book is *An Environmental History of Australian Rainforests: Fire, Rain, Settlers and Conservation* (Routledge, 2020).

Pantea Hakimian, an Assistant Professor at Shahid Beheshti University, Iran, holds a PhD in urban design, and an MSc degree in architecture and urban design. With over a decade of dedicated research in this field, she has authored papers on the characteristics of the historic context of Iranian cities. These papers focus on exploring the socioeconomic and heritage potential of historic bazaars, highlighting their morphological and physical significance. The research in this book delves into the prospect of transforming these bazaars into hubs for shopping and retail tourism, while also demonstrating the factors crucial to the successful revitalization of historic Iranian bazaars through retail-led urban regeneration.

Haipeng Jin is an Associate Professor in the School of Humanities at Southeast University, China. He received his PhD degree from James

Cook University Australia under the supervision of Professor Gianna Moscardo and Associate Professor Laurie Murphy. He then worked as a post-doctoral research fellow in the School of Sociology and Anthropology at Sun Yat-sen University, China, before taking up his current position. He has published research articles in journals such as *Tourism Management*, *Journal of Travel Research* and *Current Issues in Tourism*. His research interests include tourist shopping, cultural tourism, the sociology of tourism and the sociology of consumption.

Erdogan Koc is Professor of marketing at Bahcesehir University, Faculty of Economics and Administrative, and Social Sciences, Turkey. He has published extensively in top-tier tourism, hospitality, and services marketing and management journals. He has authored/edited several books (with publishers such as Routledge and CABI) and book chapters. He serves on the editorial boards of several journals (such as *Current Psychology*, *Journal of Hospitality Marketing and Management* and *International Journal of Intercultural Relations*). His research primarily focuses on the human element (both as consumers and employees) in tourism, hospitality, services marketing and management.

Azadeh Lak is an architect and urban designer, and an Associate Professor in the Faculty of Architecture and Urban Planning at Shahid Beheshti University, Tehran, Iran. Her research and publications concentrate on urban regeneration, urban tourism, sustainable public spaces and quality of life in cities, particularly historic cities. She also tries to employ the physical characteristics of old cities to boost urban regeneration process in different approaches, e.g. sustainable urban regeneration and retail-led urban regeneration.

Roger Marjavaara is Associate Professor in human geography with a speciality in tourism and geographical information systems (GIS), in the Department of Geography and the Center for Regional Science (CERUM), at Umeå University, Sweden. The main focus of Marjavaara's research is human mobility of different kinds. This includes tourism geography, especially second-home tourism, and its impacts and consequences. Further, the production and reproduction of places, and post-mortal mobility is of interest, besides the nexus between retail and tourism.

Paulo Martins is Assistant Professor at the Instituto Superior Manuel Teixeira Gomes (ISMAT). He holds a PhD in community psychology and a master's in counselling psychology from ISPA. His work focuses on applied psychology, particularly within community settings. He has also published academic papers and remains active in research and professional activities related to his field.

Gianna Moscardo is a Professor of Tourism in the College of Business, Law and Governance at James Cook University, Australia. Her qualifications in psychology and sociology support her research interests in understanding how communities perceive, plan for and manage tourism development opportunities, how tourists learn about and from their travel experiences and how to design more sustainable tourism experiences. She has published extensively on tourism and related areas with more than 250 refereed papers and book chapters. Her recent project areas include evaluating tourism as a tool for economic development in rural regions, tourist experience analysis and tourist storytelling.

Laurie Murphy is an Adjunct Associate Professor in the College of Business, Law and Governance at James Cook University, Australia. Her research interests focus on improving tourism's contribution to regional communities with an emphasis on tourism marketing, destination image and choice, destination branding at tourism's contribution to community wellbeing. Laurie has served on the editorial board of both the *Journal of Travel Research* and the *Journal of Travel and Tourism Marketing*, and the Tourism and Events Strategic Advisory Committee for Townsville Enterprise Ltd.

Elin Nilsson is an Associate Professor at Umeå School of Business, Economics and Statistics at Umeå University and a research fellow at CERUM (Centre for Regional Science), Umeå University, specializing in marketing. Her research focuses on consumer behaviour, retailing, digital interaction and sustainability. Her research also includes servicescape, service dominant logic and co-creation. Her work has been published in scholarly journals such as the *Journal of Services Marketing*, *Service Marketing Quarterly*, *International Journal of Retail and Distribution Management* and *The International Review of Retail, Distribution and Consumer Research*.

Sangeeta Peter is an Assistant Professor in Business Administration at Loyola Academy, in Hyderabad, India. Her primary areas of interest are marketing, retail and tourism, with a focus on integrating these domains. She has studied shopping festivals with an emphasis on tourist shopping behaviour. She leads the BBA Tourism Department at Loyola Academy.

Chiara Rabbiosi is Associate Professor at the University of Padova in Economic and Political Geography. She is also the coordinator of the Master in Local Development at the same university. Her research interests deal with the social and spatial dimensions of tourism mobilities, including the critical geographies of consumer culture, cultural heritage and place branding. She has published more than 40 articles in academic journals including *Annals of Tourism Research*; *Tourist Studies*; *Cultural Geographies*; *Gender, Place and Culture* and the *Journal of Consumer Culture*.

Matthew J. Stone, PhD, is Associate Professor of Marketing at California State University, Chico. His research focuses on consumer behaviour in tourism with a specialization in food tourism (culinary tourism) research. He serves as lead research advisor for the World Food Travel Association and was a Fulbright Scholar at Haaga-Helia University of Applied Sciences in Finland. Dr Stone is a native of Central Illinois (USA), and he completed his studies at Texas A&M University (PhD), University of Houston (MHM) and University of Illinois (BS). His passion is travel.

Magnus Strömgren is an associate professor of human geography in the Department of Geography, Umeå University. He earned his PhD in 2004 from Umeå University with a dissertation about the spatial diffusion of telemedicine in Sweden. A specialist in geographical information systems (GIS), Dr Strömgren is an experienced educator in the field and has contributed to a broad range of research endeavours spanning medical geography, population studies and transportation geography.

Dallen J. Timothy is Professor of Community Resources and Development at Arizona State University, USA, and Senior Global Futures Scientist at the Julie Anne Wrigley Global Futures Laboratory. He is also visiting professor at Ningxia University, China; University of Girona, Spain; and Senior Research Associate at the University of Johannesburg, South Africa. He serves on the editorial boards of 23 international journals and is commissioning editor for three book series with Channel View Publications and Routledge. He has ongoing research projects in Europe, the Middle East, North America, the Pacific islands and Asia.

Mohamad Fadhili Yahaya, PhD, is an Associate Professor at the Academy of Language Studies, Universiti Teknologi Mara, Perlis Branch. Being a member of the humanities and social science cluster of the National Professor Council of Malaysia, he has become an active and versatile social scientist working on numerous interdisciplinary projects mostly within the scope of humanities and social sciences. His work can be traced to education, language learning, applied linguistics, ubiquitous learning, assessment, artificial intelligence, tourism, accountancy and finance.

1 Shopping, Tourism and Consumer Societies

Dallen J. Timothy

Shopping is among the most common and popular tourist activities before and during the travel experience. Some tourists spend as much money (sometimes more) on shopping as they do on lodging, food services or transportation, and retail is one of the most visible elements of the tourism landscapes of destinations (Ben Said *et al.*, 2024; Timothy, 2005). Shopping is the main motive for many people's travel choices, and the tourism economies of some destinations are bolstered overwhelmingly by tourist shopping. Retail in this book refers to non-transportation, non-lodging and non-food service sales – e.g. shopping. Tourists' shopping is extremely desirable in destinations for the employment and tax revenues it generates (Henderson *et al.*, 2011). It also provides opportunities for sharing local cultures (i.e. handicrafts, souvenirs and shopping festivals) and empowering communities economically as it stimulates entrepreneurial activity.

Utilitarian, leisure and tourist shopping today is quite different from what it was 25 years ago. In many developed regions of the Global North, shopping malls have been declining in popularity, whereas in much of the Global South, they continue to be built and function as foci of community leisure and tourism (Ferreira & Paiva, 2017; Kiriri, 2021). Globalization has affected consumer products and the leisure shopping experience through the standardization of products and retail spaces. Increased media and internet technology, and enhanced ICT, have drastically changed the face of shopping, including retail activities in the context of tourism. These issues and others are the focus of this edited volume.

In the contemporary world of guaranteed change, shopping remains a constant in the tourism experience, even though its forms, measures, functions and manifestations are continuously evolving (Coles, 2004; Michalkó, 2004; Timothy, 2014). Building upon the success of the editor's book, *Shopping Tourism, Retailing and Leisure* (2005, Channel View Publications), this edited volume reviews traditional perspectives on shopping and tourism, although its primary aim is to update current thinking in relation to experiences, internal and external forces that affect retail change and shopping behavior, changing geopolitics and the social environment, and how changes in travel and technology create new retail spaces, non-place virtual spaces and retailscapes in tourism contexts.

Most tourists are passionate shoppers; even people who loathe shopping at home may let down their guard and participate in some style of retail shopping while on holiday. The phenomenon of people traveling away from home for the primary purpose of shopping is known as shopping tourism (or retail tourism). People engaging in shopping activities while in a destination for other purposes is tourist shopping (Coles, 2004; Timothy, 2005, 2018). Differentiating between these two may be important for destination management organizations as they try to identify their markets and cater to the needs of different retail consumers. For example, in general, shopping tourists choose their vacation destination in large part because of its retail opportunities and how known it is for certain products or its retail landscape (Moscardo, 2004). Thus, this segment tends to spend more and buy larger-ticket or luxury items. Conversely, tourist shoppers are less inclined to decide on a destination based on its shopping opportunities. For them, shopping is an auxiliary activity that ranks second or third only to sightseeing, participating in a recreational pursuit, or other primary motive. Tourist shoppers, therefore, are more likely to buy less expensive products, such as souvenirs that memorialize their holiday experience, and many might be satisfied with only minimal engagement with shopping.

Regardless of whether someone is a shopping tourist or a tourist shopper, retail activities are an extremely important part of the overall travel experience, and a lucrative commercial endeavor for a destination to pursue (McIntyre, 2012b; Moscardo, 2004; Williams, 2012). Retail activities can make or break a vacation, leave a positive or negative impression in one's mind about a destination, support a destination through increased income and employment (Albayrak *et al.*, 2016; Jin *et al.*, 2017), or contribute to the deterioration of a destination if overcommercialization is allowed to overshadow other attractions or destination values.

With the realization that shopping is one of the most impactful elements of the travel experience that nearly everyone participates in while traveling, and the research on shopping and tourism is 'piecemeal and fragmented' (Jin *et al.*, 2017: 120), this book aims to delve into the multifaceted phenomenon of shopping and tourism. It provides an overview of various trends and patterns, provides recommendations for destination planners and management organizations, and elucidates important factors of success from empirical cases around the world, which other localities can emulate or draw from to achieve their own successes. This introductory chapter provides a definitional and conceptual foundation for the studies that follow, and outlines the specific contributions of each of the chapters herein.

Global Shopping Trends and Patterns

Since the Second World War, Western societies have become more affluent, mobile and globally minded. These, among other variables, have stimulated the growth of shopping tourism in many corners of the globe. In 2019, just prior to the COVID-19 outbreak, shopping accounted for approximately 6% ($178 billion USD) of all tourism expenditures (World Travel & Tourism Council, 2023). Likewise, in 2023, approximately 18% of travelers worldwide claimed shopping to be their primary reason for traveling away from home (YouGov, 2023). Although this number has teetered off and on in 2023 and 2024 with reduced consumer confidence, high fuel prices and generally high inflation rates (Erdly, 2024), people are still willing to travel for the sake of shopping, and it is those people who are generally least affected by inflated costs of living.

Although almost everyone engages in shopping during a trip, certain nationalities far surpass others in their propensity to spend on retail goods. Nationals of several Asian countries are the most enthusiastic shopping tourists, while Europeans appear to be the least motivated by shopping (Kim *et al*., 2011). Although Singaporeans are not the largest shopping market overall, they appear to be the most shopping-oriented per capita, with 36% of that nationality claiming that shopping was their most important purpose of travel in 2023. This was followed by residents of Malaysia (28%), Indonesia (27%), the Philippines (26%), Saudi Arabia (26%), Japan (25%), Thailand (24%), India (23%), the United Arab Emirates (23%) and China (20%) (Statista, 2023). The Saudi and Emirati positions in this top-10 category reflects those nationalities' growing practice of traveling for luxury shopping to other parts of the Middle East and Europe (Alshammari *et al*., in press; Marciniak & Mohsen, 2014), and the Chinese are increasingly demanding luxury retail opportunities in their outbound travel as well (Correia *et al*., 2018; Jin *et al*., 2020; Li *et al*., 2021).

Singaporeans' preferred shopping destinations are in the general region of East and Southeast Asia: Japan (57%), Malaysia (41%), Taiwan (41%), South Korea (38%) and Thailand (36%). Shoppers from the United Arab Emirates prefer the Maldives (43%), Singapore (40%), the United Kingdom (40%), Switzerland (39%) and Saudi Arabia (38%) for their key shopping destinations, and Indians spend most of their retail rupees in the Maldives (44%), the UAE (44%), Singapore (39%), Switzerland (35%) and Indonesia (34%) (YouGov, 2023). Thus, there is a mix of regional preferences and long-distance options – all reflecting the growing middle and wealthier classes in all those countries.

Although nationality and cultural background are deeply influential variables in determining retail behavior, food consumption and lodging preferences (Correia *et al*., 2011; Kim *et al*., 2011; Koc & Ayyildiz, 2021; Lin *et al*., 2020; Lojo, 2020), other characteristics also contribute to determining shoppers' behaviors and intentions. Different age groups have different expectations of what they want to buy in the travel destination. Baby boomers, Gen Z consumers, millennials and other age groupings have different travel needs and interests, which includes their varying inclinations to shop while on holiday (Amin *et al.*, 2021; Azmi *et al.*, 2020b; Zackariya, 2023). Millennials, for example, tend to spend less on transportation but more on accommodations and shopping than their baby boomer counterparts (Kim & Park, 2020), the latter preferring to spend more on comfortable transportation.

Some Research Orientations

Possessing such a high position in the tourism industry, it is little wonder shopping has received a great deal of research attention in recent years. Much research has been carried out on many aspects of shopping over the past 50 years with three general areas being the crux of investigations: the tourist, the merchant and the retail destination.

The shopping tourist/tourist shopper

Much of the work from the tourist's perspective has focused on shopping as a leisure versus a utilitarian activity (Albayrak *et al.*, 2016; Azmi *et al.*, 2020a). For decades, general shopping studies from a retail management perspective focused on the utilitarian elements of shopping, with the recreational or ludic perspective only receiving notable attention in the 1980s–1990s. The retail binary of leisure versus utilitarian shopping is now seen as outdated, as shopping is overwhelmingly accepted as being driven by a mix of hedonic and practical motivations – a needful activity with leisure undertones or a leisure activity with utilitarian elements. This has led to a greater appreciation of shopping tourism and tourist shopping, with a recognition that what might be considered mundane or routine at home can don a more enjoyable cloak in a tourist destination (Murphy *et al.*, 2011b).

Tourists' experiences with shopping is another significant theme that has permeated much of the literature to date. Shopping tourists' satisfaction with the experience and what variables create enjoyable and memorable experiences have been at the forefront of this work (e.g. Egresi & Polat, 2016; Murphy *et al.*, 2011a; Suhartanto, 2018). Factors such as varied merchandise, a shop's internal arrangement and its surrounding environment, service quality and the friendliness of staff, reasonable prices and special offers are significant factors in creating enjoyable and memorable retail encounters (Wong *et al.*, 2024). Similarly, recent work on co-created tourist experiences has been expanded to include the shopping context where tourists are active participants in creating their own retail experience with the retail merchants and other destination stakeholders (McIntyre, 2012a; Sthapit *et al.*, 2018). Co-creation helps create more memorable experiences that are more meaningful and transformational for everyone. The rise in customized souvenirs, which individual tourist customers can design and create for themselves based on their own impressions of their destination experience, is a prominent example (Anastasiadou & Vettese, 2019).

For foreign retail guests, the language abilities of staff members, attractive packaging, accepting foreign credit cards or currency, shopkeepers' cultural sensitivities and staff knowledge of foreign customers' needs have been shown to increase shopper satisfaction (Choi *et al.*, 2016b; Parasakul, 2020; Reisinger & Turner, 2002; Timothy, 2005; Tosun *et al.*, 2007).

Souvenir studies have also played a prominent role in understanding tourists' exposure to shopping. Years of research in the area of souvenirs have attempted to demonstrate what tourists consider to be memorable, durable, authentic and representative of the places they visit (He & Timothy, 2024a; Littrell *et al.*, 1993; Sthapit *et al.*, 2024; Swanson & Timothy, 2012). The nature of souvenirs has changed through time as tourists' tastes, interests and practical considerations (e.g. size and portability) have changed (Collins-Kreiner & Zins, 2011; Swanson & Timothy, 2012). The growing importance of food products as souvenirs (Buczkowska, 2014; Lin & Mao, 2015; Stone, in press) and a movement towards useful merchandise (e.g. aprons, cooking utensils and mouse pads) rather than only something lifeless to sit on the shelf (He & Timothy, 2024a, 2024b) is indicative of these changes.

Tourism is typically defined as the sum of all activities, services and experiences that derive from someone's travels away from home. All of these elements can be found in three geographic spheres: the home/origin, the transit space and the destination. Shopping plays an important role in all of these three spaces. Before leaving home, people buy things they wish to use while on vacation. In transit, people often purchase duty-free items or goods they might have unintentionally left at home. Despite the importance of pre-trip and in-transit retail activities, most academic research focuses on tourists' expenditures in the destination. Nonetheless, travelers may also have satisfying or unsatisfactory shopping experiences as they prepare for a journey or between home and the destination. For various reasons, many tourists are inclined to make impulse purchases in airports or other transit spaces (Liang & Yu, 2024). Duty-free shops and other retailers rely heavily on this touristic behavior and encourage spontaneous spending by convincing travelers of their need for something 'in-between'.

The merchant

Research on the merchants' perspective mirrors that of the shoppers' perspective, except that it often comes in the form of recommendations for maximizing sales and co-creating positive experiences for consumers (Hurst *et al.*, 2009; Journée & Weber, 2017). Creating in-shop retail environments and offering special deals that induce buying have been the focus of much work to date (Egresi & Polat, 2016; Murphy *et al.*, 2011a; Reisinger, 2006; Wong & Wan, 2013). Recent research has begun to focus on the benefits of shop or company loyalty programs that can keep shoppers coming back because of various reward systems, trust-building and general customer satisfaction. This includes current efforts to cater to the individual with personalized loyalty programs, which retailers can develop through artificial intelligence and data analytics that make it easier for sellers to cater to individual shoppers' needs (Erdly, 2024; Lam & Wong, 2020). Similarly, personalized communication uses the same sorts of algorithms to target individual shoppers through emails and social media based on their interests, behaviors and past experiences (Erdly, 2024). These sorts of 'personal touches' have the potential to get individuals to buy more often as messages are catered to their specific needs and buying behaviors.

Currently, scholars and business owners are examining the value of offering both a physical shop and an online shop simultaneously to increase sales, build a public image, establish a brand and increase the store's visibility (Al-Sulaiti, 2022; Kacen *et al.*, 2013; Moes & van Vliet, 2017). The results seem to indicate that people who do not like to shop in stores are more willing to shop online (Rajamma & Neeley, 2005; Törőcsik *et al.*, 2023), which can have significant outcomes for tourist destinations where retailers can offer alternatives for people to buy souvenirs and mementos of the destination after they return home (Yuan *et al.*, 2022). Some souvenir sellers are going to the online option with email reminders or social media messages to capture customers who have been to the destination but might have forgotten to buy a certain item, did not have room to carry it home or decided afterwards that they wanted a specific item after all. Similarly, an emerging market is people who have never been to the destination but have in interest in acquiring a souvenir of that locality for various reasons.

The destination

Destinations have significant control over how tourists perceive them and the quality of experiences tourists have while there (Irimiás *et al.*, 2017; Michalkó *et al.*, 2015). Systematic branding is among the most effective means of accomplishing this. A destination's image is something that typically evolves organically through time as the place becomes associated with certain values and opportunities. Branding refers mostly to efforts to raise a destination's image intentionally through marketing campaigns, hosting events and other ways of setting itself apart from its competitors (Azmi *et al.*, 2019). Certain localities have evolved naturally as shopping destinations (e.g. London, New York, Singapore and Paris), whereas others have undertaken massive branding work to cement themselves as high-end retail destinations, including Doha, Dubai and Abu Dhabi in recent years, although many of the traditional shopping destinations have augmented their traditional retail positions by marketing efforts to increase their notoriety as shopping destinations. Today's shopping branding efforts tend to focus on luxury or other high-end items, such as fashion wear, vehicles, jewelry, makeup and perfumes (Azmi *et al.*, 2019), although a shopping image can emerge from any sort of retail offerings, including flea markets, outlet malls and farmers markets.

From the destination's perspective, building destination loyalty through various frequent-visit programs, and follow-up reminders via email or social media messaging help keep would-be shopping visitors informed about what is happening in the destination and what new retail opportunities abound. In common with individual merchants, this sort of personalized communication from the destination to the consumer, as well as personalized loyalty programs, are currently receiving considerable attention by the industry as a way for destinations to become part of the everyday lexicon of past consumers, current consumers, or future would-be consumers through loyalty and trust building (Choi *et al.*, 2016a; Erdly, 2024).

The connection between shopping and various types of tourism is important, as each type of travel experience has different needs with regard to retail. This will be discussed in greater detail in the conclusion, but one prevalent theme in this book that deserves attention at the outset is shopping's relationship with heritage. Heritage tourism is one of the biggest and most pervasive subsectors of tourism. It is part of almost every tour package (even many cruises), and living and built cultural heritage is among the most prominent assets communities use to build their tourism economies (Timothy, 2021). The close relationship between heritage and retail is unmistakable (Moscardo *et al.*, 2020; Murphy *et al.*, 2013). Many tourists buy the heritage of places in the form of souvenirs, handicrafts and traditional food items. Many commercial exchanges take place in heritage spaces, such as traditional markets or historic shopping centers, and in some cases, shopping itself is a manifestation of a destination's heritage and history. The creation and maintenance of a sense of place is crucial in this regard (Miller *et al.*, 1998; Timothy, 2024).

Heritage is also a crucial factor in shopping-led urban regeneration. Urban center and waterfront development are clear examples of how cities utilize heritage and shopping to revitalize districts that face decline and decay. Bringing retail tourists back to city centers, historic marketplaces and even suburban retail agglomerations is a pervasive goal in older cities throughout the world. A significant part of this is the mix of leisure shopping and entertainment. The notion of 'shoppertainment' has existed for decades, as urban and suburban shopping malls have re-invented themselves into leisureplexes where entertainment and recreation are often more important than the shops. Recent manifestations of this same phenomenon, 'retailtainment', occur at the city or district scale (Escudero-Gómez, 2024; Zaidan, 2016).

Creating a retail brand is at the forefront of many destinations' marketing plans. Establishing shopping festivals or other retail events, increasing the prominent place of shopping on their websites and other marketing means are among the most common ways of building a global image of a shopping destination (Goldblatt, 2012; Henderson *et al.*, 2011; Peter & Anandkumar, 2016).

There is also a growing research trend on the transformation of place through shopping development. Rabbiosi (2011, 2016a, 2016b) discusses the trend in Rimini, Italy. Traditionally known for its beaches and waterfront, Rimini has undergone a significant evolution from emphasizing the seaside to becoming a renowned shopping destination, with the tourism landscape changes that go with such a place evolution. Certain parts of Dubai, Singapore, Macau and Hong Kong have undergone dramatic transformations in their urban forms and functions from something else to thriving retail metropolises. Although the traditional sense of place and people's place attachment harkens back to what made these retail localities popular for many generations, the meaning of place is changing, and shopping has an important role to play in those changes.

Contents of the Book

This book aims to address many of the shopping and tourism-related issues raised here and elsewhere in other literature. It also aims to raise questions that require additional research attention. This volume comprises 15 chapters, a mix of descriptive and analytical contributions, each of which addresses a timely theme in tourism studies in relation to retail and shopping. Following the introduction, the chapters are divided into four parts based upon the thematic content of the book: consumer behavior and cultural aspects; economic, urban heritage and spatial perspectives; experiential and niche aspects; and the concluding section on current, past and future trends.

Chapter 2 (Antónia Correia and Paulo Martins) examines loyalty in shopping tourism. Correia and Martins examine loyalty programs as promotional tools that lend a strong competitive advantage for individual shops, retail corporations and shopping destinations. They outline many different types of loyalty programs (e.g. credit card loyalty and frequent shopper programs) and the benefits and parameters of each one. With regard to place, Chapter 2 particularly highlights how destinations can also build shopping loyalty through branding awareness-building, stimulating emotional responses and providing high-quality products and services. All of these have salient planning and marketing implications.

In Chapter 3, Erdogan Koc and Ahu Yazici Ayyildiz shed important light on intercultural perspectives in retail and tourism. They examine how different cultures perceive shopping in tourism contexts and highlight specific nationalities and their unique shopping behaviors. The authors also eruditely discuss how cross-cultural perspectives manifest in the destination marketing mix (i.e. product, price, place, promotion, process and people). The implications of nationality and culture in relation to shopping are unique in each one of these Ps and have significant consequences for destination marketing and management, which industry leaders and DMOs need to address in their efforts to capture markets that are expanding rapidly throughout the world.

The focus of Chapter 4 (Gianna Moscardo, Haipeng Jin and Laurie Murphy) is the ever-changing global market demand for shopping experiences in tourism. The authors examine the evolving demand for tourism from individual market perspectives, highlighting several critical themes that are influencing current demand, which they gleaned from a review of current literature. These include growing consumer concerns about sustainability, growing consumer desires for personalized experiences, globalization and diversity of demand and supply, the growing use of digital technology and the after-effects of the COVID-19 pandemic. Moscardo, Jin and Murphy also suggest that, to remain competitive, planners and marketers must pay particular attention to changing and growing retail markets, such as Chinese and Indian, as well as younger generations of consumers.

Cemile Ece and Efnan Ezenel scrutinize luxury shopping in Chapter 5. They highlight the importance of luxury shopping in today's retail marketplace, spurred by growing affluence, globalization processes and increasing human mobility. Social media plays a role in this growing phenomenon, as consumers need something to boast about, and luxury shopping-based tourism gives them an opportunity to 'keep up with the Joneses'. Luxury retail is commonly used as a branding tool, making some places 'premier' shopping destinations. The authors provide several examples of luxury retail in the marketing and branding strategies of places worldwide.

Azila Azmi, Azrul Abdullah and Mohammad Fadhili Yahaya (Chapter 6) examine the complex phenomenon of tax-free and duty-free shopping. They explain the meanings and parameters of each of these and how they manifest in tourism. Tax-free and duty-free programs are designed and implemented by national governments as a means of stimulating increased retail expenditures on usually high-taxed merchandise. There is a lot of confusion among tourists about what duty-free means and how it operates. This chapter addresses this issue, highlights the role of supranational trade blocs (e.g. the European Union) on tax-free shopping, and argues that duty-free shopping can be a major impetus for choosing a particular holiday destination, especially among travelers of more shopping-inclined nationalities.

Chapter 7 (Dallen J. Timothy and Gülsel Çiftci) naturally follows Chapter 6 as it describes the phenomenon of cross-border shopping. This phenomenon had its roots in serving the utilitarian needs of borderlanders but has now expanded to include leisure motives, typically involving a mix of need satisfaction and spending leisure time abroad. Timothy and Çiftci argue that certain conditions need to exist at national borders for cross-border shopping to commence and grow, including consumer awareness of what lies on the other side of the border, contrasting economic and fiscal situations (i.e. taxes and exchange rates), and freedom of movement with open enough borders to allow shopping-based day-trips abroad. Retail activities in border areas have significant economic development potential for regions that are typically marginalized and largely ignored by central state authorities, but careful planning is required.

The crossover between shopping tourism and heritage is of primary concern in Chapter 8. Here, Azadeh Lak and Pantea Hakimian examine how shopping-led urban renewal efforts can reinvigorate urban retail zones and arrest urban decline, including historic marketplaces, through the resurrection of shopping tourism in historic cities. They present a data-based empirical study of municipal workers in several historic cities in Iran to illustrate ways in which declining shopping districts can be regenerated. Although their focus is historical markets, and the findings focus on the characteristics and perceptions of traditional markets, the tenets of their findings and discussion may be applicable beyond only historic marketplaces to other urban retail development contexts in other parts of the world.

Roger Marjavaara, Elin Nilsson and Magnus Strömgren (Chapter 9) carry on the heritage theme by analyzing a unique situation in which certain retail centers have reached a high level of commercial success, despite their not adhering to conventional factors of success for malls and other shopping centers. The main curiosity is the idea that people will travel further to these backwoods locations to buy goods that can be acquired closer to home for the same or lower costs. The authors investigate this unique situation with examples from Sweden, where several somewhat isolated stores have become day-trip destinations on their own right owing to non-conventional variables such as people wanting to get away from the city, continuing family traditions (nostalgia) and the heritagization of shopping experiences.

In Chapter 10, Chiara Rabbiosi continues the thread of heritage and retail by examining the relationships between shopping tourism and heritage in European settings. She explores the notion of 'shopping for heritage' in venues such as historic department stores, museum shops, creative industry sites and urban food halls. The notion of 'place' plays a prominent part of her chapter on heritagization, including place-making through modern heritage creations such as 'fantasy cities'. The underlying ethos of Chiara's chapter is how tourism and retail are connected through geographical associations.

In Chapter 11, Tim Coles provides a strong conceptual foundation for understanding the role of shopping in tourists' satisfaction with their chosen destinations and with their retail experiences directly. He describes the important role of shopping in the destination product mix and points out several research lacunae in the area of shopping tourism and customer satisfaction. The author also raises critical issues and concerns regarding the future of shopping tourism satisfaction in relation to digital technology and other current trends.

Whereas several chapters focus on shopping tourism, Chapter 12, by Jennifer Frost and Warwick Frost, focuses on tourist shopping in which shopping is a secondary or ancillary activity for people who visit a destination for other purposes. The authors describe several different types of tourism (e.g. wine tourism, food tourism, sport tourism, wellness tourism and heritage tourism) and how retail manifests in each one of these to create a significant part of the destination product and essential, albeit secondary, element of the destination experience. Frost and Frost rightfully acknowledge that more work is needed to bring to light the key relationships between shopping and various forms of tourism that have not been well examined in the retail context, such as food tourism, cruises and sport tourism.

In Chapter 13, Sangeeta Peter and Victor Anandkumar provide a review of shopping events and festivals as an important part of the retail landscape of destinations. Shopping festivals are planned and carried out to enhance a destination's retail image and to brand itself as a shopping destination. These retail-based festivals typically take place in localities that are already well-known shopping destinations. Peter and Anandkumar focus on several key shopping festivals in the Middle East, Asia and Europe and describe the unique characteristics of them, including their global reach, the need for safety and security, and the role of the government in policymaking and overseeing their functions, often in collaboration with non-profit organizations comprising retail industry participants. The authors also point out the main factors of success of retail festivals, underscored by private–public collaborative planning.

In Chapter 14, Matthew J. Stone interrogates the value of food as a travel souvenir. Although food and dining are among the most enjoyable and meaningful parts of many journeys, tourists also often buy food souvenirs to give away or keep for themselves. Stone details the concept of souvenirs and what they mean in the context of tourism. He then introduces the idea of food into the mix and highlights the meanings and value of food as a memory of a time, place or experience. The author proposes a typology of food souvenirs in relation to tourists' motives for buying.

The concluding chapter (Chapter 15, Dallen J. Timothy) summarizes many of the conceptual threads and practical applications of the discussion points brought out in the previous chapters. The author also takes a forward-looking view to suggest areas that remain under-researched in the shopping tourism literature.

The information presented on shopping and tourism at the outset of this chapter sets the tone for the chapters that follow. The next 14 chapters present the research, knowledge and wisdom of their authors, providing a strong conceptual foundation for future research and highlighting important management implications for retailers and destination management organizations that desire to increase their tourism impact through shopping and retail development.

References

Albayrak, T., Caber, M. and Çömen, N. (2016) Tourist shopping: The relationships among shopping attributes, shopping value, and behavioral intention. *Tourism Management Perspectives* 18, 98–106.

Alshammari, B., South, R.B. and Raleigh, K. (in press) Saudi Arabia outbound tourism: An analysis of trends and destinations. *Journal of Policy Research in Tourism Leisure and Events*, 1–23.

Al-Sulaiti, I. (2022) Mega shopping malls technology-enabled facilities, destination image, tourists' behavior and revisit intentions: Implications of the SOR theory. *Frontiers in Environmental Science* 10, 965642.

Amin, M., Ryu, K., Cobanoglu, C., Rezaei, S. and Wulan, M.M. (2021) Examining the effect of shopping mall attributes in predicting tourist shopping satisfaction and behavioral intentions: Variation across generation X and Y. *Journal of Quality Assurance in Hospitality and Tourism* 22 (3), 367–394.

Anastasiadou, C. and Vettese, S. (2019) From souvenirs to 3D printed souvenirs: Exploring the capabilities of additive manufacturing technologies in (re)-framing tourist souvenirs. *Tourism Management* 71, 428–442.

Azmi, A., Abdullah, A., Nurhidayati, S.E. and Shaw, G. (2020a) Shopping and tourism: A state-of-the-art review. *PalArch's Journal of Archaeology of Egypt/Egyptology* 17 (5), 1220–1239.

Azmi, A., Zubir, H.A. and Shaw, G. (2020b) Mature tourist shopping behaviour in Kuching, Sarawak. *Asian Journal of Arts, Culture and Tourism* 2 (3), 1–13.

Azmi, A., Ahmad, M.A., Abdullah, A. and Shaw, G. (2019) Shopping in Malaysia: A study of international tourists' experience and expectation. *International Journal of Innovation, Creativity and Change* 8 (4), 199–212.

Ben Said, F., Meyer, N., Bahri-Ammari, N. and Soliman, M. (2024) Shopping tourism: A bibliometric review from 1979 to 2021. *Journal of Tourism and Services* 15, 61–88.

Buczkowska, K. (2014) Local food and beverage products as important tourist souvenirs. *Turystyka Kulturowa* 1 (1), 47–57.

Choi, M., Law, R. and Heo, C.Y. (2016a) Shopping destinations and trust–tourist attitudes: Scale development and validation. *Tourism Management* 54, 490–501.

Choi, M.J., Heo, C.Y. and Law, R. (2016b) Progress in shopping tourism. *Journal of Travel and Tourism Marketing* 33 (1), 1–24.

Coles, T. (2004) Tourism, shopping, and retailing: An axiomatic relationship? In A.A. Lew, C.M. Hall and A.M. Williams (eds) *A Companion to Tourism* (pp. 360–373). Blackwell.

Collins-Kreiner, N. and Zins, Y. (2011) Tourists and souvenirs: Changes through time, space and meaning. *Journal of Heritage Tourism* 6 (1), 17–27.

Correia, A., Kozak, M. and Ferradeira, J. (2011) Impact of culture on tourist decision-making styles. *International Journal of Tourism Research* 13, 433–446

Correia, A., Kozak, M. and Kim, S. (2018) Luxury shopping orientations of mainland Chinese tourists in Hong Kong: Their shopping destination. *Tourism Economics* 24 (1), 92–108.

Egresi, I. and Polat, D. (2016) Assessing tourists' satisfaction with their shopping experience in Istanbul. *GeoJournal of Tourism and Geosites* 9 (2), 172–186.

Erdly, C. (2024) Four major trends that will shape retail in 2024. *Forbes*, online: https://www.forbes.com/sites/catherineerdly/2024/01/26/four-major-trends-that-will-shape-retail-in-2024/?sh=1c03a4aa64a5

Escudero-Gómez, L.A. (2024) Shopping centers challenging decline: Competitive strategies in three case studies from Madrid's urban area. *Journal of Retailing and Consumer Services* 79, 103826.

Ferreira, D. and Paiva, D. (2017) The death and life of shopping malls: An empirical investigation on the dead malls in Greater Lisbon. *The International Review of Retail, Distribution and Consumer Research* 27 (4), 317–333.

Goldblatt, J. (2012) Creating the sensual temporal marketplace experience and the role of planned events in tourist retail sustainable development. In C. McIntyre (ed.) *Tourism and Retail: The Psychogeography of Liminal Consumption* (pp. 51–62). Routledge.

He, L. and Timothy, D.J. (2024a) Tourists' perceptions of 'cultural and creative souvenir' products and their relationship with place. *Journal of Tourism and Cultural Change* 22 (2), 143–163.

He, L. and Timothy, D.J. (2024b) Understanding souvenirs from a place-product perspective: Territorialization, deterritorialization and reterritorialization. *Tourism Review International* 28 (1), 25–48.

Henderson, J.C., Chee, L., Mun, C.N. and Lee, C. (2011) Shopping, tourism and retailing in Singapore. *Managing Leisure* 16 (1) 36–48.

Hurst, J.L., Niehm, L.S. and Littrell, M.A. (2009) Retail service dynamics in a rural tourism community: implications for customer relationship management. *Managing Service Quality* 19 (5), 511–540.

Irimiás, A., Michalkó, G., Timothy, D.J. and Franch, M. (2017) Disappointment in tourism and hospitality: The influence of films on destinations. In E. Koc (ed.) *Service Failures and Recovery in Tourism and Hospitality: A Practical Manual* (pp. 214–227). CABI.

Jin, H., Moscardo, G. and Murphy, L. (2017) Making sense of tourist shopping research: A critical review. *Tourism Management* 62, 120–134.

Jin, H., Moscardo, G. and Murphy, L. (2020) Exploring Chinese outbound tourist shopping: A social practice framework. *Journal of Travel Research* 59 (1), 156–172.

Journée, R. and Weber, M. (2017) Co-creation of experiences in retail: Opportunity to innovate in retail business. In J. Bellemare, S. Carrier, K. Nielsen and F.T. Piller (eds) *Managing Complexity: Proceedings of the 8th World Conference on Mass Customization, Personalization, and Co-Creation* (MCPC 2015), Montreal, Canada, October 20th–22th 2015 (pp. 391–404). Springer International Publishing.

Kacen, J.J., Hess, J.D. and Chiang, W.Y.K. (2013) Bricks or clicks? Consumer attitudes toward traditional stores and online stores. *Global Economics and Management Review* 18 (1), 12–21.

Kim, D.Y. and Park, S. (2020) Rethinking millennials: How are they shaping the tourism industry? *Asia Pacific Journal of Tourism Research* 25 (1), 1–2.

Kim, S.S., Timothy, D.J. and Huang, J. (2011) Understanding Japanese tourists' shopping preferences using the Decision Tree Analysis method. *Tourism Management* 32 (3), 544–554.

Kiriri, P.N. (2021) Determinants of shopping mall attractiveness: The case of shopping malls in Nairobi, Kenya. *European Journal of Economics and Business* 7 (2), 112–130.

Koc, E. and Ayyildiz, A.Y. (2021) Culture's influence on the design and delivery of the marketing mix elements in tourism and hospitality. *Sustainability* 13 (21), 11630.

Lam, I.K.V. and Wong, I.A. (2020) The role of relationship quality and loyalty program in tourism shopping: A multilevel investigation. *Journal of Travel and Tourism Marketing* 37 (1), 92–111.

Li, C.S., Zhang, C.X., Chen, X. and Wu, M.S.S. (2021) Luxury shopping tourism: Views from Chinese post-1990s female tourists. *Tourism Review* 76 (2), 427–438.

Liang, C.C. and Yu, A. (2024) Customer impulse shopping in airports. *International Journal of Retail and Distribution Management* 52 (3), 372–385.

Lin, J., Guia Julve, J., Xu, H. and Cui, Q. (2020) Food habits and tourist food consumption: An exploratory study on dining behaviours of Chinese outbound tourists in Spain. *Journal of Policy Research in Tourism, Leisure and Events* 12 (1), 82–99.

Lin, L. and Mao, P.C. (2015) Food for memories and culture: A content analysis study of food specialties and souvenirs. *Journal of Hospitality and Tourism Management* 22, 19–29.

Littrell, M.A., Anderson, L.F. and Brown, P.J. (1993) What makes a craft souvenir authentic? *Annals of Tourism Research* 20 (1), 197–215.

Lojo, A. (2020) Young Chinese in Europe: Travel behavior and new trends based on evidence from Spain. *Tourism* 68 (1), 7–20.

Marciniak, R. and Mohsen, M.G. (2014) Homogeneity in luxury fashion consumption: An exploration of Arab women. *The Business and Management Review* 5 (1), 32–41.

McIntyre, C. (2012a) Retail tourists as co-creators of tourist retail place and space. In C. McIntyre (ed.) *Tourism and Retail: The Psychogeography of Liminal Consumption* (pp. 63–89). Routledge.

McIntyre, C. (2012b) *Tourism and Retail: The Psychogeography of Liminal Consumption*. Routledge.

Michalkó, G. (2004) *A Bevásárlóturizmus (Shopping Tourism)*. S: Kodolányi János Főiskola.

Michalkó, G., Irimiás, A. and Timothy, D.J. (2015) Disappointment in tourism: Perspectives on tourism destination management. *Tourism Management Perspectives* 16, 85–91.

Miller, D., Jackson, P., Thrift, N., Holbrook, B. and Rowlands, M. (1998) *Shopping, Place and Identity*. Routledge.

Moes, A. and van Vliet, H. (2017) The online appeal of the physical shop: How a physical store can benefit from a virtual representation. *Heliyon* 3, e00336

Moscardo, G. (2004) Shopping as a destination attraction: An empirical examination of the role of shopping in tourists' destination choice and experience. *Journal of Vacation Marketing* 10 (4), 294–307.

Moscardo, G., Murphy, L., Hughes, K. and Benckendorff, P. (2020) Shopping on the edge: Identifying factors contributing to tourist retail development in heritage villages. In I. Yeoman and U. McMahon-Beattie (eds) *The Future Past of Tourism: Historical Perspectives and Future Evolutions* (pp. 188–200). Channel View Publications.

Murphy, L., Moscardo, G., Benckendorff, P. and Pearce, P. (2011a) Evaluating tourist satisfaction with the retail experience in a typical tourist shopping village. *Journal of Retailing and Consumer Services* 18 (4), 302–310.

Murphy, L., Benckendorff, P., Moscardo, G. and Pearce, P.L. (2011b) *Tourist Shopping Villages: Forms and Functions*. Routledge.

Murphy, L., Moscardo, G. and Benckendorff, P. (2013) Understanding tourist shopping village experiences on the margins. In J. Cave, L. Jolliffe and T. Baum (eds) *Tourism and Souvenirs: Glocal Perspectives from the Margins* (pp. 132–146). Channel View Publications.

Parasakul, L. (2020) Assessing Chinese tourists' satisfaction with their shopping experiences in Bangkok Metropolis. *Journal of China Tourism Research* 16 (2), 279–296.

Peter, S. and Anandkumar, V. (2016) Deconstructing the shopping experience of tourists to the Dubai Shopping Festival. *Cogent Business and Management* 3 (1), 1199109.

Rabbiosi, C. (2011) The invention of shopping tourism. The discursive repositioning of landscape in an Italian retail-led case. *Journal of Tourism and Cultural Change* 9 (2), 70–86.

Rabbiosi, C. (2016a) Itineraries of consumption: Co-producing leisure shopping sites in Rimini. *Journal of Consumer Culture* 16 (2), 412–431.

Rabbiosi, C. (2016b) Place branding performances in tourist local food shops. *Annals of Tourism Research* 60, 154–168.

Rajamma, R.K. and Neeley, C.R. (2005) Antecedents to shopping online: A shopping preference perspective. *Journal of Internet Commerce* 4 (1), 63–78.

Reisinger, Y. (2006) Shopping and tourism. In D. Buhalis and C. Costa (eds) *Tourism Business Frontiers: Consumers, Products and Industry* (pp. 127–136). Routledge.

Reisinger, Y. and Turner, L.W. (2002) The determination of shopping satisfaction of Japanese tourists visiting Hawaii and the Gold Coast compared. *Journal of Travel Research* 41 (2), 167–176.

Statista (2023) Share of shopping tourists in selected countries worldwide as of June 2023. Online: https://www.statista.com/statistics/1401364/shopping-tourists-by-country-worldwide/

Sthapit, E., Björk, P. and Rasoolimanesh, S.M. (2024) Toward a better understanding of memorable souvenir shopping experiences. *International Journal of Hospitality and Tourism Administration* 25 (1), 59–91.

Sthapit, E., Coudounaris, D.N. and Björk, P. (2018) The memorable souvenir-shopping experience: Antecedents and outcomes. *Leisure Studies* 37 (5), 628–643.

Stone, M.J. (in press) Not just another trinket: Defining unique attributes of food souvenirs. *Tourism Recreation Research* 1–6.

Suhartanto, D. (2018) Tourist satisfaction with souvenir shopping: Evidence from Indonesian domestic tourists. *Current Issues in Tourism* 21 (6), 663–679.

Swanson, K.K. and Timothy, D.J. (2012) Souvenirs: Icons of meaning, commercialization and commoditization. *Tourism Management* 33 (3), 489–499.

Timothy, D.J. (2005) *Shopping Tourism, Retailing and Leisure*. Channel View Publications.

Timothy, D.J. (2014) Trends in tourism, shopping, and retailing. In A.A. Lew, C.M. Hall and A.M. Williams (eds) *The Wiley Blackwell Companion to Tourism* (pp. 378–388). Wiley Blackwell.

Timothy, D.J. (2018) Shopping tourism. In S. Agarwal, G. Busby and R. Huang (eds) *Special Interest Tourism: Concepts, Contexts and Cases* (pp. 134–144). CAB International.

Timothy, D.J. (2021) *Cultural Heritage and Tourism: An Introduction* (2nd edn). Channel View Publications.

Timothy, D.J. (2024) Tourism, shopping and connotations of place. In C.M. Hall (ed.) *The Wiley Blackwell Companion to Tourism* (2nd edn) (pp. 501–514). Wiley.

Törőcsik, M., Csóka, L., Nemeth, P. and Neulinger, A. (2023) Changes in the attitudes of Hungarian shoppers in times of crisis. *Financial and Economic Review* 22 (4), 82–106.

Tosun, C., Temizkan, S.P., Timothy, D.J. and Fyall, A. (2007) Tourist shopping experiences and satisfaction. *International Journal of Tourism Research* 9, 87–102.

Williams, C. (2012) Re-theorising the role of retail shopping centres as tourist attractions in economic development. In C. McIntyre (ed.) *Tourism and Retail: The Psychogeography of Liminal Consumption* (pp. 11–28). Routledge.

Wong, I.A., Huang, G.I. and Li, Z.C. (2024) Axiology of tourism shopping: A cross-level investigation of value-in-the-experience (VALEX). *Journal of Hospitality and Tourism Research* 48 (3), 549–563.

Wong, I.A. and Wan, Y.K.P. (2013) A systematic approach to scale development in tourist shopping satisfaction: Linking destination attributes and shopping experience. *Journal of Travel Research* 52 (1), 29–41.

World Travel and Tourism Council (2023) WTTC unveils the power of retail tourism. Online: https://wttc.org/news-article/wttc-unveils-the-power-of-retail-tourism

YouGov (2023) Retail therapy and travel: Exploring insights into shopping tourists. Online: https://business.yougov.com/content/46958-retail-therapy-and-travel-exploring-insights-into-shopping-tourists-research

Yuan, X., Xie, Y., Li, S. and Shen, Y. (2022) When souvenirs meet online shopping–the effect of food souvenir types on online sales. *International Journal of Tourism Research* 24 (1), 58–70.

Zackariya, S. (2023) *Leading Travel and Tourism Retail: How Businesses Can Sustainably Capture New Profits in Shopping Tourism*. Kogan Page.

Zaidan, E.A. (2016) Tourism shopping and new urban entertainment: A case study of Dubai. *Journal of Vacation Marketing* 22 (1), 29–41.

Part 1

Consumer Behavior and Cultural Aspects of Shopping Tourism

2 Shopping Tourism and Consumer Emotional Loyalty: Destination and Programmatic Perspectives

Antónia Correia and Paulo Martins

Introduction

Shopping tourism is widely recognised as a highly profitable sector within the tourism industry, drawing in a diverse range of consumers who seek captivating shopping destinations and immersive experiences (Lin & Chen, 2013; Muro-Rodríguez et al., 2020). This rapidly growing market has collected significant attention, appealing to discerning travellers and solidifying its position as a lucrative segment. Scholars such as Frederick (1929) and Benson (2000) have expanded upon Veblen's (1899) seminal work on consumption, making notable contributions to the study of shopping. Retailing strategy has also emerged as a distinct area within marketing (Clarke & Belk, 1979; McGoldrick, 1990). Further insights into consumer behaviour by Hirschman (1992) have developed our understanding of the intricate dynamics of shopping. Therefore, it is vital to recognise and comprehend the interplay between shopping and the tourist experience to harness the potential of shopping tourism (Correia & Kozak, 2016). By unravelling this relationship, marketers and businesses can devise tailored strategies for immersive and memorable shopping experiences, enhancing destination appeal and maximising economic benefits.

Understanding the factors influencing tourist satisfaction is crucial for unravelling the dynamics that shape a destination's local economy (Esfandiar et al., 2024). By exploring several dimensions of consumer behaviour and destination loyalty, researchers and stakeholders gain insights to guide strategic decision-making (Chen & Chen, 2010), resulting in positive outcomes for both the destination and its visitors (Narangajavana et al., 2017). This scientific approach empowers stakeholders to optimise the shopping experience (Wu et al., 2014), enhance tourist satisfaction (Heung & Cheng, 2000) and foster economic prosperity (Sangkaew & Zhu, 2022). Adopting a holistic approach enables the creation of strategies to cater to travellers, curating unforgettable experiences that evoke emotions, spark curiosity and forge lasting connections (Zeinali *et al.*, 2014). This mutually beneficial relationship unlocks the full potential of retail enterprises and destinations, cultivating loyalty that goes beyond mere transactions (Pimpão *et al.*, 2014). The growth of shopping tourism presents a unique opportunity to enhance the tourist experience and drive economic development by understanding its connection to the broader tourism landscape (Pimpão *et al.*, 2014). Stakeholders can harness this potential to create memorable journeys for global travellers.

Previous studies (Bigné Alcañiz *et al.*, 2009; Moutinho, 1987) have focused on the cognitive aspects of tourism while neglecting the emotional dimension. However, it is essential to recognise that tourists can develop profound emotional connections to places such as Florence, Italy, when driven by their affinity for luxury brands like Gucci. These emotional attachments go beyond product quality evaluations and highlight the critical role of cognitive elements, such as personal beliefs, in shaping a place image (Qu *et al.*, 2011). The limited attention given to the emotional relationship between brands and places hinders our understanding of how tourists' emotional connections to brands influence their perception of a destination. To bridge this knowledge gap, destination marketers must acknowledge the reciprocal influence between emotional brand connections and cognitive destination evaluations. Marketers can design personalised strategies that evoke emotional responses, shaping the brand's image within the tourist experience. By integrating both affective and cognitive dimensions, marketers can create immersive experiences that deeply resonate with visitors' emotions and leave lifelong memories (Kim & Perdue, 2013).

To grasp the dynamic relationship between browsing and shopping, it is vital to acknowledge their interconnected nature and impact on consumer behaviour, as highlighted by Correia and Kozak (2019). Recognising the profound influence of browsing on shopping behaviour is indispensable for retailers and shopping destinations aiming to optimise their marketing strategies and create captivating experiences for discerning consumers (Bloch & Richins, 1983). Furthermore, the long-term success of retail businesses and shopping destinations relies heavily on shopper loyalty. Loyal customers not only make repeat purchases and spend more money but also serve as brand advocates, positively influencing the reputation and profitability of retail establishments. By understanding the drivers of shopper loyalty, including factors such as customer satisfaction, personalised experiences and destination attractiveness, retailers can implement effective strategies to foster enduring customer relationships (Correia *et al.*, 2017).

This chapter examines relevant studies on the intricate relationship between shopping and shopper loyalty. By synthesising this knowledge, we aim to comprehensively understand the factors driving shopper loyalty and the role of personalisation and technology in shaping the shopping experience. Through this exploration, retailers and shopping destinations can gain valuable insights to enhance their offerings, attract and retain loyal shoppers, and thrive in the ever-evolving landscape of shopping tourism.

Understanding Retail/Shopping Loyalty Programmes

Retail and shopping loyalty programmes have become essential strategies industries employ to foster customer loyalty amidst intense competition in the retail landscape (Pimpão et al., 2014). Therefore, it is crucial to analyse diverse loyalty programmes, including credit card loyalty, travel point programmes, store loyalty, cash rewards programs and frequent-shopper programmes, to evaluate their effectiveness and impact on customer loyalty (Sharp & Sharp, 1997). The tourism industry has proactively implemented loyalty programs to encourage repeat purchases and establish personalised connections with individual customers, gaining valuable insights into their preferences and needs (Buhalis & Law, 2008). Customer Relationship Management (CRM) systems, such as those discussed by Buhalis and Law (2008), have emerged as focal platforms that integrate loyalty programmes, facilitating interoperability, personalisation and continuous networking to enhance the customer experience.

The willingness of guests to rely on loyalty programmes assumes an essential role in mediating the connection between the tangible attributes of a website and their behavioural intentions (Herrero et al., 2015). The level of trust established in the online domain significantly influences purchasing behaviour, whereas commitment reflects the propensity to maintain that behaviour over an extended period (Pimpão et al., 2018). Trust and commitment represent critical factors in decisions to enhance customer-business relationships. However, for the relationship to thrive, satisfaction emerges as the pivotal variable (Cugelman et al., 2009). Therefore, a wide-ranging understanding of the underlying mechanisms embedded in loyalty programmes, encompassing the constructs of commitment-trust and satisfaction as complementary facets, assumes paramount importance in anticipating behavioural intentions (Garbarino & Johnson, 1999). These meticulously designed programmes are tailored to reward customers for recurring purchases and unwavering brand loyalty, nurturing a permanent bond.

Credit card loyalty and travel point programmes have revolutionised shopping tourism by integrating with travel rewards, significantly impacting consumer behaviour and destination choices (Liu, 2007). These programmes engage tourists with points and rewards for credit card purchases, driving them to select destinations where they can maximise benefits. Redeeming travel points for discounted flights, hotel stays and other perks incentivises tourists to choose destinations aligned with their shopping preferences (Choi *et al.*, 2015). Moreover, credit card loyalty programmes shape consumer spending behaviour, as tourists increase expenditures to earn more points and unlock additional benefits, benefiting businesses and fostering economic growth (Ongaya & Muathe, 2022). In addition, these programmes shape consumer preferences for shopping destinations through exclusive discounts, VIP access and personalised offers, enhancing the overall shopping experience. Integrating these programmes into tourism strategies provides a competitive advantage, attracting tourists seeking rewarding shopping experiences. However, managing complex point systems, ensuring customer engagement and programme sustainability present ongoing challenges (Cai *et al.*, 2021). Striking a balance between desirable rewards, seamless redemption processes and transparent terms and conditions is crucial for success. By embracing and leveraging these programmes, destinations can enhance the shopping tourism experience (Chen *et al.*, 2021).

Retailers widely adopt store loyalty and cash rewards schemes to enhance customer loyalty and promote repeat purchases (Cambra-Fierro *et al.*, 2019). These strategies provide exclusive rewards and incentives that foster a strong connection with the brand. Store loyalty programmes offer personalised offers and discounts, creating a sense of value and exclusivity (Correia & Kozak, 2023). By accumulating rewards specific to a particular store, customers are more inclined to remain loyal and develop a deep affinity for the brand. Conversely, cash rewards programmes deliver immediate gratification through monetary benefits like cash back or credits, increasing overall satisfaction and reinforcing brand commitment. Regular shopper programmes significantly acknowledge and reward customers based on their purchase frequency (Sharp & Sharp, 1997). By offering exclusive privileges, these programmes foster a sense of belonging and motivation for customers to maintain or elevate their status. Psychological factors, including reward ownership and a sense of progress, further nurture customer loyalty (Kandampully *et al.*, 2015). These programmes are indispensable for retailers in cultivating unwavering customer loyalty and necessitate meticulous design to grasp their profound impact on consumer behaviour. By using store loyalty and cash rewards tactics, retailers can forge robust relationships with customers, drive repeat purchases and establish a sustainable competitive advantage in the marketplace (Palmatier & Crecelius, 2019)

In conclusion, loyalty programmes are vital in retail, encompassing credit card strategies, travel points, store loyalty, cash rewards and frequent-shopper options. These programmes incentivise customer engagement through enticing rewards and personalised experiences, nurturing loyalty. They also influence shopping tourism and consumer behaviour, providing valuable data for decision-making and marketing strategies. Effective programme design and customer satisfaction are crucial for lasting loyalty. Through loyalty, retailers gain a competitive edge, drive repeat purchases and build relationships. Therefore, loyalty programmes play a pivotal role in retail, fostering loyalty, driving engagement and establishing dominance.

Shopping's Contribution to Destination Loyalty among Tourists

Shopping experiences influence tourists' emotional attachment to a destination and foster loyalty (Correia *et al.*, 2017). Shopping plays a central role in shaping travellers' trip experiences. Beyond its economic implications, it enhances emotional well-being and facilitates cultural exchanges, contributing to satisfaction and perception of the destination (Sirakaya-Turk *et al.*, 2015). Therefore, positive shopping experiences contribute to destination satisfaction, increasing the likelihood of return visits and enthusiastic recommendations, fostering destination loyalty (Zeinali *et al.*, 2014). Browsing, a recreational activity driven by curiosity, adds to the enjoyment of the shopping experience and allows tourists to acquire expertise about products and market prices. By engaging in browsing activities, tourists can enhance their overall trip experiences, satisfy their curiosity and foster a sense of loyalty towards the destination. These experiences evoke a sense of thrill, fulfilment and camaraderie, creating possibilities for immersing in the local culture and embarking on explorations through interactions with vibrant street markets, bustling bazaars and like-minded shoppers (Correia & Kozak, 2016, 2019) (see Chapter 8).

The satisfaction derived from shopping experiences inevitably influences perceptions of the destination. Shopping opportunities and retail environments position the destination as highly desirable. A diverse range of high-quality shopping options reflects the destination's attractiveness and competitiveness, showcasing a commitment to delivering unforgettable experiences (Hudson & Ritchie, 2009). In addition, Lee *et al.* (2011) report that positive shopping experiences foster long-term loyalty. Satisfied tourists develop an emotional connection with the destination, driven by positive emotions and cherished memories associated with their shopping encounters (Yüksel, 2007). These loyal visitors become advocates, promoting word-of-mouth and sharing their positive experiences. To harness the potential of shopping experiences in nurturing destination loyalty, destination marketers and businesses can craft shopping environments. Huang and Hsu (2009) reinforce that detailed attention to service quality, product variety, authentic cultural representations and engaging retail environments elevates the shopping experience. The importance of shopping in enhancing visitor satisfaction and shaping the destination experience has been consistently emphasised in prior research (Alrawadieh *et al.*, 2019; Heung & Cheng, 2000; Kumar *et al.*, 2020). These findings guide destination marketers and businesses in shaping exceptional shopping experiences that amplify visitor satisfaction and cultivate loyalty.

Case studies are essential for exploring the relationship between shopping experiences and destination loyalty in the tourism industry. By providing valuable insights, these studies enhance our understanding of this connection and emphasise the significant influence of shopping on tourists' loyalty toward a destination. In a case study developed by Myagmarsuren and Chen (2011) in Mongolia, the authors applied the customer-based brand equity (CBBE) model to examine the connection between destination brand equity, tourist satisfaction and loyalty. The findings emphasised the impact of tourists' destination awareness on their perception of the destination's image, subsequently influencing their expectations and actual experiences of service quality. Similarly, Correia and Kozak (2016) conducted a cross-national study focusing on tourists' attitudes towards street markets in Algarve, Portugal, and Bodrum, Turkey. This study highlighted the significance of satisfaction with tourism quality in fostering destination loyalty. The authors identified factors such as price consciousness and perceived utility as influential elements in the marketplace, with higher satisfaction levels positively influencing tourists' intentions to revisit and recommend street markets in both destinations. Another case study by Liu *et al.* (2022) examined the categorisation of cultural tourism attractions in Hong Kong based on the preferences of Japanese and Thai tourists. The findings revealed distinct inclinations, emphasising the need for tailored marketing strategies in cultural tourism to cater to the specific tastes of different tourist segments. Consistently, these studies highlight the link between positive shopping experiences and destination loyalty. Recognizing this relationship enables destination marketers to enhance shopping experiences and loyalty by aligning destination image, personal preferences and shopping motivations. Furthermore, Rajesh (2013) and Chi and Qu (2008) emphasise the significance of tourists' strong inclination towards shopping and their preference for unique local products in developing loyalty, particularly when coupled with positive shopping experiences. Conversely, negative experiences between tourists' expectations and available shopping options can harm loyalty.

In summary, shopping experiences are crucial in fostering destination loyalty among tourists. Extensive research and case studies have shed light on the factors that shape this relationship. By understanding the impact of shopping on destination loyalty, marketers can develop targeted strategies to improve shopping experiences, leading to increased satisfaction and loyalty.

Emotional Loyalty and its Influence on Shopping Tourism

The emotional bond between consumers and brands/destinations has received considerable focus in the field of shopping tourism (Huang *et al.*,

2017). Destination marketers and businesses need to recognise the significance of emotional loyalty and its impact on perceptions and experiences. According to Khoo-Lattimore (2011), this connection transcends rational decision-making processes and fosters a strong bond, trust, and positive associations with a particular brand or destination. In shopping tourism, emotional loyalty can consist of a profound attachment and positive emotions that consumers establish toward a destination owing to their shopping experiences.

Numerous studies (Han et al., 2019; Lam & Wong, 2020; Williams & Soutar, 2009) have explored the impact of emotional loyalty on the overall perceptions and experiences of shopping tourism. Abdulsalam and Dahanan (2022) have provided evidence that establishing an emotional connection with a destination leads to more positive perceptions of the shopping experience. Consequently, emotional loyalty positively influences various aspects, including satisfaction, trust and willingness to recommend the destination to others. Moreover, it enriches the experience by fostering a sense of belonging and personal fulfilment (Timothy, 2005). To promote emotional loyalty in the context of shopping tourism, destination marketers and businesses can implement several strategies:

Creating authentic and memorable experiences

Creating authentic and memorable shopping experiences is essential for establishing emotional connections with tourists (Mehta et al., 2013). Showcasing local culture, personalising services and curating exclusive products or events enable destinations to leave a lasting impact on visitors (Kozak & Rimmington, 2000). Emotional connections are enhanced by incorporating authentic cultural elements, offering personalised recommendations and services and providing exclusive experiences (Ramseook-Munhurrun et al., 2015). According to Chen et al. (2021), leveraging innovative technologies like AR and VR further enhances the authenticity and memorability of the shopping experience (see Chapter 15). Therefore, prioritising authenticity and creating unique moments allows shopping destinations to stand out and cultivate strong emotional loyalty among tourists.

Building trust and relationships

To foster emotional loyalty in the tourism industry, destination marketers should emphasise trust through effective communication, personalised interactions and transparent practices (Rather et al., 2019). This involves ensuring clear and timely communication, providing relevant information and engaging in tailored interactions (Chi & Qu, 2008). Transparent pricing and policies further demonstrate a commitment to care and honesty. Sparks and Browning (2011) reinforce that marketers can actively cultivate trust and nurture emotional loyalty among tourists by prioritizing communication, interaction and transparency. Sustaining trust requires consistently delivering exceptional experiences.

Engaging emotional triggers

Understanding tourists' emotional triggers and aligning them with the shopping experience is essential for fostering emotional loyalty (Hosany & Gilbert, 2009). Storytelling, sensory marketing and strategic campaigns are vital in establishing these emotional connections (Yang *et al.*, 2021). The authors emphasise that captivating narratives enable brands and destinations to engage with consumers, leaving a memorable impact. Sensory elements such as music and visuals enhance the emotional experience, while strategic campaigns evoke specific emotions. By prioritising emotional triggers, destination marketers can cultivate loyalty and encourage advocacy among tourists.

Encouraging co-creation and customer engagement

By involving tourists in co-creating shopping experiences, destination marketers can cultivate a sense of ownership and emotional attachment (Gallarza *et al.*, 2002). This approach surpasses passive consumption and promotes active customer participation. Engaging with tourists through social media facilitates real-time interaction and fosters community (Jaya & Prianthara, 2020). Feedback mechanisms like surveys and reviews allow customers to express their opinions and contribute to future experiences. In addition, loyalty programmes incentivise and reward customer engagement, reinforcing emotional loyalty (Pimpão *et al.*, 2014). Embracing co-creation strategies enables destination marketers to enhance the shopping experience, strengthen emotional connections and build long-lasting relationships.

Providing exceptional customer service

Providing exceptional customer service is crucial for fostering emotional loyalty among tourists. Training staff to offer personalised assistance is vital in creating memorable experiences. When customers receive individual attention and their specific needs are addressed, it generates a sense of being valued and cared for (Kim & Cha, 2002). Promptly addressing customer concerns is also essential, as it showcases attentiveness and dedication to resolving issues. Going above and beyond to exceed expectations can leave a lasting positive impact on tourists, making them feel valued and appreciated (Parasuraman *et al.*, 1985). By prioritising exceptional customer service, destination stakeholders can cultivate emotional loyalty and foster loyal advocates for their brand or destination.

In conclusion, emotional loyalty is central to shopping tourism, impacting tourists' perceptions, experiences and overall satisfaction. Destination marketers and businesses must recognise the significance of emotional connections and implement effective strategies to cultivate emotional loyalty. Destination marketers can foster emotional bonds between tourists and shopping destinations by creating authentic experiences, establishing trust, appealing to emotional triggers, boosting customer engagement and providing exceptional customer service. As a result, it leads to heightened satisfaction, increased loyalty and positive word-of-mouth, ultimately contributing to the long-term success of shopping tourism destinations.

Implications for Destination Marketers and Programmatic Strategies

Destination marketing plays a crucial role in shopping tourism, offering practical implications that guide marketers in maximising their efforts. Marketers must devise effective strategies to optimise shopping experiences and foster customer loyalty (Kotler *et al.*, 2021). By implementing innovative approaches, marketers can create exceptional experiences that encourage repeat visits and generate positive word-of-mouth. Integrating planning and programmatic destination marketing is a valuable strategy, enhancing marketing efforts and effectively reaching the target audience. Programmatic marketing utilises automated technologies and algorithms to deliver real-time, targeted advertisements. These topics can be explored in depth by examining the following points:

- *Tailored marketing approaches*: Understanding tourists' preferences, behaviours and motivations is instrumental in developing marketing strategies for destination marketers. By gaining insights into their target audience's unique characteristics, including demographics, interests and cultural backgrounds, marketers can customise their approaches to resonate with and engage their customers (Kim *et al.*, 2019). This involves crafting targeted campaigns that align with the specific needs and desires of the audience, ensuring that the marketing efforts are relevant, impactful and memorable.
- *Elevating customer experiences*: Valuable insights can empower destination marketers to enhance customer experiences. According to Yachin (2018), marketers can concentrate on optimising these interactions to pinpoint critical touchpoints to deliver exceptional and unforgettable experiences. This may involve investments in infrastructure, amenities and training to improve overall customer service. In addition, personalised recommendations and tailored assistance can enhance customer satisfaction, and loyalty fosters emotional connections with the destination (Kim & Kim, 2022).

- *Leveraging technology*: Valuable insights can shed light on technology in destination marketing. Marketers can leverage innovative tools like augmented reality (AR), virtual reality (VR) and artificial intelligence (AI) applications to elevate the visitor experience (Chen *et al.*, 2022; Oncioiu & Priescu, 2022). Implementing interactive digital maps, offering virtual tours and using AR for enhanced attraction information are just a few examples of how technology can be harnessed to create immersive and engaging experiences.
- *Collaborating with stakeholders*: The significance of collaboration with local stakeholders, such as businesses, community organizations and government agencies, becomes apparent. Destination marketers can foster partnerships with these stakeholders to develop joint marketing campaigns, establish alliances for exclusive offerings and promote local cultural events or festivals (Morgan *et al.*, 2004). Through collaboration, destination marketers and local stakeholders can create a unified and genuine destination experience that benefits tourists and the local community.
- *Measuring and evaluating success*: Destination marketers can utilise research findings to assess and evaluate the effectiveness of their marketing endeavours. By employing appropriate metrics and analytics, marketers can measure the impact of their strategies, identify opportunities and make well-informed decisions based on data (Sakas *et al.*, 2022). In addition, destination marketers must seek feedback and attentively listen to customers. By gathering and analysing customer feedback, marketers can identify areas that require enhancement and implement the necessary adjustments (Heung & Cheng, 2000). This encompasses monitoring visitor satisfaction, tracking social media engagement and analysing visitors' spending patterns.
- *Emotional engagement*: Building emotional connections with customers is critical to fostering loyalty (Myagmarsuren & Chen, 2011). Marketers can achieve this by incorporating storytelling elements into the shopping experience. Sharing the brand's story, highlighting local culture and heritage and creating meaningful interactions, contribute to a sense of belonging and attachment (Zhong *et al.*, 2017). Engaging customers on an emotional level creates lasting memories and strengthens the bond between the destination and the customer.
- *Rewards and incentives*: Encouraging repeat visits and nurturing customer loyalty can be achieved by implementing a loyalty programme or providing tips and incentives (Pimpão *et al.*, 2014). Marketers can offer exclusive discounts, special promotions or loyalty points that customers can collect and use. These initiatives motivate customers to return and make them feel valued and appreciated for their continued patronage.
- *Social media engagement*: Engaging customers and building community through social media is crucial. Marketers should interact with

customers, address inquiries and share captivating shopping-related content (Yu *et al.*, 2024). User-generated content and collaborations with influencers amplify campaign impact. Programmatic marketing leverages data and analytics for a precise targeting audience (Maurer, 2021). Analysing demographics, online behaviour and interests enable focused campaigns. This data-driven approach maximises effectiveness and minimises resource waste.

- *Real-time bidding*: Programmatic advertising empowers marketers to engage in real-time bidding auctions, allowing them to secure ad placements swiftly. Through this process, marketers can bid for ad space in milliseconds, using predefined criteria and targeting parameters (Maurer, 2021). Real-time bidding significantly improves the likelihood of delivering ads to the most relevant audience, enhancing engagement and conversion rates. This technology can be particularly advantageous in loyalty tourism, where precise targeting and personalised messaging are crucial to fostering customer loyalty.
- *Dynamic ad content*: Programmatic marketing empowers destination marketers to deliver dynamic and personalised ad content that adjusts to individual user preferences (Lacarcel & Huete, 2023). By utilising dynamic creative optimisation (DCO), marketers can customise messages, offers and visuals based on user behaviour and preferences, resulting in an enhanced user experience and heightened engagement and loyalty.
- *Multi-channel reach*: By utilising programmatic technology, destination marketers can establish connections with their target audience through various channels, maximising their visibility and nurturing loyalty (Lacarcel & Huete, 2023; Maurer, 2021). Through personalised ad delivery and seamless integration, programmatic marketing enhances the user experience, drives engagement and fosters loyalty in the context of shopping tourism.
- *Performance tracking and optimisation*: By leveraging real-time tracking and measuring campaign performance, marketers gain valuable insights into impressions, clicks, conversions and engagement metrics. This data-driven approach allows for continuous monitoring and optimisation of campaigns, ensuring that loyalty-focused marketing efforts are effective and impactful (Kotler *et al.*, 2021). Armed with these insights, marketers can make informed decisions to enhance customer loyalty and improve the overall effectiveness of their marketing strategies.
- *Cost efficiency*: Programmatic marketing offers cost efficiencies by automating the ad-buying process and removing the need for manual negotiations. It allows them to set budget limits, strategically bid and optimise campaigns based on performance data (Maurer, 2021). This automated approach ensures that marketers can maximise their advertising budget, maximising their return on investment (ROI).

- *Enhanced audience insights*: Programmatic marketing leverages data analysis and tracking to uncover audience insights, enabling marketers to understand their target audience's behaviour, preferences and engagement patterns (Palmatier & Crecelius, 2019). This knowledge serves as a foundation for marketing strategies, creating captivating content and delivering personalised products and experiences, all of which contribute to fostering customer loyalty.

To reiterate, these marketing practices can enhance shopping tourism experiences and foster customer loyalty, destination marketers can implement tailored marketing approaches, elevate customer experiences, leverage technology, collaborate with stakeholders and measure success. By understanding the preferences and motivations of tourists, marketers can customise their strategies to engage their target audience. Enhancing customer experiences involves optimising key touchpoints and providing personalised recommendations. Integrating technology, such as augmented reality and virtual tours, can create immersive and captivating experiences. Collaborating with local stakeholders can contribute to a cohesive and authentic destination experience. Measuring success through metrics and analytics allows marketers to make data-driven decisions and continuously improve their strategies. These strategies encompass audience understanding, experience enhancement, technology utilisation, collaboration and performance measurement.

Conclusion

This chapter focuses on destination marketers' effective strategies to foster emotional loyalty among tourists. It explores the significance of creating authentic and memorable experiences, building trustful relationships, engaging emotional triggers, encouraging co-creation and customer engagement and providing exceptional customer service. The chapter draws upon the insights of Mehta *et al.* (2013), Sparks and Browning (2011) and Gallarza *et al.* (2002), who have extensively studied these strategies and their impact on customer loyalty. Their research highlights the critical role of emotional connections and provides practical guidance for destination marketers in implementing these strategies successfully. Through unique experiences that reflect the destination's culture and heritage, marketers leave a lasting impression on tourists. Transparent communication and a focus on customer satisfaction help build trustful relationships, while emotional triggers such as resonant messages and branding deepen the connection between tourists and the destination. Involving customers in decision-making and providing exceptional customer service further enhance loyalty. This chapter presents a comprehensive framework that integrates these strategies, offering valuable insights and actionable recommendations for destination marketers seeking to cultivate emotional loyalty among tourists.

The research findings emphasise the pivotal role of emotional loyalty in establishing strong and enduring bonds between consumers and their preferred shopping destinations. Positive emotional experiences are crucial in cultivating consumer loyalty and shaping attitudes and behaviours. To establish meaningful connections, destination marketers should prioritise strategies that foster emotional engagement. Iin addition, the research underscores the substantial impact of emotional loyalty on consumer satisfaction in shopping tourism. It is evident that emotional loyalty significantly influences consumers' satisfaction with their shopping experiences, reinforcing the importance of eliciting positive emotions to shape consumer perceptions and evaluations. By creating emotionally compelling shopping experiences, destination marketers can enhance customer satisfaction and cultivate long-term loyalty. Throughout this chapter, we have explored a range of perspectives and insights from various studies, all contributing to our understanding of emotional loyalty in shopping tourism. The collective body of research reinforces the significance of emotional connections between consumers and shopping destinations, providing valuable guidance for destination marketers aiming to deliver exceptional shopping experiences and foster customer loyalty.

Future research in shopping tourism and consumer loyalty is essential. There is a need for deeper exploration into the underlying mechanisms of emotional commitment, including the influence of culture, personalisation and technology on connections and subsequent consumer behaviour. Furthermore, investigating the impact of social media and digital marketing in nurturing and affecting loyalty presents an intriguing avenue for future studies. It is crucial to emphasise the role of destination and programmatic perspectives. Destination marketers must acknowledge the significance of emotional connections and implement effective strategies to foster loyalty among tourists. By customising marketing approaches, enhancing customer experiences, harnessing technology, collaborating with stakeholders and measuring success, destination marketers can create exceptional shopping experiences, maximise their marketing endeavours and encourage long-term customer loyalty. In conclusion, this chapter has underscored the importance of emotional fidelity in shopping tourism and provided valuable insights into successful strategies for cultivating emotional connections between consumers and shopping destinations.

In summary, emotional loyalty is critical in shopping tourism, fostering strong bonds between consumers and shopping destinations. This chapter highlights the significance of emotional loyalty and provides practical insights into effective strategies for nurturing these connections. Moreover, it opens exciting prospects for future research in shopping tourism and consumer loyalty. To deepen our understanding of emotional loyalty, forthcoming studies can delve into the underlying mechanisms, including cultural influences, personalisation strategies and the impact of technology. In addition, exploring emerging trends such as social media and digital marketing in shaping emotional connections offers intriguing avenues for investigation. By delving into these areas, researchers can contribute valuable insights into the complex interplay of emotions, consumer behaviour and the shopping tourism industry. This knowledge will empower destination marketers to devise innovative strategies that align with evolving consumer preferences. Finally, this chapter catalyses future research, unravelling the intricacies of emotional loyalty and exploring emerging trends to enhance consumer loyalty in the shopping tourism industry.

References

Abdulsalam, M. and Dahanan, W.D. (2022) Influence of product involvement on tourist shopping behavior: The mediating role of place attachment, satisfaction, and visit frequency. *Journal of International Consumer Marketing* 34 (5), 467–482.

Alrawadieh, Z., Alrawadieh, Z. and Kozak, M. (2019) Exploring the impact of tourist harassment on destination image, tourist expenditure, and destination loyalty. *Tourism Management* 73, 13–20.

Benson, A.L. (2000) *I Shop, Therefore I Am: Compulsive Buying and the Search for Self.* Aronson.

Bigné Alcañiz, E., Sánchez García, I. and Sanz Blas, S. (2009) The functional-psychological continuum in the cognitive image of a destination: A confirmatory analysis. *Tourism Management* 30 (5), 715–723.

Bloch, P.H. and Richins, M.L. (1983) Shopping without purchase: An investigation of consumer browsing behavior. In R.P. Bagozzi and A.M. Tybout (eds) *Advances in Consumer Research* (Vol. 10, pp. 389–393). Association for Consumer Research.

Buhalis, D. and Law, R. (2008) Progress in information technology and tourism management: 20 years on and 10 years after the internet – The state of eTourism research. *Tourism Management* 29 (4), 609–623.

Cai, G., Xu, L. and Gao, W. (2021) The green BandB promotion strategies for tourist loyalty: Surveying the restart of Chinese national holiday travel after COVID-19. *International Journal of Hospitality Management* 94, 102704.

Cambra-Fierro, J., Gao, L.X., Melero-Polo, I. and Sese, F.J. (2019) What drives consumers' active participation in the online channel? Customer equity, experience quality, and relationship proneness. *Electronic Commerce Research and Applications* 35, 100855.

Chen, C.F. and Chen, F.S. (2010) Experience quality, perceived value, satisfaction, and behavioral intentions for heritage tourists. *Tourism Management* 31 (1), 29–35.

Chen, Y., Mandler, T. and Meyer-Waarden, L. (2021) Three decades of research on loyalty programs: A literature review and future research agenda. *Journal of Business Research* 124, 179–197.

Chen, R., Perry, P., Boardman, R. and McCormick, H. (2022) Augmented reality in retail: A systematic review of research foci and future research agenda. *International Journal of Retail and Distribution Management* 50 (4), 498–518.

Chi, C.G.-Q. and Qu, H. (2008) Examining the structural relationships of destination image, tourist satisfaction and destination loyalty: An integrated approach. *Tourism Management* 29 (4), 624–636.

Choi, M.J., Heo, C.Y. and Law, R. (2015) Progress in shopping tourism. *Journal of Travel and Tourism Marketing* 33 (1), 1–24.

Clarke, K. and Belk, R.W. (1979) The effects of product involvement and task definition on anticipated consumer effort. *ACR North American Advances, Association for Consumer Research* 313–318.

Correia, A., Águas, P., Ferradeira, J. and Portugal, J. (2017) Emotional loyalty: The case of Algarve. In M. Kozak, A. Artal-Tur and N. Kozak (eds) *Proceedings of the 9th World Conference for Graduate Research in Tourism, Hospitality and Leisure* (pp. 71–74). Anatolia and Technical University of Cartagena.

Correia, A. and Kozak, M. (2016) Tourists' shopping experiences at street markets: Cross-country research. *Tourism Management* 56, 85–95.

Correia, A. and Kozak, M. (2019) Browsing and shopping. In P. Pearce (ed.) *Tourist Behaviour: The Essential Companion* (pp. 175–198). Edward Elgar.

Correia, A. and Kozak, M. (2023) Luxury product and brand purchasing behavior: Introduction to a Journal of Global Scholars of Marketing Science (JGSMS) special section. *Journal of Global Scholars of Marketing Science* 33 (3), 327–331.

Cugelman, B., Thelwall, M. and Dawes, P. (2009) The dimensions of website credibility and their relation to active trust and behavioural impact. *Communications of the Association for Information Systems* 24, 455–472.

Esfandiar, K., Rahmani Seryasat, M. and Kozak, M. (2024) To shop or not to shop while traveling? Exploring the influence of shopping mall attributes on overall tourist shopping satisfaction. *Tourism Recreation Research* 49 (6), 1411–1426.

Frederick, C. (1929) *Selling Mrs. Consumer*. The Business Bourse.

Gallarza, M.G., Saura, I.G. and García, H.C. (2002) Destination image: Towards a conceptual framework. *Annals of Tourism Research* 29 (1), 56–78.

Garbarino, E. and Johnson, M.S. (1999) The different roles of satisfaction, trust, and commitment in customer relationships. *Journal of Marketing* 63 (2), 70–87.

Han, H., Moon, H. and Kim, W. (2019) The influence of international tourists' self-image congruity with a shopping place on their shopping experiences. *Journal of Hospitality and Tourism Management* 41, 101–109.

Herrero, Á., San Martín, H. and Hernández, J.M. (2015) How online search behavior is influenced by user-generated content on review websites and interactive hotel websites. *International Journal of Contemporary Hospitality Management* 27 (7), 1573–1597.

Heung, V.C. and Cheng, E. (2000) Assessing tourists' satisfaction with shopping in the Hong Kong special administrative region of China. *Journal of Travel Research* 38 (4), 396–404.

Hirschman, E.C. (1992) The consciousness of addiction: A general theory of compulsive consumption. *Journal of Consumer Research* 19, 155–179.

Hosany, S. and Gilbert, D. (2009) Measuring tourists' emotional experiences toward hedonic holiday destinations. *Journal of Travel Research* 49 (4), 513–526.

Huang, S. and Hsu, C.H.C. (2009) Effects of travel motivation, past experience, perceived constraint, and attitude on revisit intention. *Journal of Travel Research* 48 (1), 29–44.

Huang, Z. (Joy), Zhang, C. and Hu, J. (2017) Destination brand personality and destination brand attachment – the involvement of self-congruence. *Journal of Travel and Tourism Marketing* 34 (9), 1198–1210.

Hudson, S. and Ritchie, J.R.B. (2009) Branding a memorable destination experience: The case of Brand Canada. *International Journal of Tourism Research* 11 (2), 217–228.

Jaya, I.P.G.I.T. and Prianthara, I.B.T. (2020) Role of social media influencers in tourism destination image: How does digital marketing affect purchase intention? In *Proceedings of the 3rd International Conference on Vocational Higher Education (ICVHE 2018)*, (pp. 114–120).

Kandampully, J., Zhang, T. (Christina) and Bilgihan, A. (2015) Customer loyalty: A review and future directions with a special focus on the hospitality industry. *International Journal of Contemporary Hospitality Management* 27 (3), 379–414.

Khoo-Lattimore, C.S.C. (2011) The tourism and leisure experience: Consumer and managerial perspectives. *Annals of Tourism Research* 38 (3), 1205–1206.

Kim, D. and Perdue, R.R. (2013) The effects of cognitive, affective, and sensory attributes on hotel choice. *International Journal of Hospitality Management* 35, 246–257.

Kim, H., Kim, J.J. and Asif, M. (2019) The antecedents and consequences of travelers' well-being perceptions: Focusing on Chinese tourist shopping at a duty free. *International Journal of Environmental Research and Public Health* 16 (24), 5081.

Kim, W.G. and Cha, Y. (2002) Antecedents and consequences of relationship quality in hotel industry. *International Journal of Hospitality Management* 21 (4), 321–338.

Kim, Y.-J. and Kim, H.-S. (2022) The impact of hotel customer experience on customer satisfaction through online reviews. *Sustainability* 14 (2), 848.

Kotler, P., Bowen, J., Makens, J. and Baloglu, S. (2021) *Marketing for Hospitality and Tourism* (8th edn). Pearson.

Kozak, M. and Rimmington, M. (2000) Tourist satisfaction with Mallorca, Spain, as an off-season holiday destination. *Journal of Travel Research* 38 (3), 260–269.

Kumar M.D., Govindarajo, N.S. and Khen, M.H.S. (2020) Effect of service quality on visitor satisfaction, destination image and destination loyalty – practical, theoretical and policy implications to avitourism. *International Journal of Culture, Tourism and Hospitality Research* 14 (1), 83–101.

Lacarcel, F.J. and Huete, R. (2023) Digital communication strategies used by private companies, entrepreneurs, and public entities to attract long-stay tourists: A review. *International Entrepreneurship and Management Journal* 19, 691–708.

Lam, I.K.V. and Wong, I.A. (2020) The role of relationship quality and loyalty program in tourism shopping: A multilevel investigation. *Journal of Travel and Tourism Marketing* 37 (1), 92–111.

Lee, S., Jeon, S. and Kim, D. (2011) The impact of tour quality and tourist satisfaction on tourist loyalty: The case of Chinese tourists in Korea. *Tourism Management* 32 (5), 1115–1124.

Lin, Y.H. and Chen, C.F. (2013) Passengers' shopping motivations and commercial activities at airports: The moderating effects of time pressure and impulse buying tendency. *Tourism Management* 36, 426–434.

Liu, Y. (2007) The long-term impact of loyalty programs on consumer purchase behavior and loyalty. *Journal of Marketing* 71 (4), 19–35.

Liu, Z., Wang, A., Weber, K., Chan, E.H.W. and Shi, W. (2022) Categorisation of cultural tourism attractions by tourist preference using location-based social network data: The case of central, Hong Kong. *Tourism Management* 90, 104488.

Maurer, C. (2021) Digital marketing in tourism. In Z. Xiang, M. Fuchs, U. Gretzel and W. Höpken (eds) *Handbook of e-Tourism* (pp. 1–17). Springer.

Myagmarsuren, O. and Chen, C.-F. (2011) Exploring relationships between destination brand equity, satisfaction, and destination loyalty: A case study of Mongolia. *Journal of Tourism, Hospitality and Culinary Arts* 3 (2), 81–94.

McGoldrick, P.J. (1990) *Retail Marketing*. McGraw-Hill.

Mehta, R., Sharma, N.K. and Swami, S. (2013) The impact of perceived crowding on consumers' store patronage intentions: Role of optimal stimulation level and shopping motivation. *Journal of Marketing Management* 29 (7–8), 812–835.

Morgan, N., Pritchard, A. and Pride, R. (2004) *Destination Branding: Creating the Unique Destination Proposition* (2nd edn). Elsevier.

Moutinho, L. (1987) Consumer behaviour in tourism. *European Journal of Marketing* 21 (10), 5–44.

Muro-Rodríguez, A.I., Pérez-Jiménez, I.R. and Sánchez-Araque, J.A. (2020) Impact of shopping tourism for the retail trade as a strategy for the local development of cities. *Frontiers in Psychology* 11, 67.

Narangajavana, Y., Callarisa Fiol, L.J., Moliner Tena, M.Á., Rodríguez Artola, R.M. and Sánchez García, J. (2017) The influence of social media in creating expectations. An empirical study for a tourist destination. *Annals of Tourism Research* 65, 60–70.

Oncioiu, I. and Priescu, I. (2022) The use of virtual reality in tourism destinations as a tool to develop tourist behavior perspective. *Sustainability* 14 (7), 4191.

Ongaya, L. and Muathe, S.M. (2022) Do customer loyalty programs enhance customers buying behaviour in supermarkets in Kenya? *International Journal of Research in Business and Social Science* 11 (3), 1–15.

Palmatier, R.W. and Crecelius, A.T. (2019) The 'first principles' of marketing strategy. *AMS Review* 9, 5–26.

Parasuraman, A., Zeithaml, V.A. and Berry, L.L. (1985) A conceptual model of service quality and its implications for future research. *Journal of Marketing* 49 (4), 41–50.

Pimpão, P., Correia, A., Duque, J. and Zorrinho, J.C. (2014) Exploring effects of hotel chain loyalty program. *International Journal of Culture, Tourism and Hospitality Research* 8 (4), 375–387.

Pimpão, P., Correia, A., Duque, J. and Zorrinho, J.C. (2018) Social diffusion and loyalty programs: a path to succeed. *International Journal of Contemporary Hospitality Management* 30 (1), 475–494.

Qu, H., Kim, L.H. and Im, H.H. (2011) A model of destination branding: Integrating the concepts of the branding and destination image. *Tourism Management* 32 (3), 465–476.

Rajesh, R., (2013) Impact of tourist perceptions, destination image and tourist satisfaction on destination loyalty: A conceptual model. *PASOS: Revista de Turismo y Patrimonio Cultural* 11 (3), 67–78.

Ramseook-Munhurrun, P., Seebaluck, V.N. and Naidoo, P. (2015) Examining the structural relationships of destination image, perceived value, tourist satisfaction and loyalty: Case of Mauritius. *Procedia – Social and Behavioral Sciences* 175, 252–259.

Rather, R.A., Tehseen, S., Itoo, M.H. and Parrey, S.H. (2019) Customer brand identification, affective commitment, customer satisfaction, and brand trust as antecedents of customer behavioral intention of loyalty: An empirical study in the hospitality sector. *Journal of Global Scholars of Marketing Science* 29 (2), 196–217.

Sakas, D.P., Reklitis, D.P., Terzi, M.C. and Vassilakis, C. (2022) Multichannel digital marketing optimizations through big data analytics in the tourism and hospitality industry. *Journal of Theoretical and Applied Electronic Commerce Research* 17 (4), 1383–1408.

Sangkaew, N. and Zhu, H. (2022) Understanding tourists' experiences at local markets in Phuket: An analysis of trip advisor reviews. *Journal of Quality Assurance in Hospitality and Tourism* 23 (1), 89–114.

Sharp, B. and Sharp, A. (1997) Loyalty programs and their impact on repeat-purchase loyalty patterns. *International Journal of Research in Marketing* 14 (5), 473–486.

Sirakaya-Turk, E., Ekinci, Y. and Martin, D. (2015) The efficacy of shopping value in predicting destination loyalty. *Journal of Business Research* 68 (9), 1878–1885.

Sparks, B.A. and Browning, V. (2011) The impact of online reviews on hotel booking intentions and perception of trust. *Tourism Management* 32 (6), 1310–1323.

Timothy, D.J. (2005) *Shopping Tourism, Retailing and Leisure*. Channel View Publications.

Veblen, T. (1899) *The Theory of the Leisure Class*. Houghton Mifflin.

Williams, P. and Soutar, G.N. (2009) Value, satisfaction and behavioral intentions in an adventure tourism context. *Annals of Tourism Research* 36 (3), 413–438.

Wu, M.-Y., Wall, G. and Pearce, P.L. (2014) Shopping experiences: International tourists in Beijing's Silk market. *Tourism Management* 41, 96–106.

Yachin, J.M. (2018) The 'customer journey': Learning from customers in tourism experience encounters. *Tourism Management Perspectives* 28, 201–210.

Yang, F., Huang, A. and Huang, J. (2021) Influence of sensory experiences on tourists' emotions, destination memories, and loyalty. *Social Behavior and Personality: An International Journal* 49 (4), 1–13.

Yüksel, A. (2007) Tourist shopping habitat: Effects on emotions, shopping value and behaviours. *Tourism Management* 28 (1), 58–69.

Yu, W.-J., Hung, S.-Y., Yu, A.P.-I. and Hung, Y.-L. (2024) Understanding consumers' continuance intention of social shopping and social media participation: The perspective of friends on social media. *Information and Management* 61 (4), 103808.

Zeinali, B., Jafarpour, M., Hessam, A., Shojaeivand, B. and Zolghadr, H. (2014) Tourists' satisfaction with and loyalty to shopping experience: A socio-demographic analysis. *International Journal of Academic Research in Business and Social Sciences* 4 (6), 67–84.

Zhong, Y.Y.S., Busser, J. and Baloglu, S. (2017) A model of memorable tourism experience: The effects on satisfaction, affective commitment, and storytelling. *Tourism Analysis* 22 (2), 201–217. https://doi.org/10.3727/108354217X14888192562366

3 Intercultural Perspectives on Shopping and Tourism

Erdogan Koc and Ahu Yazici Ayyildiz

Introduction

With the internationalisation of tourism activities as increasing numbers of customers and suppliers derive from different parts of the world (Koc, 2021a; Mihalič & Fennell, 2015), understanding shopping and tourism from a cross-cultural perspective is necessary. Several studies show that culture may significantly influence tourists' expectations and perceptions (Koc & Ayyildiz, 2021, 2022; Zhang *et al.*, 2015). In addition, as tourism and shopping activities may involve relatively intense and frequent social interactions, mostly intercultural between the customers and the service providers (Cooper *et al.*, 2021; Koc & Ayyildiz, 2021, 2022), the need to understand the implications of cross-cultural factors for tourism and shopping has increased significantly.

Given this context, this chapter explains and discusses the influence of culture on tourism and shopping experiences from a dyadic perspective, or from the demand perspective (the cultures of tourist-shoppers), and the supply perspective (the cultures of destination retailers). The explanations and discussions in the chapter are organised around the seven marketing mix elements (7Ps), namely product, price, place, promotion, physical evidence, people and process (Ayyildiz & Ayyildiz, 2021). Contrary to what some may believe, the marketing mix framework not only represents the demand – the customer and marketing perspective – but also the supply, organisational behaviour and management perspective as more evident, particularly with regard to the elements of *people*, *process* and *physical evidence*.

Culture, Shopping and Tourism

Culture has a pervasive influence on people's perceptions, beliefs, attitudes and behaviours (Hofstede *et al.*, 2010; Koc, 2021a) and hence may influence almost all aspects of tourists' shopping experiences, their interactions and decisions. Thus, decision makers need to understand how each cultural characteristic may influence certain elements of their operations, activities and processes. Litvin *et al.* (2004) and Litvin and Kar (2003) demonstrate that tourists' purchasing and consumer activities are congruent with their self-image, largely shaped by their cultural characteristics. Likewise, Crotts and Erdmann (2000) argue that tourists tend to think and behave in accordance with their mental programming which, again, is shaped by their cultural background.

The differences among customers in terms of their cultural or national characteristics have been noted by several researchers. For instance, Su *et al.* (2018) show that tourists from masculine cultures (e.g. Japan, USA, UK and Germany) and high power-distance cultures (e.g. Russia, China, India and Japan) are more likely to allocate relatively more money to shopping while on holiday. Likewise, studies such as the one by Kim *et al.* (2011) demonstrate that, whereas American, Canadian and European tourists are more likely to be interested in cultural attractions, the Japanese are more inclined to spend more time and money on shopping (Jang *et al.*, 2004; Kim *et al.*, 2011; Rosenbaum & Spears, 2006).

Against this background, the chapter explains and discusses the influence of culture on retail behaviour from the perspective of cultural types developed by Hall (1977) and Hofstede *et al.* (2010) (see Table 3.1). This is because the cultural dimensions developed by these researchers are among the most frequently studied cultural dimensions in the social sciences (Cardon, 2008; Kumar & Dhir, 2021; Nam *et al.*, 2014; Portugal Ferreira *et al.*, 2014). For instance, Portugal Ferreira *et al.* (2014) studied the frequency of cross-cultural dimensions used in top-tier journals and discovered that more than 75% of the studies were based on Hofstede *et al.*'s. (2010) and Hall's (1977) concepts. Their work forms the essence of the conceptual framework of this chapter, but from time to time, references to other cultural characteristics such as humane orientation and performance orientation (House *et al.*, 2004) are made in this chapter.

The Influence of National Culture on Consumer Behaviour, Shopping and Tourism

Although tourists may purchase a wide variety of products including things like luxury goods, as bought frequently by the Japanese (Correia *et al.*, 2019; Rosenbaum *et al.*, 2005; Spears & Rosenbaum, 2012), Dimanche (2003) argues that tourist shopping is related mainly to the purchase of souvenirs or items that may not be easily available or expensive in a tourist's home country. As much as 78% of souvenir purchases are bought for family members and friends (Amaro *et al.*, 2020) and often constitute a substantial proportion of tourists' expenditures, comprising a significant portion of tourism-induced income to destinations (Decrop & Masset, 2014; Swanson & Timothy, 2012; Timothy, 2005).

In addition, souvenirs are important in helping tourists identify with the destination (Collins-Kreiner & Zins, 2011; Swanson & Timothy,

Table 3.1 Characteristics and dimensions of national cultures*

Cultural dimension	Abbreviation	Characteristics	Prominent examples
High-context	HC	Behavioural norms, rules and values are communicated not only by words but also by contextual cues such as tone of voice, body language, facial expressions and extant knowledge.	Brazil, Taiwan, France, Spain
Low-context	LC	Straightforward and more explicit in communicating, using words and clear expressions.	Canada, USA, Germany, Sweden
Power-distance (high)	PDI	Social distances are maintained between leaders and subordinates, and class distinctions are often emphasised.	Saudi Arabia, Indonesia, France, China
Individualism	IDV	People are more self-oriented and are concerned foremost with their own well-being.	USA, Netherlands, Germany, Australia
Collectivism	COL	People are more concerned about how they fit in with others, their role in society and how they care for the greater social good.	Japan, Turkey, Mexico, China
Uncertainty avoidance (high)	UA	Predictability and control are key in high UA cultures. Uncertainties and controlled situations are generally avoided.	Russia, Finland, South Korea, Greece
Indulgence	IND	Societies spend more money on luxury items and time in leisure pursuits. Having friends and portraying a sense of happiness and success are common.	Canada, USA, Nigeria, Chile
Restraint	RES	Restrained societies tend to control personal desires and withhold pleasures to align more with societal norms. Positive emotions are less freely expressed, and leisure and luxury are not as valued.	Russia, China, Japan, Pakistan
Masculinity	MAS	Masculine cultures emphasise gender roles, assertiveness, personal ambitions and egos and the pursuit of wealth.	Austria, India, Mexico, Hungary
Femininity	FEM	Feminine cultures demonstrate nurturing and caring behaviours, gender equality, personal relationships, environmental concern and quality of life.	Sweden, Denmark, Thailand, Costa Rica
Long time-orientation	LTO	Virtues oriented toward the future, including thrift, perseverance and preparedness.	Finland, New Zealand, Argentina, Malaysia

* Note: These types are not mutually exclusive, and the list is not exhaustive.
Source: Based on Hall (1977, 2000) and Hofstede et al. (2010).

2012), increase the popularity and memorability of a destination (Ansari *et al.*, 2019; Dumbrovská & Fialová, 2020), and form revisit intentions (Sthapit & Björk, 2019). Souvenir purchases are also important because they provide the first impetus for tourists to initiate shopping more generally. After first buying traditional souvenirs, tourists may turn to purchasing other items, including clothes, jewellery, books, arts and crafts, duty-free goods and electronics (García-Milon *et al.*, 2021; Josiam *et al.*, 2005; Timothy & Butler, 1995). Hence, the cross-cultural discussion below starts with souvenirs and gift-buying behaviour, by using the marketing mix perspective noted previously.

The product

Tourists' self-images (Litvin & Kar, 2003; Litvin *et al.*, 2004) and their mental programming (Crotts & Erdman, 2000) are both largely shaped by their cultural characteristics, including national culture, and these influence the types of products and services they purchase, and how, when and why they buy. The cultural distance between tourists' cultures and the destination culture is likely to increase tourists' interest in purchasing souvenirs that demonstrate local uniqueness (Su *et al.*, 2018). Conversely, when tourists visit similar cultures, as in the case of mainland Chinese tourists visiting Hong Kong, they tend to shop for utilitarian products such as food, groceries and clothes (Timothy & Butler, 1995), and for many Chinese travellers in Hong Kong and elsewhere, brand significance is a key motivator in buying daily use merchandise and souvenir items (Hung *et al.*, 2021).

Research shows that because of their group orientations, customers from COL cultures (e.g. China, Russia and Japan) and LTO cultures (e.g. Japan, China and Singapore) tend to buy more gifts, have more gift-giving obligations and allocate more money for gift buying than customers from IDV cultures (e.g. USA, Australia, the UK, the Netherlands and Sweden) (Lu *et al.*, 2016; Manrai & Manrai, 2011; Reisinger, 2009; Timothy, 2005). Customers from IDV cultures are less inclined to buy and give gifts and have fewer gift-giving obligations, owing to the low level of pressure they feel to reciprocate (Park, 1998).

Tourists from high PDI cultures (e.g. Russia, China, India and Japan) consider tangible objects to be more significant than intangible memories (Mou *et al.*, 2019; Nath *et al.*, 2016). The grandeur of tangible items satisfies the needs of customers from high PDI cultures for status and hierarchy as they may give the impression of high status, luxury, exclusivity and elitism. Hence, destinations catering to tourist-shoppers from high PDI cultures ought to pay more attention to tangible merchandise and retail environment, ranging from the design of products to the physical atmosphere in and around the retail outlets.

Coming from a high PDI culture, the Japanese tend to place greater importance on product availability and relative value (Keown, 1989) and prefer duty-free shops (Rosenbaum & Spears, 2006). Perhaps because of their high PDI and RES cultural backgrounds, the Japanese tend to attach more value to the products they buy – what they can own, particularly with regard to famous brands, rather than the shopping experience (Hobson & Christensen, 2001). People from high PDI cultures also appear to have higher product expectations. Hence, they may be more critical of products and retail experiences and may be more difficult to please (Ladhari *et al.*, 2011; Mou *et al.*, 2019; Nath *et al.*, 2016). To this end, businesses that offer basic products and services with only must-have elements may appeal far less to customers from high PDI cultures (Nath *et al.*, 2016).

Regarding product design and the physical environment in and around retail outlets, it is also important to note that customers from HC cultures, including Central Europeans, Southeast Asians, Arabs and people from the Mediterranean, Latin America and Africa – in fact, almost 70% of the world (Tung, 2002), experience more sensory involvement in dining, entertainment, socialising and shopping (Becker, 2000). The tangibility of objects and environments is also important in reducing customers' risk perceptions (Koc, 2013a). This is especially the case for consumers from high UA cultures (e.g. Russia, Japan, France, Turkey and Italy). UA culture significantly influences shopping and tourism, as tourists from uncertainly avoidance cultures usually prefer all-inclusive package holidays. During these sorts of vacations, especially sun–sea–sand holidays, customers may lack shopping opportunities, depending on how isolated their resorts are, or the inclination to leave the resort for shopping purposes. It should also be noted that in holiday package preferences, customers from high UA cultures tend to prefer low-risk activities such as staged events and shopping (Manrai & Manrai, 2011; Reisinger & Crotts, 2010). As uncertainty avoidance may influence shopping in tourism positively or negatively, retail products and the shopping environment should be designed carefully, and the risk reduction features of the merchandise need to be made appropriately more salient through marketing communications.

While customers from high UA cultures and COL cultures may tend to prefer prepaid package holidays, customers from IDV cultures prefer individual holidays and tend to be more interested in discovering the destination more deeply (Koc, 2007), being exposed more to shopping environments and therefore more likely to engage in shopping activities. Moreover, whereas customers from IDV cultures tend to pursue variety and innovativeness in their purchase and consumption preferences, people from COL and high PDI cultures may express more loyalty and seek value stability (Manrai & Manrai, 2011).

Based on the fact that customers from IDV cultures expect more personalisation (Muller et al., 2003; Staus & Meng, 1999) and prefer variety-seeking and innovation (Manrai & Manrai, 2011), tourism and retail practitioners should ensure personalized shopping opportunities, not only in terms of the products and services, and their varieties alone, but also in terms of the treatment of and interaction with these customers.

The price

Price is an important element of the marketing mix as it determines how many of a product or service, if any, will be purchased by shoppers. This includes in shopping tourism settings, as prices may deter people from buying a souvenir or purchasing many to give away at home. Many research findings show how customers' cross-cultural characteristics influence their perceptions, attitudes and behaviours towards certain aspects of pricing. For instance, Correia et al. (2011) conclude that tourist-consumers from COL cultures are more likely to be price-conscious and make their decisions based on brand, price and number of alternatives, while IDV customers emphasise factors such as novelty and individuality. In a similar study, Jeong et al. (2019) compared consumers from COL and HC cultures with IDV and LC cultures (e.g. Western Europeans, Scandinavians and North Americans) and found that customers from the former group placed more importance on the price–quality relationship than customers in the latter group. Similarly, tourists from future-oriented cultures (e.g. Singapore, Switzerland, the Netherlands, Sweden, the UK and USA) are more likely to do a mental cost–benefit analysis (value for money) when they purchase products and services (Bergadaa, 1990; Lu et al., 2016).

Research relating to consumers' reactions towards price also shows that those from high PDI cultures have a greater tendency to use price to judge the quality of products and services (Lalwani & Forcum, 2016). Similarly, Nath et al. (2016) discovered that Chinese and Malaysian tourists, both from PDI cultures, were more likely to believe that higher prices mean higher quality. Interestingly, as prices get higher, these shoppers' positive expectations also increased, and they became happier (Lalwani & Forcum, 2016). As higher prices are sometimes indicative of superior quality, customers from high PDI cultures tend to be happier with their display of social status (Nath et al., 2016). Customers from high PDI cultures usually refrain from bargain hunting or haggling, as such a behaviour may be unworthy of their aspirations for status and prestige (Bathaee, 2011).

Similar to consumers from high PDI cultures, consumers from high UA cultures are also comfortable with displays of socioeconomic superiority as reflected in high prices (Sabiote-Ortiz et al., 2016). This is because a high level of risk aversion correlates with a desire for order and structure, which they believe may be established with more expensive merchandise (Sabiote-Ortiz et al., 2016). Likewise, Mattila and Choi (2006) discovered that the display of price information had a positive impact on South Koreans' (a high UA culture) level of satisfaction, though the satisfaction level of North Americans (low UA) remained stable upon exposure to the same information. Koc (2013b) argues that the provision of information (e.g. regarding price details) satisfies customers' need for control and leads to success in service businesses, especially those that target customers from high UA cultures. Furthermore, as customers from LTO cultures are less likely to make impulse purchases (Kwak *et al.*, 2004), are more prudent and are more inclined to delay gratification when shopping (Park *et al.*, 2013), price discounts and merchandising activities may be less effective on these customers.

The place and the physical environment

The place refers to the location where products and services are sold or made available to consumers (Koc, 2021a). In tourist destinations, this can include a wide range of settings, such as malls, markets, souvenir vendors, museum and heritage site shops and similar localities. Decisions regarding where products and services are made available for purchase must consider the physical environment as well (Koc, 2021b). For this reason, place and environment-related cross-cultural characteristics are discussed here together. For instance, Zein (2015) finds that parents in low UA cultures are more willing to leave their children to play alone, whereas parents from high UA cultures look after their children meticulously and rarely let them out of their sight. Retail outlets might take this into account and organise their retail space in a way that allows parents to watch over their children while shopping, having the peace of mind to concentrate on their retail activities.

As previously noted, the opulence and luxuriousness of products and retail environments appeal to the social status aspirations of people from high PDI cultures (Correia *et al.*, 2011). Thus, the level of luxury and brand pervasiveness may also be important factors in influencing the satisfaction of high PDI consumers. Likewise, customers from high UA cultures place greater importance on the tangible elements of shopping as well (Hsieh & Tsai, 2009). However, unlike customers from high PDI cultures who are interested in status and prestige, high UA cultures see luxurious merchandise and lavish environments as an assurance of reduced risk and higher quality (Hsieh & Tsai, 2009). Hence, retailers ought to consider developing strategies to appease risk-averse customers' risk perceptions. For instance, they might offer three types of control to reduce the risk perceptions of customers from high UA cultures (Koc, 2013a): cognitive control (provision of more relevant and high-quality information – e.g. prices and construction material), decisional control (giving customers plenty of alternatives and opportunities to make choices) and behavioural control (e.g. offering customers opportunities to revoke transactions and return or replace an item easily) (Koc, 2021a). In addition, retail spaces should be designed to allow customers to feel the products with as many senses as possible to reduce their misgivings and risk perceptions. Koc (2021a) argues that customers from UA cultures prefer retail experiences that allow the use of as many senses as possible.

The design of retail outlets from the perspective of proxemics, or the amount of space or physical distance people feel is necessary to have between themselves and others, is also important. According to Hall (1977), people from LC cultures tend to value proxemics and desire to maintain a relatively greater physical distance with strangers. Conversely, people from HC cultures are more relaxed and casual regarding the use and 'infringement' of social space. Hence, stores catering to tourists from Western low-context cultures should be designed to allow customers to feel they have enough space between themselves, other customers and retail staff.

Koc (2020) suggests that, different to tourist-customers from RES cultures, customers from IND cultures collect and digest more information about their leisure activities and tend to be more interested in the 'authenticity' of the destination (Koc, 2021a). This means that customers from IND cultures are more likely to shop in traditional, local, and 'authentic' places, such as the Grand Bazaar in Istanbul, Turkey. The fact that Japanese shoppers, a RES culture-based market, prefer shopping in duty-free shops (Rosenbaum & Spears, 2006) supports Koc's (2020) statement above. Customers from RES cultures, such as Japanese and Russians, tend to concentrate on acquiring luxury brands, trademarks and well-known images to reinforce their social standing. In doing this, they may lose the value of the immersive shopping experience and the heritage opportunities often associated with it. Conversely, customers from IND cultures may enjoy the shopping experience more, so that shopping at a duty-free outlet may be viewed as dull and unexciting.

Promotion

Promotion is probably the most salient element of the marketing mix that reflects cross-cultural differences. In tourism settings, these differences are critical to understand to cater to the needs of different tourist markets. However, the explanations, findings and examples presented below should not be thought of only from the perspective of promotion, as they may express customers' preferences pertaining to other marketing mix elements as well. For instance, the individuality, independence and distinctiveness themes in advertisements resonates positively with the customers from IDV cultures (Pollay & Gallagher, 1990), but these do not influence the promotion and marketing communications element alone. The preference for individuality, independence and distinctiveness in IDV cultures may be reflected in consumers' preferences for products, prices, places, physical environments, processes and people.

As reflected in Pollay and Gallagher's (1990) findings, Albers-Miller and Gelb (1996) also discovered that customers from IDV cultures seem to value independence and distinctiveness cues in a positive manner, while they evaluated cues such as affiliation, family and community in a less positive manner. Research on advertising themes has revealed that customers from high UA cultures view safety, tameness and durability in a positive manner (Albers-Miller & Gelb, 1996; Cheong et al., 2000). However, shoppers from FEM cultures consider characteristics such as natural, frail and modest more positively, customers from MAS cultures consider advertisement cues such as effective, convenient and productive more positively (Cheong et al., 2000).

Time orientation reflects itself in terms of customers' responses towards the types of sales promotion campaigns as well. Consumers from IDV and short-term-oriented cultures, such as Australia, the US, Norway and Denmark, may have a greater proclivity to respond positively to heavy price discounts, whereas customers from LTO cultures may be less content with price discounts (McNeill, 2006; McNeill et al., 2014). Research also shows that, while people from FEM cultures do not respond positively towards monetary-based sales promotions, customers from MAS cultures may seek sales promotions (McNeill, 2006; McNeill et al., 2014).

Swaminathan's (2012) study finds that, compared with low PDI American advertisements, high PDI Chinese advertisements more frequently contain the advertising cues of ornamental, status and costly. In India, a high PDI culture, Parker (2003) found that the cues of ornamental, vain, expensive and high status were seen positively by Indian consumers, whereas the themes of humility, nurturing and ordinary/plain were viewed negatively. Nath et al. (2016) compared high PDI culture advertisements (Chinese and Malaysian) with low PDI culture advertisements (the British) and concluded that symbols denoting expensiveness were more common and successfully used in the high PDI context because these symbols reflect status, prestige and power.

Various other symbols or cues may be relevant for other cultures. For instance, Rojas-Méndez et al.'s (2017) study finds that the cues of altruism, benevolence, kindness, love, generosity, affiliation, belongingness and personal and family relationships in advertisements educed a positive assessment of an advertisement among consumers humane cultures. Conversely, cues related to achievement, urgency, results and control elicited positive evaluations of an advertisement by both high- and low-performance-oriented cultures (Rojas-Méndez et al., 2017). However, the difference between high- and low-performance-oriented cultures was that, for shoppers from high-performance-oriented cultures, much stronger cues were necessary for them to view the advertisements as performance oriented (Rojas-Méndez et al., 2017).

Some studies on time orientation suggest that cues like health and nutrition are considered more important by LTO customers than short-term oriented cultures (Diehl *et al.*, 2016; Parker, 2003). Mattila (1999) compared the UK, Saudi Arabia, Thailand and Chile and demonstrated that customers from past-oriented cultures were more likely to avoid advertisements. Present- and future-oriented customers, however, appear to have a more positive attitude towards advertisements. However, present- and future-oriented customers expressed different reasons for having positive attitudes towards advertisements. Future-oriented consumer attitudes were largely utilitarian, as they thought the advertisements may be beneficial for them to learn about products and some of the questions they had (Mattila, 1999). Present-oriented customers held positive attitudes towards advertisements for hedonic reasons, such as just enjoying the advertisements' entertainment value (Mattila, 1999). Regarding hedonism, customers from low PDI cultures tend to assign significant value to the hedonic aspects of consumption, compared with customers from high PDI cultures (De Mooij, 2010; Koc & Ayyildiz, 2022).

In line with a hedonistic orientation, humour was more frequent in low PDI culture ads compared to those shown in high PDI cultures, as they are more likely to parody themselves (Humprey, 2018). Likewise, particularly well aligned with the 'people' category below, smiling at strangers, as in the case of shop assistants smiling at customers, may be seen as suspicious in high PDI cultures (Hofstede *et al.*, 2010). Likewise, people in RES cultures are far less comfortable with smiling at people they do not know (Hamid, 2016).

When McDonald's opened up its first restaurant in Moscow in 1992, the staff were trained to smile at the customers. However, the smiles attracted suspicion among the high PDI- and RES-oriented Russian customers (Koc, 2021a). Zhao's (2017) study of Chinese webpages also found relatively few smiles, as most faces bore serious expressions, which is in line with their RES and high PDI orientations. A study of Japanese advertisements, another high PDI and RES culture, discovered that the commercials contained a several fear, rules-oriented and formality cues rather than expressions of joy (Huang *et al.*, 2021).

Similar to refraining from humorous expressions in high PDI cultures, promotions in MAS cultures tend to refrain from nudity and sexually suggestive content, as these societies generally have stricter taboos related to sex and nudity (Nelson & Paek, 2005). Koc (2020) suggests that refraining from nudity and sexual public behaviour in advertising may be common in RES cultures as well, as the social tendencies there tend to restrict public expressions of pleasure, enjoyment, sexual gratification, spending and consumption.

Hall's (1977) HC–LC framework is probably the most relevant paradigm for the promotion element, as it primarily has to do with communication and interaction. For instance, Würtz (2005) suggests that advertisements in LC cultures are rather direct and informative, while in HC cultures they tend to be less direct and informative. In HC cultures, information and communication of meaning are likely to be implied rather than stated overtly (Hofstede *et al.*, 2010; Koc, 2021c). In relation to context and advertisements, Le Pair and Van Mulken (2008) discovered that HC French and Spanish customers enjoy complex advertisement messages, as opposed to the LC Dutch customers, who did not. Similarly, Hornikx and Le Pair (2017) compared Dutch consumers (LC) with Belgian consumers (HC) and discovered that the former had much lower preferences for visual metaphors than their Belgian neighbours, who were more able to understand indirectly expressed meanings.

Studies on how information is used by customers are important as marketing communications play an important role in determining the ability of a business to reach its customers efficiently and effectively. In one such study, Seo *et al.* (2018) discovered that, while American customers (low UA) resorted to actual knowledge and relevant attributes, South Korean consumers (high UA) depend more on their own past experience to reduce risk and make purchasing decisions. This attitude of high risk-averse South Koreans may be caused by an illusion of control, a tendency to believe that an individual's own decisions and preferences may result in more positive outcomes (Koc, 2013a). Illusion of control may occur from decisional control, or the ability or opportunity of a customer to make free choices regarding a product or service (Ayyildiz *et al.*, 2024)

The process

The process element in shopping tourism may relate to sequential activities such as arriving at the shop, browsing, viewing product displays, interacting with shop assistants, queuing and paying. In each stage, customers' perceptions, beliefs, attitudes and behaviours may be influenced by their cultural backgrounds and their touristic surroundings. For instance, a comparative study of customers from a high PDI culture and a low PDI culture demonstrates that in the former group, consumers were unhappy about the service providers' attempts to initiate interaction (Lee, 2015). This was primarily a result of these customers viewing themselves as socioeconomically superior to the employees. However, customers in the latter group, from the low PDI society, perceived themselves to be essentially equal to the service providers and did not seem to mind employees' attempts to initiate interaction (Lee, 2015). This finding aligns with Hofstede's (2011) assertion that high UA and high PDI nationals expect to maintain greater social distance and more formal customer–merchant interactions.

Swanson *et al.*'s (2011) study finds that customers from LTO societies value responsiveness and empathy in commercial interactions, whereas customers from IDV, MAS and LC societies tend to value technical quality more (Mattila, 2000; Yüksel & Yüksel, 2001). Customers from LC cultures attach more value to efficiency and task completion, while customers from

HC societies value to the quality of interaction more (Mattila, 2000; Yüksel & Yüksel, 2001). As consumers from LC cultures value technical quality of service, such as efficiency and time savings, they may be more sensitive towards being kept waiting during the shopping process. Bilgili *et al.* (2020) and Ozkul *et al.* (2020) show that factors in the service environment, including the colour of lighting, may affect customers' perceptions of wait times and their overall assessments of service quality.

After a service failure, the way customers respond to service recovery attempts may vary according to their cultural characteristics. Customers from RES cultures have a greater proclivity to remember negative events, whereas consumers from IND cultures have a greater tendency to remember positive emotions and events (Koc *et al.*, 2020). Koc *et al.* (2020) discovered that customers from one RES culture (i.e. Russia) were not likely to experience the service recovery paradox – when customer satisfaction is higher after a service failure and subsequent genuine apology and compensation than if it had not happened at all (McCollough, 2009) – because they were more inclined to remember negative events. The expectations of customers in service recovery also vary according to culture. For instance, when an apology is made to ensure interactional justice, customers from high PDI cultures expect the apology to come from someone of authority, such as a manager or owner rather than a frontline employee (Mueller *et al.*, 2003; Patterson *et al.*, 2006).

The people

Doyle (2008) suggests that dissatisfaction with product quality may account for 14% of all customers changing product use, whereas problems arising from interactions between buyers and service staff may account for as much as 67% of all customer switching to other service providers. Shopping tourism activities may involve intense and frequent interactions between tourists and retailers. Likewise, retail marketing and human resource management activities may be significantly intertwined (Pellegrini *et al.*, 2018). Hence, the people element, or human resource management activities in tourism retailing may be more important than it is in other sectors, particularly with regard to satisfaction and first impressions.

It was noted earlier that people from LC cultures prefer to maintain greater physical distance from strangers. For instance, if a shop clerk from an HC culture (roughly 70% of the world) (Tung, 2002) is not aware of this cultural fact, he or she may stand too close to a customer from an LC culture while explaining the details of a product. This situation may be off-putting and make an LC customer uncomfortable, even causing the shopper to leave the store without making a purchase.

Perceptions of time also vary between cultures. People from sequential time cultures (i.e. monochronic cultures, such as western Europeans and

North Americans) prefer order and precision. They value being on time as they consider time to be a dear commodity not to be 'wasted'. People from monochronic cultures prefer to plan ahead and expect others to adhere to these plans and deadlines, and tasks are handled one by one sequentially. Monochronic people visiting polychronic cultures where time constraints are less relevant often have a hard time coping with the relaxed nature of time accounting, such as northern Europeans visiting the Mediterranean region.

Hall (2000) asserts that a culture's time orientation, whether monochronic or polychronic, may determine how people behave in their social and work lives. People from a polychronic culture (e.g. Southern Europe, South America and many parts of Asia) may perceive the future, present and the past as interconnected, and they may be dexterous enough to work on several projects simultaneously. Several studies reveal certain advantages of employee polychronicity or multitasking, ranging from service orientations to service quality in various service sectors including retail and tourism (Arndt et al., 2006; Carraher et al., 2009; Daskin et al., 2015; Liu et al., 2021).

Wen et al. (2020) argue that, owing to work requirements, service employees are expected to behave in a polychronic manner – multitasking. In this work environment, monochronic people will often experience stress and burnout at work, resulting in poor and inefficient work (Wen et al., 2020). Hence, retail businesses, when appropriate, may recruit staff from polychronic cultures, or specifically train monochronic staff to cope better with the polychronic nature of the retail business. Similar to polychronicity and moonchronicity, Koc (2020) asserts that as people in RES cultures assign less importance to fun, leisure and hedonic activities (Hoffstede et al., 2010), they may find it difficult to internalise the specific requests and aspirations of customers and may perceive them as bizarre. Hence, managers in RES societies may wish to establish training programmes to enable their employees to internalise tourist-shoppers' requests and aspirations. Also, as people from RES cultures may be more cynical and pessimistic (Hofstede et al., 2010), retail managers can focus on recruiting, appraising and training appropriate staff.

Furthermore, the empowerment of frontline employees in retailing is crucial in responding to the specific needs of customers and solving problems that may arise on a daily basis (van Esch et al., 2020; Wilder et al., 2014). From a cultural perspective, Magnini et al. (2013) find that employees in COL societies are less comfortable with individual empowerment. Likewise, Koc's (2013b) study of the communication styles between service employees and their superiors in IDV–COL and high and low PDI cultures shows marked differences between the two groups in terms of reporting service failures to their superiors. In IDV and low PDI cultures (e.g. the UK), employees readily communicate service failures directly to their managers, but in COL and high PDI societies (e.g. Turkey) employees avoid communicating directly, instead using mitigated speech to communicate service failures to their managers. This means that in the case of the latter group delays may occur in managers becoming aware of problems. Consequently, in COL and high PDI cultures, owing to delays in responding to service failures, problems often arise related to customer satisfaction, customer loyalty and maintaining service quality (Koc, 2013b).

From a marketing perspective, several studies show the importance of corporate social responsibility in establishing a competitive advantage (Hanaysha, 2020; Schill & Godefroit-Winkel, 2022; Vlachos *et al.*, 2010). Corporate social responsibility appears to be determined by the cultural orientations of staff. For instance, Filimonau *et al.* (2018) and Kang *et al.* (2016) show that people from MAS and IDV cultures have less interest in the environment and corporate social responsibility.

Research about the influence of culture on the service employees' behaviour is expansive (Cooper *et al.*, 2021). It is difficult to include all perspectives in one book chapter, but the elements of the marketing mix may be looked at from the perspective of what service employees need to know in retailing and may be used to develop training programmes. For instance, as people from RES cultures are less likely to smile and accept overly friendly behaviours on the part of employees (Zhao, 2017), service staff should be trained to be personable with some customers but not all, including not smiling too much at customers from RES cultures, who may feel a sense of suspicion.

Conclusions

This chapter explains and discusses the ways in which cross-cultural characteristics influence people's consumer behaviour. The influence of culture has been examined both from demand and supply perspectives. This essay shows the importance of developing cultural sensitivities and cultural competence among service employees and managers so that they can successfully provide high-quality retail experiences for their international customers. However, as Koc's (2021b, 2021c) study shows, self-reported scales used to identify people's intercultural competence in various human resource roles, ranging from recruitment to training, may prove to be futile. This is because, in general, people tend to inflate self-efficacy and end up exaggerating their abilities in self-reported measurements (Koc, 2021c). Hence, these measurement exercises may not be reliable and useful.

The main message here is the need for continuity of research to discover different aspects of culture that influence the ways in which tourists shop. Hence, in addition to for tourism and retail practitioners, and government officials, the chapter provides several recommendations for future academic research as well. For instance, as previously noted, customers from LTO cultures are more likely to view negatively retail employees' efforts to interact and to be friendly (Lee, 2015). Thus, future research should investigate how people from different cultures react to consumer participation (e.g. co-creation), as these activities may necessitate higher levels of interaction with service providers (Koc & Ayyildiz, 2022). Also, as certain customers view interaction with staff members negatively, their attitudes towards service robots may differ.

Future research also ought to investigate issues relating to proxemics. Customers from HC cultures may wish to maintain a greater physical distance with others (Hall, 1977). Researchers may explore whether both HC and LC customers' preferences for greater personal space have been extended after the COVID-19 pandemic. Finally, researchers should fruitfully investigate whether the low probability of the occurrence of the service recovery paradox in RES cultures is dependent on the type of recovery, or the compensation used. Customers from RES cultures may expect distributive justice and excessive compensation, and in the absence of these, adequate service recovery may not occur.

References

Albers-Miller, N. and Gelb, B. (1996) Business advertising appeals as a mirror of cultural dimensions: A study of eleven countries. *Journal of Advertising* 25, 57–70.
Amaro, S., Morgado Ferreira, B. and Henriques, C. (2020) Towards a deeper understanding of the purchase of souvenirs. *Tourism and Hospitality Research* 20 (2), 223–236.
Ansari, F., Jeong, Y., Putri, I.A. and Kim, S.I. (2019) Sociopsychological aspects of butterfly souvenir purchasing behavior at Bantimurung Bulusaraung National Park in Indonesia. *Sustainability* 11 (6), 1789.
Arndt, A., Arnold, T.J. and Landry, T.D. (2006) The effects of polychronic-orientation upon retail employee satisfaction and turnover. *Journal of Retailing* 82 (4), 319–330.
Ayyıldız, A.Y. and Ayyıldız, T. (2021) *Postmodern Pazarlama ve Turizm*. Detay Yayıncılık.
Ayyıldız, T., Ayyıldız, A.Y. and Koc, E. (2024) Illusion of control in service failure situations: Customer satisfaction/dissatisfaction, complaints, and behavioural intentions. *Current Psychology* 43, 515–530.
Bathaee, A. (2011) *Culture Affects Consumer Behavior: Theoretical Reflections and an Illustrative Example with Germany and Iran*. Universität Greifswald, Rechts- und Staatswissenschaftliche Fakultät.
Becker, C. (2000) Service recovery strategies: The impact of cultural differences. *Journal of Hospitality & Tourism Research* 24 (4), 526–538.
Bergadaa, M. (1990) The role of time in the action of the consumer. *Journal of Consumer Research* 17 (3), 289–302.
Bilgili, B., Ozkul, E. and Koc, E. (2020) The influence of colour of lighting on customers' waiting time perceptions. *Total Quality Management & Business Excellence* 31 (9–10), 1098–1111.
Cardon, P.W. (2008) A critique of Hall's contexting model: A meta-analysis of literature on intercultural business and technical communication. *Journal of Business, Communication & Technology* 22, 399–428.
Carraher, S.M., Parnell, J. and Spillan, J. (2009) Customer service-orientation of small retail business owners in Austria, the Czech Republic, Hungary, Latvia, Slovakia, and Slovenia. *Baltic Journal of Management* 4 (3), 251–268.

Cheong, Y., Kim, K. and Zheng, L. (2000) Advertising appeals as a reflection of culture: A cross-cultural analysis of food advertising appeals in China and the US. *Asian Journal of Cummunication* 20, 1–16.

Collins-Kreiner, N. and Zins, Y. (2011) Tourists and souvenirs: Changes through time, space and meaning. *Journal of Heritage Tourism* 6 (1), 17–27.

Cooper, M.A., Camprubí, R., Koc, E. and Buckley, R. (2021) Digital destination matching: Practices, priorities and predictions. *Sustainability* 13 (19), 10540.

Correia, A., Kozak, M. and Ferradeira, J. (2011) Impact of culture on tourist decision-making styles. *International Journal of Tourism Research* 13 (5), 433–446.

Correia, A., Kozak, M. and Kim, S. (2019) Investigation of luxury values in shopping tourism using a fuzzy-set approach. *Journal of Travel Research* 58 (1), 77–91.

Crotts, J.C. and Erdmann, R. (2000) Does national culture influence consumers' evaluation of travel services? A test of Hofstede's model of cross-cultural differences. *Managing Service Quality* 10 (6) 410–419.

Daskin, M., Arasli, H. and Kasim, A. (2015) The impact of management commitment to service quality, intrinsic motivation and nepotism on front-line employees' affective work outcomes. *International Journal of Management Practice* 8 (4), 269–295.

de Mooij, M. (2010) *Consumer Behavior and Culture: Consequences for Global Marketing and Advertising*. Sage.

Decrop, A. and Masset, J. (2014) 'This is a piece of coral received from Captain Bob': Meanings and functions of tourist souvenirs. *International Journal of Culture, Tourism and Hospitality Research* 8 (1), 22–34.

Diehl, S., Terlutter, R. and Mueller, B. (2016) Doing good matters to consumers: The effectiveness of humane-oriented CSR appeals in cross-cultural standardized advertising campaigns *International Journal of Advertising* 35, 730–757.

Dimanche, F. (2003) The Louisiana tax free shopping program for international visitors: A case study. *Journal of Travel Research* 41 (2), 311–314.

Doyle, P. (2008) *Value-Based Marketing: Marketing Strategies for Corporate Growth and Shareholder Value* (2nd edn). Wiley.

Dumbrovská, V. and Fialová, D. (2020) The city of one hundred souvenir shops: Authenticity of souvenirs in Prague. *Journal of Tourism and Cultural Change* 18 (2), 187–200.

García-Milon, A., Olarte-Pascual, C., Juaneda-Ayensa, E. and Pelegrín-Borondo, J. (2021) Tourist purchases in a destination: What leads them to seek information from digital sources? *European Journal of Management and Business Economics* 30 (2), 243–260.

Portugal Ferreira, M., Li, D., Rosa Reis, N. and Ribeiro Serra, F. (2014) Culture in international business research: A bibliometric study in four top IB journals. *Management Research* 12 (1), 68–91.

Filimonau, V., Matute, J., Mika, M. and Faracik, R. (2018) National culture as a driver of pro-environmental attitudes and behavioural intentions in tourism. *Journal of Sustainable Tourism* 26, 1804–1825.

Hall, E.T. (1977) *Beyond Culture*. Anchor Press.

Hall, E.T. (2000) Monochronic and polychronic time. In L.A. Samovar and R.E. Porter (eds) *Intercultural Communication: A Reader* (pp. 280–286). Belmont, CA: Wadsworth.

Hamid, M.A. (2016) Does culture impact choice of pictures for websites: An analysis of Chinese cultural dimensions on websites of Chinese universities. *New Media and Mass Communication* 45, 34–45.

Hanaysha, J.R. (2020) Marketing mix elements and corporate social responsibility: Do they really matter to store image? *Jindal Journal of Business Research* 9 (1), 56–71.

Hobson, J.P. and Christensen, M. (2001) Cultural and structural issues affecting Japanese tourist shopping behaviour. *Asia Pacific Journal of Tourism Research* 6 (1), 37–45.

Hofstede, G. (2011) Dimensionalizing cultures: The Hofstede Model in context. *Online Readings in Psychology and Culture* 2 (1), 2307–0919.

Hofstede, G., Hofstede, G.J. and Minkov, M. (2010) *Cultures and Organizations: Software of the Mind* (3rd edn). McGraw-Hill.

Hornikx, J. and Le Pair, R. (2017) The influence of high-/low-context culture on perceived ad complexity and liking. *Journal of Global Marketing* 30 (4), 228–237.

House, R.J., Hanges P.J., Javidan M., Dorfman, P.W. and Gupta, V. (2004) *Culture, Leadership, and Organizations: The GLOBE Study of 62 Societies*. Sage.

Hsieh, A.T. and Tsai, C.W. (2009) Does national culture really matter? Hotel service perceptions by Taiwan and American tourists. *International Journal of Culture, Tourism and Hospitality Research* 3 (1), 54–69.

Huang, Z., Huang, S., Yang, Y., Tang, Z., Yang, Y. and Zhou, Y. (2021) In pursuit of happiness: Impact of the happiness level of a destination country on Chinese tourists' outbound travel choices. *International Journal of Tourism Research* 23 (5), 713–725.

Humphrey, C. (2018) To smile and not to smile: Mythic gesture at the Russia-China border. *Social Analysis* 62 (1), 31–5.

Hung, K., Ren, L. and Qiu, H. (2021) Luxury shopping abroad: What do Chinese tourists look for? *Tourism Management* 82, 104182.

Jang, S., Bai, B., Hong, G. and O'Leary, J.T. (2004) Understanding travel expenditures: A study of Japanese pleasure travellers to the United States by income level. *Tourism Management* 25 (3), 331–341.

Jeong, J.Y., Crompton, J.L. and Hyun, S.S. (2019) What makes you select a higher price option? Price–quality heuristics, cultures, and travel group compositions. *International Journal of Tourism Research* 21 (1), 1–10.

Josiam, B.M., Kinley, T.R. and Kim, Y.K. (2005) Involvement and the tourist shopper: Using the involvement construct to segment the American tourist shopper at the mall. *Journal of Vacation Marketing* 11 (2), 135–154.

Kang K.H., Lee, S. and Yoo, C. (2016) The effect of national culture on corporate social responsibility in the hospitality and tourism industry. *International Journal of Contemporary Hospitality Management* 28 (8), 1728–1758.

Keown, C.F. (1989) A model of tourists' propensity to buy: The case of Japanese visitors to Hawaii. *Journal of Travel Research* 27 (3), 31–34.

Kim, S.S., Timothy, D.J. and Hwang, J. (2011) Understanding Japanese tourists' shopping preferences using the Decision Tree Analysis method. *Tourism Management* 32 (3), 544–554.

Koc, E. (2007) Total quality management and business excellence in services: The implications of all-inclusive pricing system on internal and external customer satisfaction in the Turkish tourism market. *Total Quality Management & Business Excellence* 17 (7), 857–877.

Koc, E. (2013a) Inversionary and liminoidal consumption: Gluttony on holidays and obesity. *Journal of Travel & Tourism Marketing* 30 (8), 825–838.

Koc, E. (2013b) Power distance and its implications for upward communication and empowerment: Crisis management and recovery in hospitality services. *The International Journal of Human Resource Management* 24 (19), 3681–3696.

Koc, E. (2021a) *Cross-cultural Aspects of Tourism and Hospitality: A Services Marketing and Management Perspective*. Routledge.

Koc, E. (2021b) *Hizmet Pazarlaması ve Yönetimi*. Ankara: Baskı, Seçkin Yayıncılık.

Koc, E. (2021c) Intercultural competence in tourism and hospitality: Self-efficacy beliefs and the Dunning Kruger effect. *International Journal of Intercultural Relations* 82, 175–184.

Koc, E., Yilmaz, O. and Boz, H. (2020) Service recovery paradox in restraint cultures: An implementation in tourism sector. *Journal of Empirical Economics and Social Sciences* 2 (1), 72–91.

Koc, E. and Ayyildiz, A.Y. (2021) Culture's influence on the design and delivery of the marketing mix elements in tourism and hospitality. *Sustainability* 13 (21), 11630.

Koc, E. and Ayyildiz, A.Y. (2022) An overview of tourism and hospitality scales: Discussion and recommendations. *Journal of Hospitality and Tourism Insights* 5 (5), 927–949.

Kumar, S. and Dhir, A. (2021) Associations between travel and tourism competitiveness and culture. *Journal of Destination Marketing and Management* 18, 100501.

Kwak, H., Zinkhan, G.M. and Roushanzamir, E.P.L. (2004) Compulsive comorbidity and its psychological antecedents: A cross cultural comparison between the US and South Korea. *Journal of Consumer Marketing* 21 (6), 418–434.

Kwok, S. and Uncles, M. (2005) Sales promotion effectiveness: The impact of consumer differences at an ethnic-group level. *Journal of Product and Brand Management* 14, 170–186.

Ladhari, R., Pons, F., Bressolles, G. and Zins, M. (2011) Culture and personal values: How they influence perceived service quality. *Journal of Business Research* 64 (9), 951–957.

Lalwani, A.K. and Forcum, L. (2016) Does a dollar get you a dollar's worth of merchandise? The impact of power distance belief on price-quality judgments. *Journal of Consumer Research* 43 (2), 317–333.

Lee, H.E. (2015) Does a server's attentiveness matter? Understanding intercultural service encounters in restaurants. *International Journal of Hospitality Management* 50, 134–144.

Le Pair, R. and Van Mulken, M. (2008) Perceived complexity and appreciation of visual metaphors by consumers with different cultural backgrounds. In F. Costa Pereira, J. Veríssimo and P. Neijens (eds) *New Trends in Advertising Research* (pp. 279–290). Sílabo.

Litvin, S.W. and Kar, G.H. (2003) Individualism/collectivism as a moderating factor to the self-image congruity concept. *Journal of Vacation Marketing* 10 (1), 23–42.

Litvin, S.W., Crotts, J.C. and Hefner, F.L. (2004) Cross-cultural tourist behaviour: A replication and extension involving Hofstede's uncertainty avoidance dimension. *International Journal of Tourism Research* 6 (1), 29–37.

Liu, J., Cho, S., Yang, S. and Xue, C. (2021) How and when does multitasking affect customer orientation of hotel employees? *Journal of Hospitality and Tourism Management* 47, 335–342.

Lu, J., Hung, K., Wang, L., Schuett, M.A. and Hu, L. (2016) Do perceptions of time affect outbound-travel motivations and intention? An investigation among Chinese seniors. *Tourism Management* 53, 1–12.

Magnini, V.P., Hyun, S.S., Kim, B.P. and Uysal, M. (2013) The influences of collectivism in hospitality work settings. *International Journal of Contemporary Hospitality Management* 25 (6), 844–864.

Manrai, L.A. and Manrai, A. (2011) Hofstede's cultural dimensions and tourist behaviors: A review and conceptual framework. *Journal of Economics, Finance & Administrative Science* 16 (31), 23.

Mattila, A.S. (1999) Culture in evaluating services. *Journal of Service Research* 1, 250–261.

Mattila, A.S. (2000) The impact of culture and gender on customer evaluations of service encounters. *Journal of Hospitality and Tourism Research* 24, 263–273.

Mattila, A.S. and Choi, S. (2006) A cross-cultural comparison of perceived fairness and satisfaction in the context of hotel room pricing. *International Journal of Hospitality Management* 25 (1) 146–153.

McCollough, M.A. (2009) The recovery paradox: The effect of recovery performance and service sailure severity on postrecovery customer satisfaction. *Academy of Marketing Studies Journal* 13 (1), 89–104.

McNeill, L. (2006) The influence of culture on retail sales promotion use in Chinese supermarkets. *Australas: Australasian Marketing Journal* 14, 34–46.

McNeill, L.S., Fam, K.S. and Chung, K. (2014) Applying transaction utility theory to sales promotion–the impact of culture on consumer satisfaction. *The International Review of Retail, Distribution and Consumer Research* 24, 166–185.

Mihalič, T. and Fennell, D. (2015) In pursuit of a more just international tourism: The concept of trading tourism rights. *Journal of Sustainable Tourism* 23 (2), 188–206.

Mou, X., Gao, L. and Yang, W. (2019) The joint effects of need for status and mental imagery perspective on luxury hospitality consumption in China. *Journal of Travel & Tourism Marketing* 36 (9), 1050–1060.

Mueller, R.D., Palmer, A., Mack, R. and McMullan, R. (2003) Service in the restaurant industry: An American and Irish comparison of service failures and recovery strategies. *International Journal of Hospitality Management* 22, 395–418.

Nam, K-A., Cho, Y. and Lee, M. (2014) West meets East? Identifying the gap in current cross-cultural training research. *Human Resource Development Review* 13 (1), 36–57.

Nath, P., Devlin, J. and Reid, V. (2016) Expectation formation in the case of newer hotels: The role of advertising, price, and culture. *Journal of Travel Research* 55 (2), 261–275.

Nelson, M.R. and Paek, H.J. (2005) Cross-cultural differences in sexual advertising content in a transnational women's magazine. *Sex Roles* 53, 371–383.

Ozkul, E., Bilgili, B. and Koç, E. (2020) The influence of the color of light on the customers' perception of service quality and satisfaction in the restaurant. *Color Research & Application* 45 (6), 1217–1240.

Park, S.B., Chung, N. and Woo, S.C. (2013) Do reward programs build loyalty to restaurants? The moderating effect of long-term orientation on the timing and types of rewards. *Managing Service Quality* 23 (3), 225–244.

Park, S.Y. (1998) A comparison of Korean and American gift-giving behaviours. *Psychology and Marketing* 15 (6), 577–593.

Parker, B.J. (2003) Food for health: The use of nutrient content, health, and structure/function claims in food advertisement. *Journal of Advertising* 32, 47–55.

Patterson, P.F., Cowley, E. and Prasongsukarn, K. (2006) Service failure recovery: The moderating impact of individual-level cultural value orientation on perceptions of justice. *International Journal of Research in Marketing* 23 (3), 263–277.

Pellegrini, C., Rizzi, F. and Frey, M. (2018) The role of sustainable human resource practices in influencing employee behavior for corporate sustainability. *Business Strategy and the Environment* 27 (8), 1221–1232.

Pollay, R.W. and Gallagher, K. (1990) Advertising and cultural values: Reflections in the distorted mirror. *International Journal of Advertising* 9 (4), 359–372.

Reisinger, Y. (2009) Cross-cultural differences in tourist behaviour. In M. Kozak and A. Decrop (eds) *Handbook of Tourist Behaviour: Theory & Practice* (pp. 237–255). Routledge.

Reisinger, Y. and Crotts, J. (2010) Applying Hofstede's national culture measures in tourism research: Illuminating issues of divergence and convergence. *Journal of Travel Research* 49 (2), 153–164.

Rojas-Méndez, J.I., Davies, G., Omer, O., Chetthamrongchai, P. and Madra, C.A. (2017) Time attitude scale for cross-cultural research. *Journal of Global Marketing* 15, 117–147.

Rosenbaum, M.S., Ostrom, A.L. and Kuntze, R. (2005) Loyalty programs and a sense of community. *Journal of Services Marketing* 19, 222–233.

Rosenbaum, M.S. and Spears, D.L. (2006) An exploration of spending behaviors among Japanese tourists. *Journal of Travel Research* 44, 467–473.

Sabiote-Ortiz, C.M., Frías-Jamilena, D.M. and Castañeda-García, J.A. (2016) Overall perceived value of a tourism service delivered via different media: A cross-cultural perspective. *Journal of Travel Research* 55 (1), 34–51.

Schill, M. and Godefroit-Winkel, D. (2022) Consumer responses to environmental corporate social responsibility and luxury. *Journal of Services Marketing* 36 (6), 769–780.

Seo, S., Kim, K. and Jang, J. (2018) Uncertainty avoidance as a moderator for influences on foreign resident dining out behaviors. *International Journal of Contemporary Hospitality Management* 30, 900–918.

Spears, D. and Rosenbaum, M. (2012) The packaged tourist: A Japanese and American perspective. *Tourismos* 7 (1), 19–40.

Stauss, B. and Mang, P. (1999) Culture shocks in inter-cultural service encounters. *Journal of Services Marketing* 13 (4/5), 329–346

Sthapit, E. and Björk, P. (2019) Relative contributions of souvenirs on memorability of a trip experience and revisit intention: A study of visitors to Rovaniemi, Finland. *Scandinavian Journal of Hospitality and Tourism* 19 (1), 1–26.

Su, N., Min, H., Chen, M.H. and Swanger, N. (2018) Cultural characteristics and tourist shopping spending. *Journal of Hospitality & Tourism Research* 42 (8), 1210–1231.

Swaminathan, F. (2012) The uncommon sense of advertising: Understanding contemporary advertising appeals in press of top advertisers in India. In Proceedings of the LCBR European Marketing Conference, August 2012.

Swanson, K.K. and Timothy, D.J. (2012) Souvenirs: Icons of meaning, commercialization and commoditization. *Tourism Management* 33, 489–499.

Swanson, S.R., Frankel, R., Sagan, M. and Johansen, D.L. (2011) Private and public voice: Exploring cultural influence. *Managing Service Quality: An International Journal* 21 (3), 216–239.

Timothy, D.J. (2005) *Shopping Tourism, Retailing and Leisure*. Channel View Publications.

Timothy, D.J. and Butler, R.W. (1995) Cross-border shopping: A North American perspective. *Annals of Tourism Research* 22 (1), 16–34.

Tung, R.L. (2002) International organizational behaviour. In *Virtual O.B. Electronic Data Base* (pp. 20–23). McGraw-Hill.

Van Esch, P., Arli, D. and Gheshlaghi, M.H. (2020) Creating an effective self-managed service climate for frontline service employees. *Journal of Retailing and Consumer Services* 57, 1–10.

Vlachos, P.A., Theotokis, A., Pramatara, K. and Vrechopoulos, A. (2010) Consumer–retailer emotional attachment: Some antecedents and the moderating role of attachment anxiety. *European Journal of Marketing* 44 (9–10), 1478–1499.

Wen, B., Zhou, X., Hu, Y. and Zhang, X. (2020) Role stress and turnover intention of front-line hotel employees: The roles of burnout and service climate. *Frontiers in Psychology* 11 (36), 1–13.

Wilder, K.M., Collier, J.E. and Barnes, D.C. (2014) Tailoring to customers' needs: Understanding how to promote an adaptive service experience with frontline employees. *Journal of Service Research* 17 (4), 446–459.

Würtz, E. (2005) Intercultural communication on web sites: A cross-cultural analysis of web sites from high-context cultures and low-context cultures. *Journal of Computer-mediated Communication* 11 (1), 274–299.

Yüksel, A. and Yüksel, F. (2001) Comparative performance analysis: Tourists' perceptions of Turkey relative to other tourist destinations. *Journal of Vacation Marketing* 7(4), 333–355.

Zein, O. (2015) *Culture and Project Management: Managing Diversity in Multicultural Projects*. Gower.

Zhang, Z., Li, H. and Law, R. (2015) Differences and similarities in perceptions of hotel experience: The role of national cultures. *Journal of Travel & Tourism Marketing* 32 (1), S2–S14.

Zhao, D. (2017) The effects of culture on international advertising appeals: A cross-cultural content analysis of US and Japanese global brands. *Professional Projects from the College of Journalism and Mass Communications* 15, 13–14.

4 The Evolving Tourism Marketplace: Changing Tourist Shopping Markets

Gianna Moscardo, Haipeng Jin and Laurie Murphy

Introduction

Global mass tourism is a relatively recent phenomenon emerging at the start of the 21st century with the rise of budget airlines, aircraft capable of longer flights with larger capacities, globalisation, decreased travel restrictions and the enthusiasm of Chinese consumers to take up opportunities for international travel (Buckley *et al.*, 2015). While tourism has a much longer history, the scale and diversity of truly global mass tourism is limited to the last 30 years. Although this is not a long time in any historical context it does mean that a large proportion of the world's human population have never lived without global tourism as either a guest or a host or both. Not surprisingly then, a question asked at a tourism academic conference in 2019 about imagining a world without international tourism resulted in an extensive pause and much consternation. Less than six months after this question was posed, the onset of the COVID-19 pandemic essentially shut down international travel and much domestic travel. This resulted in an avalanche of academic predictions on the future of tourism, many of which have not yet been subjected to any critical analysis. These discussions fell into two main categories: once the pandemic was under control tourism would/should return to its pre-pandemic levels and quickly grow beyond those, or tourism would/should return but in a much-changed form (Higgins-Desbiolles, 2020). While to date neither of these two forecasts appear to be correct, on closer inspection with more critical analysis, changing trends and emerging forces for change can be identified.

This chapter seeks to explore these changing trends and emerging forces and their impact on tourist shopping markets. For the purposes of this chapter, tourist shopping is defined as 'a recreational activity in which tourists browse, select and purchase goods to take home during their travel' (Jin *et al.*, 2017: 121). This definition includes people who take shopping-focused tours/trips, as well as tourists who buy personal items and souvenirs as part of a trip (Timothy, 2005). There is considerable overlap between leisure shopping and tourism. Both are hedonic, recreational activities that involve discretionary expenditure (Bäckström, 2011). For some leisure shopping is a prime feature of tourist activities and for others the two phenomena are substitutes for each other. After outlining key trends and forces in both the retail shopping and tourism systems, the chapter will use social practice theory from sociology as the basis for identifying and analysing changing and emerging tourist shopping markets. It will conclude the analysis with some initial implications for both research into tourist shopping and for tourism practitioners.

The Changing Leisure Shopping and Tourism Context

To understand changing and evolving markets for tourist shopping it is necessary to have an overview of the larger forces impacting on both the general tourism system and the larger phenomenon of leisure shopping. An examination of the recent industry/business consulting reports and reviews of academic research into consumers and leisure shopping and tourism reveals a set of four themes common to both areas, which are summarised in Table 4.1.

An additional theme, the impact of COVID-19, was also evident in these reviews although examined and discussed in a different way in the two areas. In leisure shopping, the COVID-19 discussion has been based much more on empirical data and focused on the extent to which the pandemic has altered consumption patterns and the likelihood of these changed consumption patterns remaining post the pandemic (cf. Eger *et al.*, 2021). In this regard, academic research is more closely aligned with business intelligence based on larger scale market survey research (cf. Angus & Westbrook, 2020). The consistent conclusion has been that the pandemic accelerated the existing trends, especially digital transformations, and that this change has become stable (Eger *et al.*, 2021; McKinsey & Company, 2020; Sheth, 2020).

Discussion of the impact of the pandemic on tourist markets in the tourism academy has been much more focused on descriptions of, and suggestions for, policies and strategies to restore pre-COVID tourism consumption patterns with the available empirical research-based papers focusing more on immediate traveller concerns and fears over health risks (Aldao *et al.*, in press; Rogerson & Rogerson, 2021). In addition, many reviews of this literature are bibliometric in style telling us who published what and where but with almost no discussion of the actual research results (cf. Viana-Lora & Nel-lo-Andreu, 2022; Yang *et al.*, 2021; Zopiatis *et al.*, 2021) offering few, if any, insights into more permanent changes in tourist consumption. By way of contrast, tourism business intelligence based on large-scale market survey research has identified several changes

Table 4.1 Key themes in reviews of research into leisure shopping and tourist consumption patterns

Theme	Implications for leisure shopping	Implications for tourism
Rising consumer use of digital technology, especially mobile online connections to both businesses and social collectives through social media	Increased power to buyers and their social collectives Increased expectations that physical stores will blend with and incorporate digital technology	Balance use of technology for mundane tasks with more human interactions Increased power to buyers and their social collectives Increased expectations that physical stores will blend with and incorporate digital technology
Globalization and diversification of both supply and demand	Sellers have potential exposure to new and diverse markets Buyers can access goods from anywhere Rise of purchasing power among younger generations is diversifying markets	Rise of purchasing power among younger generations is diversifying markets Changes in migration mean more diverse domestic tourism markets Less interest in global brands and more in local products
Increasing expectations for explicit and personalised experiential opportunities	Pressure on physical stores to provide unique, multi-sensory, social experiences Rise of themed pop-up stores and business flagship museums	Pressure on tourist retail settings to provide unique, multi-sensory, social experiences
Growing consumer concerns over sustainability issues	Increased demand for more environmentally responsible products with shorter more local supply chains More explicit details wanted on social justice and employment conditions throughout the production and supply chain Greater interest in inclusivity in stores and product design and presentation	Increased demand for sustainable/local products Less distance travelled and longer stays in destinations More explicit details wanted on social justice and employment conditions and environmental performance throughout the production and supply chain
Sources	Angus and Westbrook, 2020; De-Juan-Vigaray and Espinsosa Seguie, 2019; Eger et al., 2021; Helm et al., 2020	Aldao et al., in press; Robertson, 2021; Rogerson and Rogerson, 2021; UNWTO and ADB, 2022; WTTC, 2022

likely to remain post the pandemic. Like leisure shopping, some of these are accelerations of trends already occurring pre-pandemic, such as increasing digitalisation, concerns over sustainability and the rise of experiential travel (Rogerson & Rogerson, 2021; UNWTO & ADB, 2022; WTTC, 2022). Others are trends that emerged during or immediately after COVID-19 and appear likely to remain in place for some time. These include:

- a shift to domestic tourism from international travel in response to perceived health risks, uncertainty about cancellations, political instability, extreme weather events and the rising cost of travel;
- China emerging from travel restrictions as a dominant source of international tourists challenged by increased growth in travel out of India;
- increased preferences for smaller travel groups and independent travel options over larger package tourism;
- more interest in, and travel to, regional, rural and nature-based destinations and decreased interest in urban tourism; and
- slower tourism where tourists travel to fewer destinations in a single trip and stay in those places longer (Sources: Aldao *et al.*, in press; Rogerson & Rogerson, 2021; UNWTO & ADB, 2022; WTTC, 2022).

Experience and sustainability have been identified as key themes linked to both changing and emerging new tourist shopping markets (Deloitte, 2022; Francis & Hoefel, 2018; Sthapit *et al.*, 2018). In the case of sustainable, socially responsible and ethical products, consumer demand is 'transforming a once-niche market into a powerful mainstream audience' (IBM Institute for Business Value, 2022: 7). It no longer makes sense to see experience and sustainability as characteristics linked to specific tourist shopper markets, as these are forces changing the nature of products and shopping opportunities for all tourist shoppers.

One key difference between leisure shopping and tourism analyses of changing patterns of consumption is greater awareness in the leisure shopping literature of the diversity and impact of culture, especially national or ethnic culture and generational culture, and prevailing economic conditions on consumption patterns (cf. Lu *et al.*, 2017; Shavitt & Barnes, 2020; Thomas *et al.*, 2020; Timothy, 2005). Tourism researchers have paid much less attention to this, focusing instead on identifying markets according to their country of origin, often treating these groups as homogeneous and stable. This suggests a need to be more sophisticated in identifying markets for tourist shopping based on examining in more detail how tourists shop, as well as why they shop and not just who they are and what they buy. It requires a shift from simple demographic and travel choice-based markets to tourist shopping markets based on shared social practices.

Social Practice Theory

Social practice theory (SPT) provides a distinctive perspective on understanding the social world. It is not a unified theory, but a collection of accounts about the workings of social life that centre on social practices (see Reckwitz, 2002; Schatzki, 1996; Shove *et al.*, 2012). In the seminal work of Reckwitz (2002: 249), social practice was defined as 'a routinized type of behavior which consists of several elements, interconnected to each other: forms of bodily activities, forms of mental activities, "things" and their use, a background knowledge in the form of understanding, know-how, states of emotion and motivational knowledge'. These elements later were conceptualised by Shove *et al.* (2012) into a widely accepted analytical framework, where a social practice consists of three elements: materials, competences and meanings. Specifically, materials include 'things, technologies, tangible entities, and the stuff of which objects are made'; competences encompass 'skills, know-how and technique'; and meanings refer to 'symbolic meanings, ideas and aspirations' (Shove *et al.*, 2012: 14).

The originality of SPT mainly lies in that it treats social practices as the basic unit of analysis, with individuals being decentralised as carriers of the practices (Reckwitz, 2002; Shove *et al.*, 2012; Warde, 2005). In the words of Reckwitz (2002: 256), 'the social world is first and foremost populated by diverse social practices which are carried by agents'. SPT situates the social in practices, which is fundamentally different to both individualistic and holistic views. It provides a new level of analysis that allows for the integration of individual and social levels of causal explanation, helping to bridge the dualisms of agency and structure (Shove *et al.*, 2012).

SPT has become increasingly adopted by tourism researchers over the last decade, with two main streams of research being identified. First, SPT has been utilised to (re)theorise tourism (Bargeman & Richards, 2020; de Souza Bispo, 2016; Lamers *et al.*, 2017). Specifically, this theory can contribute to advancing tourism research in three ways: by analysing the performed tourism practices in depth, by investigating change in tourism practices over time and by unravelling the embeddedness of tourism practices in broader practice networks (Lamers *et al.*, 2017). Second, SPT has been applied to a rising number of empirical tourism studies. Tourism phenomena that have been analysed from a social practice perspective include cruise tourism (Lamers & Pashkevich, 2018), backpacking (Iaquinto & Pratt, 2020), tourist scams (Xu *et al.*, 2021), family travel (Yi *et al.*, 2023), technology use of young travellers at airports (Pant, 2022) and solo female travel (Jin & Zhang, 2023).

Of relevance to the present discussion is the work of Jin and colleagues (Jin *et al.*, 2020, 2021a, 2021b; Jin & Cai, 2022) who conducted several studies guided by SPT to investigate Chinese tourist shopping in Australia, which is especially relevant to analyses of tourist shopping markets. As Jin *et al.* (2020) stated, tourist shopping reflects both individual agency and social structure, and practice-based inquiries into this activity shift the focus from the choices and values of individual tourists to the nexus of their doings and sayings on shopping (Schatzki, 1996), contributing to a more comprehensive understanding of its complexity and dynamics.

An Expanded Social Practice Theory Model of the Evolving Tourist Shopping Marketplace

Combining an SPT approach with major conclusions identified in the literature review suggests a preliminary framework for analysing changing tourist shopping markets. Figure 4.1 presents this framework, which was based on Jin and colleagues' (2020) SPT model of pre-COVID Chinese tourist shoppers in Australia. That model set out the three core elements of social practices – competences, meaning and materials (Shove *et al.*, 2012) – and added a fourth one specific to tourism: settings (Jin *et al.*, 2020; Jin & Zhang, 2023). Meanings are the values, beliefs and expected benefits of tourist shopping (Jin *et al.*, 2020). Changes in leisure shopping suggest additional meanings to be considered including the pursuit of meaningful leisure experiences that include opportunities for social interaction, expression of social identity and personal skill development (Pantano & Gandini, 2018; von Briel, 2018). Competences include the skills and knowledge of how, where and what to shop for while travelling. The materials element refers to the products available to purchase and tools, such as mobile apps, that facilitate tourist shopping. Finally, there are settings, with changes such the rise of pop-up stores, flagship brand experience stores and museums, and shopping more closely linked to production facilities (Kahn *et al.*, 2018; Wu, 2017).

The four SPT elements are also subject to four forces – increased awareness and concern about sustainability; culture, including ethnic and generational cultures; technology; and opportunities. Each of these forces can change any or all the SPT elements. For example, sustainability concerns can influence the type of products offered or desired, place

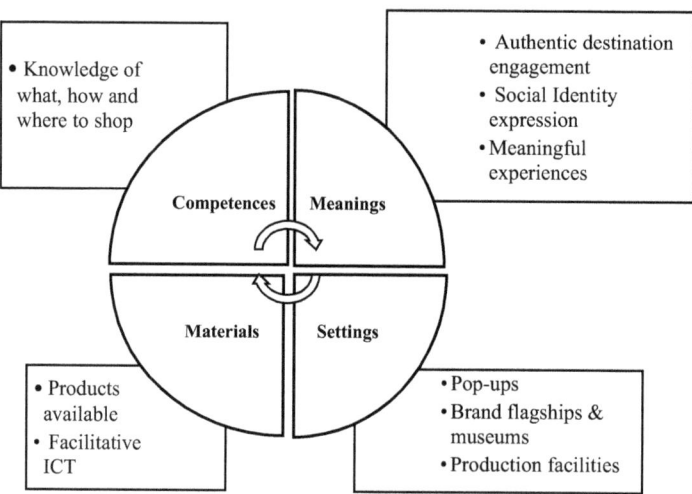

Figure 4.1 Social practice framework for tourist shopping markets

additional pressures on tourists' competences with a need to learn and use sustainability certifications and information, highlight the importance of local products in tourist shopping meanings, and encourage more direct purchases from production facilities to cut down supply chains. Similarly, changing technology creates new materials, such as apps, and the need for new competences related to using these apps. New technologies allow for new ways of buying and transporting purchases and change the meanings of tourist shopping to be less about the actual items paid for and more about the total shopping experience. This SPT framework offers a way to examine changes in tourist shopping markets. Specifically, the framework is used to examine changes in two key existing markets – domestic tourists worldwide and Chinese tourists – and two emerging markets: tourists from India and newer generations. While the inclusion of two country of origin markets may seem contrary to the previous argument for more sophistication in tourism market analysis, each presents a special case. Chinese outbound tourists are already widely treated as a specific tourist shopping market and the uniformity of their shopping practices has been established and linked to core cultural dimensions and shared circumstances (Jin *et al.*, 2021b). Indian outbound tourists have also been identified as a very specific group with common characteristics and consumption practices and are not seen as representative of the larger country.

Changing Tourist Shopping Markets – Domestic Shopping Tourists

Two key changes can be identified for domestic tourists as shoppers: increased travel closer to home and increased cultural diversity of domestic tourists in many destinations. While the lack of research attention paid to these two forces in tourism makes it difficult to map out these changes in detail in the SPT framework, it is likely that both changes will result in the emergence of new tourist shopping markets and, based on the review of trends in leisure shopping in general, it is possible to suggest some ways in which these changing markets might adapt their tourist shopping practices. Increased domestic travel, especially travel closer to home, is likely to mean repeat travel to familiar destinations which in turn may have the following consequences:

- less demand for destination souvenirs as these have been purchased on previous visits;
- greater demand for local arts, crafts, food and beverage;
- opportunities to purchase larger quantities and/or larger objects because of lower costs and greater ability to transport goods home and
- increased demand for experiential shopping (Angus & Westbrook, 2020; De-Juan-Vigaray & Espinosa Seguie, 2019; Eger *et al.*, 2021; Helm *et al.*, 2020; Moscardo, 2024; Szolnoki *et al.*, 2022).

Although there is little tourism research available on this change, there is emerging evidence to support these claims. Szolnoki *et al.*'s (2022) study of changes in tourist wineries across the USA, Australia, Germany, Hungary and Romania found an increased focus on domestic markets who sought more themed, educational, personalised experience beyond just wine tasting and purchasing. Moscardo's (2024) review and analysis of staycations also highlighted demand from local/domestic tourists for experiences linked to learning and improving skills, and spending time with family and friends.

The evidence from the leisure shopping literature overall, and from the few studies in tourism that have explored these factors (cf. Kirillova *et al.*, 2020; Su *et al.*, 2018; Weidenfeld, 2018) indicates that there is significant diversity and change among domestic tourists as many countries experience more diverse patterns of migration. Countries including the USA, Canada, Australia, the UK and many parts of Europe have had significant immigration over the last decade from an increasingly diverse range of source countries. For example, in the UK, nearly 15% of residents are recent immigrants mainly from Poland, India, Pakistan, Bangladesh, South Africa and China, similar to patterns reported for other major European countries (Sturge, 2021; The Migration Observatory at Oxford University, 2020). The main countries for immigrants to Canada are India, China, the Philippines, Pakistan, Nigeria and the USA (Singer, 2020). The USA has a similar immigration pattern to the UK with nearly 15% of the current population being recent immigrants coming from Mexico, China, India, the Philippines and El Salvador (Buddiman, 2020). These new domestic tourists are likely to share some of the characteristics of their cultural counterparts who shop as international tourists but also to be different from these traditional tourist shoppers. For example, Chinese immigrants on domestic holidays in Australia may continue to pursue gift purchases as part of their need to develop and maintain *guanxi* (personal social network), but they may no longer focus on buying utilitarian objects such as medicines seen as safer than those available in China because they have easy access to these in their new home.

Changing Tourist Shopping Markets – Chinese Tourists

Chinese tourists played a prominent role in the global tourist shopping markets before the COVID-19 pandemic. Back then, it was common to see that Chinese tourists shopped in droves no matter in the Galleries Lafayette in central Paris, in the Ginza district of Tokyo or in the supermarkets and pharmacies in Sydney, Australia. According to the China Tourism Academy (2020), there were 155 million Chinese tourists spending over USD134 billion abroad in 2019, and shopping constituted a major proportion of their travel expenditure. The COVID-19 pandemic largely suspended Chinese outbound tourism from 2020 to 2022, during which time the global tourist shopping markets were severely affected by the absence of Chinese customers (e.g. Baker, 2022; Stockdill, 2020). As China lifted its COVID-19-related travel restrictions in early 2023, it was expected that Chinese outbound tourism would revive, albeit slowly.

Some continuities and changes can be identified for Chinese tourist shoppers, based on a combination of reviews of existing research on this market (Jin *et al.*, 2020; Wen & Kozak, 2022) and that of trends in tourism consumption in China (China Tourism Academy, 2022). Figure 4.2 presents the key features of the evolving Chinese tourist shopping practices.

Firstly, Chinese tourists continue to have interests in purchasing a wide variety of products, ranging from clothes and handbags to food and supplements. As Jin *et al.* (2021b) stated, one major reason why shopping dominated the outbound travel of Chinese tourists was their consumerism, featuring the desire for global brands and pursuit of high-quality products, and these items were either not available in China or much cheaper abroad as compared to China. Although this consumerism has not changed, the competitive advantages of retail businesses abroad over Chinese counterparts in these regards are shrinking. Chinese consumers are now able to buy these products at lower prices from more sources at home through e-commerce platforms and duty-free shops (e.g. Kim *et al.*, 2021).

Given the variety of products Chinese tourist shoppers are interested in purchasing, it is not surprising that this group of customers would continue to visit many types of shopping venues. Based on previous research (Jin *et al.*, 2020; Li, 2016; Wen & Kozak, 2022), these venues include shopping malls, department stores, direct factory outlets, airports, supermarkets, tourist attractions and so on. Meanwhile, as independent travel is increasing

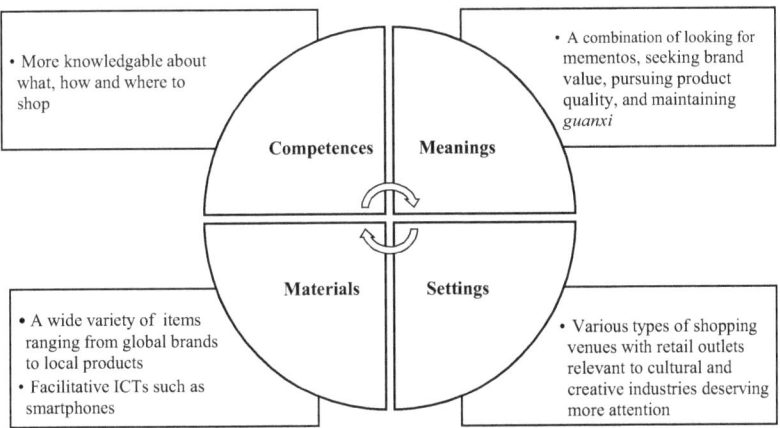

Figure 4.2 Social practice framework for Chinese outbound tourist shoppers

among Chinese tourists, it is expected that their shopping activities will expand further into retail sectors that conventionally are not included in the tourism industry. Taking pharmacies as an example, although it is not new that Chinese tourists seek healthcare products abroad (Jin et al., 2020; Xu & McGehee, 2012), the COVID-19 pandemic has increased the attention Chinese customers pay to their health. As the COVID-19 surge across China in late 2022 led to panic buying of some medicines (e.g. ibuprofen) and medical devices (e.g. pulse oximeter), pharmacies abroad are likely to be places where Chinese tourists will seek health-related products that are either not available or in short supply at home.

Another type of shopping venue that deserves particular attention are retail outlets linked to cultural and creative industries. For instance, the online shop of the British Museum in China has over 2 million followers (accessed 27 January 2023), with average sales of nearly USD 120 million per month from January to May 2022 (Mktindex, 2022). Creativity plays an increasingly important role in the popularity of some souvenirs in China. Examples include the popularity of LinaBell from the Shanghai Disney Resort and products inspired by the Palace Museum (Mktindex, 2022). Chinese tourists, especially younger generations, are likely to patronise these retail outlets if they offer unique products with cultural meaning and creativity. Souvenirs with creative designs have become more attractive to Chinese tourists than the mass-produced and cheap items that previously dominated.

Chinese tourists are also becoming more knowledgeable about what, how and where to shop while travelling abroad. This is not only because they have more travel experience, but also because they are able to obtain more information on shopping from multiple sources, especially shopping websites and social media accessed through smartphones (Jin & Cai, 2022). As Jin and Cai (2022) found, Chinese tourists used smartphones during their shopping process in various ways, including getting product information in Chinese, comparing prices between home and the destination, and doing translations. This empowerment of Chinese shoppers, on the one hand, requires retail businesses at tourist destinations to be transparent about information provision and communication, and on the other hand, provides new possibilities to improve customer engagement and mobile marketing by sending personalised greetings, coupons and notifications (Faulds et al., 2018).

As with the diversity of Chinese tourists' purchases stated above, the meanings of their shopping practices continue to be multiple. In accordance with Jin et al. (2020), these meanings can be grouped into four main categories:

- *Looking for mementos.* Chinese tourists, especially first-time travellers, commonly buy mementos to help preserve and commemorate their travel experiences. Although these mementos may only occupy a

small part of what they purchase, they are essential with the process of searching for them often involving exploration and engagement with locals.
- *Seeking brand value.* Chinese tourists, the majority of whom are middle class, prefer global or even luxury brands. Outbound travel functions as an opportunity to show their brand loyalty, with the purchase of relevant products at relatively low prices being commonly counted as good value for money.
- *Pursuing product quality and authenticity.* Chinese customers hold sceptical attitudes towards domestically produced daily commodities, owing to a series of food safety scandals. This largely explains why this group of tourists prefer utilitarian products abroad such as baby formula and supplements. It is expected that these utilitarian products with high quality abroad will continue to be favoured by Chinese tourists.
- *Maintaining and reinforcing guanxi.* Chinese tourists not only shop for themselves, but also purchase products for others at home, either as gifts or as per request. As *guanxi* functions as a cornerstone of Chinese society (Bian, 2019), Chinese tourists will continue to shop while travelling to maintain or even extend *guanxi*.

Emerging Tourist Shopping Markets – India

Discussions of growth and change in international tourism often highlight two countries – China and India, the largest countries in terms of population and economic growth in the world (World Population Review 2022a; World Bank 2022). While China has dominated tourism discussion, growth in Indian outbound travel follows very closely behind (UNWTO & ADB, 2022). Given the size of both these groups, it is surprising to see them usually presented as relatively homogenous tourism markets. It is also often the case that they are presented as culturally very similar perhaps because of their geographic closeness and similarities in the rise of their middle class and adoption of western consumption patterns (Hur *et al.*, 2015). While the assumption of homogeneity is not true for either group, it is very clear that India is a much more diverse country than China in terms of religion, ethnicity and economic disparity (Roy, 2018; Witt & Redding, 2013). It is also a much younger country in demographic terms, with a median age of 29 years compared to 38 years in China, with higher levels of English proficiency, although this is changing rapidly (EF, 2022; World Population Review, 2022b). Despite these differences, Indian outbound tourists are more homogeneous than China. The growth of an affluent middle class in India has been rapid but is still restricted to a much smaller percentage of the population than in China (Roy, 2018; Witt & Redding, 2013). There is generally a view that Indian outbound tourists are younger, more affluent, better educated, with higher levels of travel experience, and, while unlikely to travel in package tours, likely to travel with a multigenerational family group (Future Market Insights, 2022; Roy, 2018; Singh & Srivastava, 2019).

One area of commonality between the two groups is the importance of shopping while travelling. International Indian tourists have been consistently reported as very focused on shopping as a major travel activity and reason for destination choices with Indian tourists often the largest spenders on shopping at a destination (see Chapter 3). The European Travel Commission (2022) reported that 65% of Indian tourists list shopping as a major holiday activity and the US International Trade Administration (2022) reporting that shopping was their main activity while in the US, and in 2019 they spent more than USD14 billion on shopping. Although shopping is important to both groups it is important not to assume that Indian tourists shop in the same way that Chinese tourists do as analyses of cultural dimensions demonstrate that India and China are very different culturally (Muthukrishna et al., 2020). Multiple studies using both Hofstede's cultural dimensions and Schwartz's cultural values report that Indians are more feminine, egalitarian, embedded, short-term in their life orientation and individualistic as well as higher on uncertainty avoidance or risk aversion (Hur et al., 2015; Maheswaran & Shavitt, 2000; Schwartz, 2014; Tu et al., 2011; Witt & Redding, 2013). This means that they place a greater value on:

- helping others, equality, social justice and environmental preservation;
- individuals generating pragmatic, immediate and innovative solutions to problems;
- individual self-expression, independence and social recognition; and
- family security and traditions over the welfare of a larger group (Hur et al., 2011; Maheswaran & Shavitt, 2000; Schwartz, 2014; Tu et al., 2011; Witt & Redding, 2013).

These are all features that influence consumer and shopping behaviour (Hur et al., 2011; Maheswaran & Shavitt, 2000).

Despite the importance of Indian outbound tourism in general and the importance of shopping to Indian tourists specifically, there has been almost no research into how Indian tourists shop. There is, however, research into leisure shopping within India and these leisure shoppers are also those very likely to shop when traveling internationally. Figure 4.3 summarises what is known from this leisure research into the social practice framework. Arguably the widespread adoption of Euro-American shopping places and styles in India (Hiremath et al., 2023; Khare & Sarkar, 2021) provides Indian tourists with the ability to simply transfer shopping competence from home into the new tourism settings. Indian outbound tourist shoppers see shopping while on holiday as an individual, hedonistic, recreational experience that is done for fun and novelty, to exhibit personal and social identity, and to be fashionable (Hur et al.,

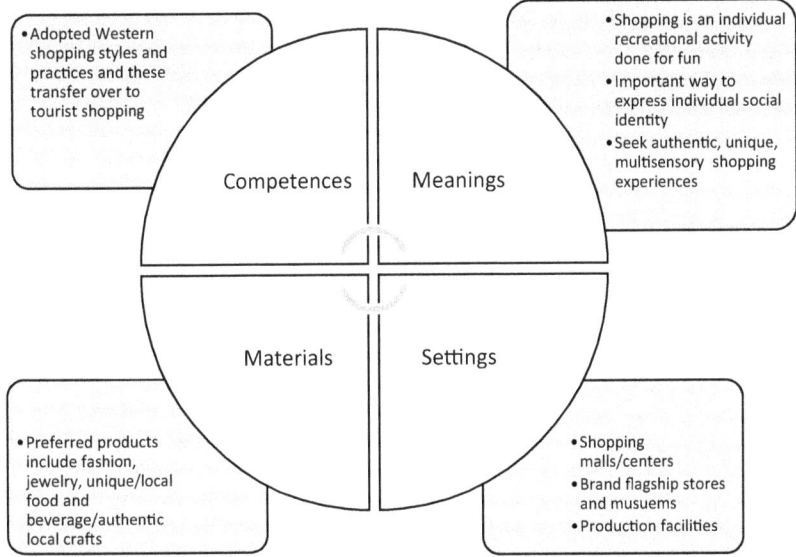

Figure 4.3 Social practice framework for Indian outbound tourist shoppers

2015; Nair & Shams, 2021; Paulose & Veliyath, 2020; Varshneya, 2022). In the context of tourist shopping, a desire for authentic local experiences has been noted (Tourism Australia, 2022; Varghese, 2018). Indian tourist shoppers are likely to visit a wide range of tourist settings to shop but seem especially attracted to shopping malls and are likely to visit brand flagship stores or museums and shops attached to production facilities for specific purchases (Hiremath *et al.*, 2023; Hur *et al.*, 2015; Khare & Sarker, 2021). Finally, there is emerging evidence that Indian tourists and shoppers are becoming more aware of, and concerned about, sustainability (Times of India, 2018; Roy, 2018) and are more likely to take personal responsibility for sustainable choices, especially among younger generations (Hanson-Rasmussen & Lauver, 2018; Schwartz, 2014).

Emerging Tourist Shopping Markets – New Generations

One critical feature of Indian outbound tourist shoppers is that they are much more likely to be younger consumers from the generational cohorts referred to as Gen Y and Gen Z. Although there is some recognition of generational culture as a key influence on leisure shopping (Eger *et al.*, 2021) and tourism (Monaco, 2018), they have not yet been specifically considered as emerging tourist shopping markets. Generational cohorts can be defined as a set of people born within a specified time who all experience the same major and significant events, such as a war, economic recession or pandemic, and who develop in similar broad socioeconomic conditions (Thangavel *et al.*, 2022). Generational cohort theory

(GCT) from sociology argues that these groups develop a distinctive shared culture based on beliefs, values and communication styles that influence their behaviours including consumption patterns (Howe & Strauss, 1992; Strauss & Howe, 2008). While definitions of specific generations in terms of birth years can vary Gen Y (also known as millennials) were born between the early 1980s and mid-1990s with Gen Z born between the mid-1990s and the late 2010s (Dimock, 2019; Monaco, 2018). These two generational cohorts are a substantial proportion of leisure shoppers in general and a large and growing proportion of tourists, both domestic and international (Ketter, 2021; Monaco, 2018), making them significant emerging tourist shopping markets.

It is important to be cautious about claims made about Gen Z, as GCT argues that defining events have to occur when most of the cohort are in their teenage or early adulthood years when values and beliefs are consolidated (Dimock, 2019). There is, however, business research focused on Gen Y and Gen Z shopping behaviours, motivations and beliefs that can inform a preliminary SPT framework for this new tourist shopping market. There is evidence that the two generations share many similar features with a few distinctive differences emerging. Figure 4.4 outlines the key elements of this framework focused on the common features of the two cohorts.

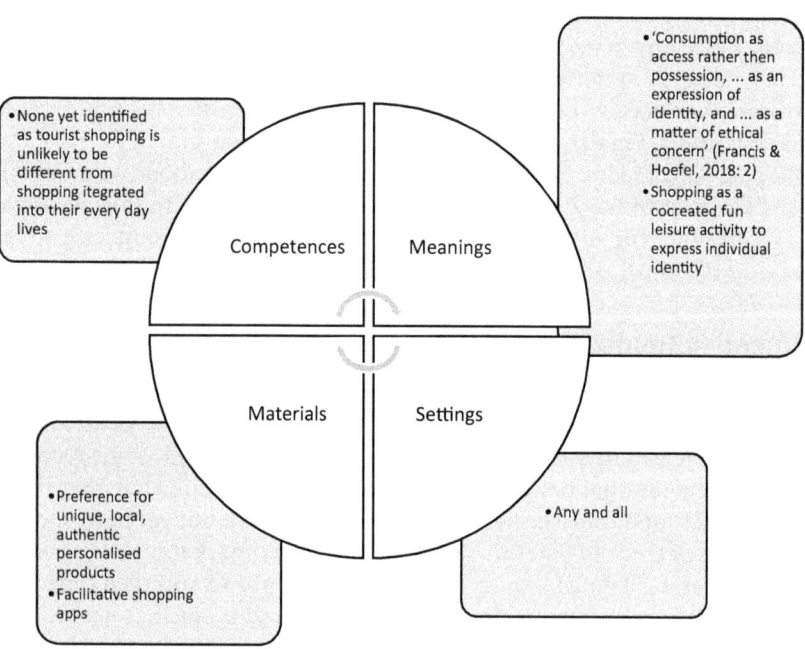

Figure 4.4 Social practice framework for Gen Y and Gen Z tourist shoppers

The available academic research and business survey-based evidence on what Gen Y and Gen Z shoppers seek can be organised around four key themes:
- the importance of technology assisted and integrated shopping where mobile apps and websites provide information on products and their availability and accessibility and in store use of augmented and virtual reality enhances physical shopping experiences (Priporas et al., 2017);
- the importance of the social collective, both online through social media platforms and reviews, and onsite in shared social shopping activity, in generating, informing and supporting purchase decisions (Thangavel et al., 2022);
- the search for experiential shopping which incorporates fun, multisensory pleasure, learning and social interaction (Grant, 2022; IBM Institute for Business Value, 2018; McKinsey & Company, 2020) and
- shopping, especially while on holiday, to express personal identity (Eger et al., 2021; Monaco, 2018).

The available evidence indicates several major characteristics for Gen Z that differ from Gen Y. Although both see shopping as a way to express personal identity, for Gen Y this is about brand purchases, for Gen Z this about unique personalised co-created products to express their unique individual identity (Francis & Hoefel, 2018; IBM Institute for Business Value, 2018; Wunderman Thompson, 2022). While both see sustainability as important, Gen Z is much more likely to search for and confirm sustainability and ethical credentials and specifically search out locally sourced, authentic products that can be linked to specific causes (McKinsey, 2020; Voyado, 2022). Gen Z are also seen as more price conscious and wanting loyalty rewarded by personalised special offers (Grant, 2022; Thangavel et al., 2022) and to expect businesses to use mobile technology to ease purchase and delivery of products (IBM institute for Business Value, 2018).

Conclusion: Implications for Practice and Research

In this chapter the authors used a SPT framework adapted to tourism to organise and critically assess the available evidence on changing tourist shopping practices with a focus on looking towards the future of this core element of tourist consumption. This framework identified two evolving or changing tourist shopping markets – Chinese tourists and Domestic tourists worldwide – and two new or emerging tourist shopping markets – Indian international travellers and new generations.

Although based on currently limited empirical evidence for the specific tourist shopping markets discussed, the more extensive evidence on tourist shopping trends in general offer some initial implications for tourism practitioners. Firstly, sustainability, technology and experience have become central to the tourist shopping experience across all consumers.

All tourist shoppers are likely to expect to be able to blend online and mobile technologies with their physical shopping to organise delivery and customisation of products. This is both a challenge for tourist shopping providers to provide this integration but also an opportunity to increase sales as visitors are less likely to be concerned about, and limited by, luggage limits when they purchase and organise independent delivery. There will be increasing pressure on tourist shopping providers to provide information on the sustainability performance of both the products they sell and their own business. Again, this is both a challenge for tourist shopping retailers, who have to date paid little attention to sustainability issues, and an opportunity to improve their positive impact on destination communities through support for locally produced goods for purchase. Finally, the rising expectation among consumers for rewarding memorable personally meaningful experiences rather than just good service supporting product purchase also offers challenges and opportunities. While tourist shopping providers will have to change and take up experience design principles, rebuilding shopping as an experience offers opportunities for more creative shopping-focused tours, for enhancing local production businesses, and for developing loyal customers beyond the visit.

The recognition that current markets are changing and that new markets are emerging also has practical implications. Tourist shopping providers need to be aware of these changes and of the emerging new markets. For example, as Chinese tourists change their interest from bulk cheap souvenirs to more personalised and creative destination mementos and replace some of their utilitarian purchasing with regionally distinctive gifts for their social networks, retail businesses in many destinations will have to change the range of products offered, the information provided about these and how they are presented. While it is easy to tell practitioners they need to change in general, there is clearly a need for more research into tourist shoppers to provide more useful detail for practice. A common feature of the discussion of the four evolving and emerging markets was the limited research into their characteristics and actions. Researchers need to pay attention to the increasing cultural diversity of domestic shoppers to identify the ways in which they differ in how, where and what they shop for, from both previous domestic tourists and international tourists. Similarly, there is very little academic research into Indian outbound tourist shopping or younger generations, especially Gen Z.

The SPT framework allows us to look at how, why and where different tourist markets shop while travelling. It expands the focus on defining markets from their sociodemographic and generic lifestyle factors to specific shared collective consumption practices. This is emerging as an approach in marketing beyond tourist shopping (cf. Moraes *et al.*, 2017; Nairn & Spotswood, 2015). Combining the SPT insights into existing tourism marketing communication and distribution systems provides a new way to identify, analyse, describe and reach the tourist shopping markets of the future.

References

Aldao, C., Blasco, D. and Espallargas, M.P. (in press) Lessons from COVID-19 for the future. *Journal of Tourism Futures.* https://doi.org/10.1108/JTF-02-2022-0059.

Angus, A. and Westbrook, G. (2020) Top 10 global consumer trends 2020. *Euromonitor International.* Online: https://go.euromonitor.com/white-paper-EC-2020-Top-10-Global-Consumer-Trends.html.

Bäckström, K. (2011) Shopping as leisure. *Journal of Retailing and Consumer Services* 18 (3), 200–209.

Baker, M. (2022, November 23) The China Syndrome: Japan department stores missing tourist shoppers. Online: https://insideretail.asia/2022/11/23/the-china-syndrome-japan-department-stores-missing-tourist-shoppers/.

Bargeman, B. and Richards, G. (2020) A new approach to understanding tourism practices. *Annals of Tourism Research* 84, 102988.

Bian, Y. (2019) *Guanxi: How China Works.* Polity.

Buckley, R., Gretzel, U., Scott, D., Weaver, D. and Becken, S. (2015) Tourism megatrends. *Tourism Recreation Research* 40 (1), 59–70.

Buddiman, A. (2020) Key findings about U.S.A. Immigrants. Pew Research Centre. Online: https://www.pewresearch.org/fact-tank/2020/08/20/key-findings-about-u-s-immigrants/.

China Tourism Academy (2020) *Annual Report of China Outbound Tourism Development 2019.* Tourism Education Press.

China Tourism Academy (2022, April 8) National cultural consumption report 2021. Online: https://mp.weixin.qq.com/s?__biz=MzA4ODM4OTgxOA==andmid=2650822515andidx=2andsn=4076b5ee81b17ad918dd85138abc41a0andchksm=8bde16a6bca99fb065bd5721b3243191a76c727bbb43228c9ea63690eade36d48bdb912adfb2andscene=27.

de Souza Bispo, M. (2016) Tourism as practice. *Annals of Tourism Research* 61, 170–179.

De-Juan-Vigaray, M.D. and Seguí, A.I.E. (2019) Retailing, consumers, and territory: Trends of an incipient circular model. *Social Sciences* 8 (11), 300.

Deloitte (2022) How consumers are embracing sustainability. Online: https://www2.deloitte.com/uk/en/pages/consumer-business/articles/sustainable-consumer.html.

Dimock, M. (2019) Defining generations. Pew Research Center. Online: https://www.pewresearch.org/fact-tank/2019/01/17/where-millennials-end-and-generation-z-begins/.

EF (2022) English Proficiency Index. Online: https://www.ef.com/wwen/epi/.

Eger, L., Komárková, L., Egerová, D. and Mičík, M. (2021) The effect of COVID-19 on consumer shopping behaviour: Generational cohort perspective. *Journal of Retailing and Consumer Services* 61, 102542.

European Travel Commission (2022) Indian Outbound Travel Market. Online: https://etc-corporate.org/reports/indian-outbound-travel-market-infographic/.

Faulds, D.J., Mangold, W.G., Raju, P.S. and Valsalan, S. (2018) The mobile shopping revolution. *Business Horizons* 61 (2), 323–338.

Francis, T. and Hoefel, F. (2018) True Gen. McKinsey. Online: https://www.mckinsey.com/~/media/McKinsey/Industries/Consumer%20Packaged%20Goods/Our%20Insights/True%20Gen%20Generation%20Z%20and%20its%20implications%20for%20companies/Generation-Z-and-its-implication-for-companies.pdf

Future Market Insights (2022) India Outbound Tourism Market Outlook 2022–2032. Online: https://www.futuremarketinsights.com/reports/india-outbound-tourism-market.

Grant, M. (2022) Millennials vs. Gen Z. *Salesforce.* Online: https://www.salesforce.com/blog/how-millennials-and-gen-z-are-different/.

Hanson-Rasmussen, N.J. and Lauver, K.J. (2018) Environmental responsibility. *Journal of Global Responsibility* 9 (1), 6–20.

Helm, S., Kim, S.H. and Van Riper, S. (2020) Navigating the 'retail apocalypse'. *Journal of Retailing and Consumer Services* 54, 101683.

Higgins-Desbiolles, F. (2020) The 'war over tourism'. *Journal of Sustainable Tourism* 29 (4), 551–569.

Hiremath, S., Panda, A., Prashantha, C. and Pasumarti, S.S. (2023) An empirical investigation of customer characteristics on retail format selection – a mediating role of store image. *Journal of Indian Business Research* 15 (1), 55–75.

Howe, N. and Strauss, W. (1992) *Generations*. Harper Collins.

Hur, W.M., Ahn, K.H. and Kim, M. (2011) Building brand loyalty through managing brand community commitment. *Management Decision* 49 (7), 1194–1213.

Hur, W.M., Kang, S. and Kim, M. (2015) The moderating role of Hofstede's cultural dimensions in the customer-brand relationship in China and India. *Cross Cultural Management* 22 (3), 487–508.

Iaquinto, B.L. and Pratt, S. (2020) Practicing sustainability as a backpacker: The role of nationality. *International Journal of Tourism Research* 22 (1), 100–107.

IBM Institute for Business Value (2018) What do Gen Z shoppers really want? Online: https://www.ibm.com/thought-leadership/institute-business-value/report/genzshoppers.

IBM Institute for Business Value (2022) Balancing sustainability and profitability. Online: https://www.ibm.com/downloads/cas/5NGR8ZW2#:~:text=Our%20February%202022%20survey%20of,see%20Methodology%20on%20page%2020.

Jin, H. and Cai, W. (2022) Understanding the smartphone usage of Chinese outbound tourists in their shopping practices. *Current Issues in Tourism* 25 (18), 2955–2968.

Jin, H., Moscardo, G. and Murphy, L. (2017) Making sense of tourist shopping research: A critical review. *Tourism Management* 62, 120–134.

Jin, H., Moscardo, G. and Murphy, L. (2020) Exploring Chinese outbound tourist shopping: A social practice framework. *Journal of Travel Research* 59 (1), 156–172.

Jin, H., Moscardo, G. and Murphy, L. (2021a) Unfolding Chinese tourist shopping practices: An observational study. *Tourism Review* 76 (2), 460–472.

Jin, H., Moscardo, G. and Murphy, L. (2021b) Unraveling the mechanisms behind Chinese outbound tourist shopping: A social practice perspective. *Journal of Hospitality and Tourism Research* 45 (4), 629–651.

Jin, W. and Zhang, W. (2023) Solo female visitors in museum tourism: An exploration based on social practice theory. *Current Issues in Tourism* 26 (12), 1941–1955.

Kahn, B.E., Inman, J.J. and Verhoef, P.C. (2018) Introduction to special issue: Consumer response to the evolving retailing landscape. *Journal of the Association for Consumer Research* 3 (3), 255–259.

Ketter, E. (2021) Millennial travel: Tourism micro-trends of European Generation Y. *Journal of Tourism Futures* 7 (2), 192–196.

Khare, A. and Sarkar, S. (2021) Does cultural value influence consumers' attitudes toward mall events? *Journal of International Consumer Marketing* 33 (5), 526–542.

Kim, A., Sawaya, A. and Straub, M. (2021, July 22) Hainan's $40 billion prize: The new battleground for global luxury. Online: https://www.mckinsey.com/cn/our-insights/our-insights/hainans-40-billion-prize-the-new-battleground-for-global-luxury.

Kirillova, K., Wang, D., Fu, X. and Lehto, X. (2020) Beyond 'culture': A comparative study of forces structuring tourism consumption. *Annals of Tourism Research* 83, 102941.

Lamers, M. and Pashkevich, A. (2018) Short-circuiting cruise tourism practices along the Russian Barents Sea coast? The case of Arkhangelsk. *Current Issues in Tourism* 21 (4), 440–454.

Lamers, M., van der Duim, R. and Spaargaren, G. (2017) The relevance of practice theories for tourism research. *Annals of Tourism Research* 62, 54–63.

Li, X. (2016) *Chinese Outbound Tourism 2.0*. Apple Academic Press.

Lu, J., Yu, C.S., Liu, C. and Wei, J. (2017) Comparison of mobile shopping continuance intention between China and USA from an espoused cultural perspective. *Computers in Human Behavior* 75, 130–146.

Maheswaran, D. and Shavitt, S. (2000) Issues and new directions in global consumer psychology. *Journal of Consumer Psychology* 9 (2), 59–66.

McKinsey and Company (2020) Consumer sentiment and behavior continue to reflect the uncertainty of the COVID-19 crisis. Online: https://www.mckinsey.com/capabilities/growth-marketing-and-sales/our-insights/a-global-view-of-how-consumer-behavior-is-changing-amid-covid-19.

Mktindex (2022) Why are Pop Mart and Bing Dwen Dwen big hits? Online: https://www.mktindex.com/research/notebook/0627ipnew-Copy1.

Monaco, S. (2018) Tourism and the new generations: Emerging trends and social implications in Italy. *Journal of Tourism Futures* 4 (1), 7–15.

Moraes, C., Carrigan, M., Bosangit, C., Ferreira, C. and McGrath, M. (2017) Understanding ethical luxury consumption through practice theories: A study of fine jewellery purchases. *Journal of Business Ethics* 145 (3), 525–543.

Moscardo, G. (2024) Staycations: A sustainable tourism solution? In T. Walker, E. Demir, G. Machnik-Kekesi and V. Kelly (eds) *Sustainable Tourism: Frameworks, Practices, and Innovative Solutions*. Springer.

Muthukrishna, M., Bell, A.V., Henrich, J., Curtin, C.M., Gedranovich, A., McInerney, J. and Thue, B. (2020) Beyond Western, Educated, Industrial, Rich, and Democratic (WEIRD) psychology: Measuring and mapping scales of cultural and psychological distance. *Psychological Science* 31 (6), 678–701.

Nair, S.J. and Shams, S.M.R. (2021) Impact of store-attributes on food and grocery shopping behavior: Insights from an emerging market context. *EuroMed Journal of Business* 16 (3), 324–343.

Nairn, A. and Spotswood, F. (2015) 'Obviously in the cool group they wear designer things': A social practice theory perspective on children's consumption. *European Journal of Marketing* 49 (9/10), 1460–1483.

Pant, P. (2022) Technology social practices by Millennials and Gen Z at airport departure terminals. *Tourism Management Perspectives* 43, 100978.

Pantano, E. and Gandini, A. (2018) Shopping as a 'networked experience': An emerging framework in the retail industry. *International Journal of Retail and Distribution Management* 46 (7), 690–704.

Paulose, D. and Veliyath, O. (2020) Shopping orientations of youth: Evidence from India. *Journal of Customer Behaviour* 19 (4), 347–374.

Priporas, C-V., Stylos, N. and Fotiadis, A.K. (2017) Generation Z consumers' expectations of interactions in smart retailing. *Computers in Human Behavior* 77, 374–381.

Reckwitz, A. (2002) Toward a theory of social practices. *European Journal of Social Theory* 5 (2) 243–263.

Robertson, S. (2021) *Temporality in Mobile Lives: Contemporary Asia-Australia Migration and Everyday Time*. Bristol University Press.

Rogerson, C.M. and Rogerson, J.M. (2021) COVID-19 and changing tourism demand: Research review and policy implications for South Africa. *African Journal of Hospitality, Tourism and Leisure* 10 (1), 1–21.

Roy, A. (2018) The middle class in India. *Association for Asian Studies* 23 (1), 32–37.

Schatzki, T. (1996) *Social Practices: A Wittgensteinean Approach to Human Activity and The Social*. Cambridge University Press.

Schwartz, S.H. (2014) National culture as value orientations: Consequences of value differences and cultural distance. In V.A. Ginsburgh and D. Throsby (eds) *Handbook of the Economics of Art and Culture*, Volume 2 (pp. 547–586). Elsevier.

Shavitt, S. and Barnes, A.J. (2020) Culture and the consumer journey. *Journal of Retailing* 96 (1), 40–54.

Sheth, J. (2020) Impact of Covid-19 on consumer behavior: Will old habits return or die. *Journal of Business Research* 117, 280–283.

Shove, E., Pantzar, M. and Watson, M. (2012) *The Dynamics of Social Practice: Everyday Life and How It Changes*. Sage.

Singer, C.R. (2020) Where will Canada's 401,000 immigrants come from in 2021? Immigration Canada. Online: https://www.immigration.ca/where-will-canadas-401 000-immigrants-come-from-in-2021.

Singh, S. and Srivastava, P. (2019) Social media for outbound leisure travel: A framework based on technology acceptance model (TAM). *Journal of Tourism Futures* 5 (1), 43–61.

Sthapit, E., Coudounaris, D.N. and Björk, P. (2018) The memorable souvenir-shopping experience: Antecedents and outcomes. *Leisure Studies* 37 (5), 628–643.

Stockdill, R. (2020, February 6) Paris starts to suffer as Chinese tourists disappear. Online: https://insideretail.asia/2020/02/06/paris-starts-to-suffer-as-chinese-tourists-disappear/.

Strauss, W. and Howe, N. (2008) *Millenials and K-12 Schools: Educational Strategies for a New Generation*. LifeCourse Associates.

Sturge, G. (2021) Migration Statistics. House of Commons Library Briefing Paper Number CBP06077. Online: https://commonslibrary.parliament.uk/research-briefings/sn06077/.

Su, N., Min, H., Chen, M.H. and Swanger, N. (2018) Cultural characteristics and tourist shopping spending. *Journal of Hospitality and Tourism Research* 42 (8), 1210–1231.

Szolnoki, G., Bail, S., Tafel, M., Feher, A. and Veith, C. (2022) A Cross-cultural comparison of new implemented sustainable wine tourism strategies during the COVID-19 crisis. *Sustainability* 14 (8), 4688.

Thangavel, P., Pathak, P. and Chandra, B. (2022) Consumer decision-making style of Gen Z: A generational cohort analysis. *Global Business Review* 23 (3), 710–728.

The Migration Observatory at Oxford University (2020) Migrants in the UK: An Overview. Online: https://migrationobservatory.ox.ac.uk/resources/briefings/migrants-in-the-uk-an-overview/.

Thomas, T.C., Epp, A.M. and Price, L.L. (2020) Journeying together: Aligning retailer and service provider roles with collective consumer practices. *Journal of Retailing* 96 (1), 9–24.

Times of India (2018) 97% Indian Travellers Eye 'Sustainable' Tourism in 2018. https://timesofindia.indiatimes.com/business/india-business/97-indians-travellers-eye-sustainable-tourism-in-2018-survey/articleshow/63867379.cms

Timothy, D.J. (2005) *Shopping Tourism, Retailing and Leisure*. Channel View Publications.

Tourism Australia (2022) India Interest in Tourist Experiences. https://www.tourism.australia.com/content/dam/assets/document/1/c/1/4/0/2240928.pdf.

Tu, Y.T., Lin, S.Y. and Chang, Y.Y. (2011) A cross-cultural comparison by individualism/collectivism among Brazil, Russia, India and China. *International Business Research* 4 (2), 175–182.

UNWTO and ADB (2022) COVID-19 and the Future of Tourism in Asia and the Pacific. Online: https://dx.doi.org/10.22617/TCS220110-2.

US International Trade Administration (2022) India – Country Commercial Guide. Online: https://www.trade.gov/country-commercial-guides/india-travel-and-tourism.

Varghese, P.N. (2018) An India Economic Strategy to 2035. DFAT. Online: https://www.dfat.gov.au/publications/trade-and-investment/india-economic-strategy/ies/index.html.

Varshneya, G. (2022) Antecedents and consequences of experiential value in fashion retailing: A study on Indian consumers. *Journal of Fashion Marketing and Management* 26 (5), 832–851.

Viana-Lora, A. and Nel-lo-Andreu, M.G. (2022) Bibliometric analysis of trends in COVID-19 and tourism. *Humanities and Social Sciences Communications* 9 (1), 1–8.

von Briel, F. (2018) The future of omnichannel retail: A four-stage Delphi study. *Technological Forecasting and Social Change* 132, 217–229.

Voyado (2022) How is Generation Z shopping? Online: https://www.apptus.com/blog/generation-z-online-shopping-habits/.

Warde, A. (2005) Consumption and theories of practice. *Journal of Consumer Culture* 5 (2), 131–153.

Weidenfeld, A. (2018) Tourism diversification and its implications for smart specialisation. *Sustainability* 10 (2), 319.

Wen, J. and Kozak, M. (2022) *Chinese Outbound Tourist Behaviour: An International Perspective*. Routledge.

Witt, M.A. and Redding, G. (2013) Asian business systems: Institutional comparison, clusters and implications for varieties of capitalism and business systems theory. *Socio-Economic Review* 11 (2), 265–300.

World Bank (2022) GDP Growth (Annual %). Online: https://data.worldbank.org/indicator/NY.GDP.MKTP.KD.ZG?locations=CN-IN.

World Population Review (2022a) Total Population by Country 2022. Online: https://worldpopulationreview.com/countries.

World Population Review (2022b) Average Age by Country 2022. Online: https://worldpopulationreview.com/country-rankings/median-age.

WTTC (2022) Travel and Tourism Economic Impact 2022 Global Trends. Online: https://wttc.org/Portals/0/Documents/Reports/2022/EIR2022-Global%20Trends.pdf.

Wu, P.C. (2017) Make a loyal visitor: A study of leisure experience at Farglory corporate museum in Taiwan. *Asia Pacific Journal of Tourism Research* 22 (5), 554–564.

Wunderman Thompson (2022) What do Gen Z shoppers really look for? Online: https://www.wundermanthompson.com/insight/what-do-gen-z-shoppers-look-for#:~:text=Gen%20Z%20are%20genuine%20omnichannel%20shoppersandtext=Factors%20that%20most%20will%20look,value%20and%20improved%20customer%20service.

Xu, D., Pearce, P.L. and Chen, T. (2021) Deconstructing tourist scams: A-social-practice-theory perspective. *Tourism Management* 82, 104186.

Xu, Y. and McGehee, N.G. (2012) Shopping behavior of Chinese tourists visiting the United States' Letting the shoppers do the talking. *Tourism Management* 33 (2), 427–430.

Yang, Y., Zhang, C.X. and Rickly, J.M. (2021) A review of early COVID-19 research in tourism: Launching the *Annals of Tourism Research*'s curated collection on coronavirus and tourism. *Annals of Tourism Research* 91, 103–312.

Yi, L., Tong, Y., Wu, M.-Y. and Fu, X. (2023) Taking aging parents on holiday: A social practice perspective. *Journal of Travel Research* 62 (8), 1722-1736.

Zopiatis, A., Pericleous, K. and Theofanous, Y. (2021) COVID-19 and hospitality and tourism research: An integrative review. *Journal of Hospitality and Tourism Management* 48, 275–279.

5 Luxury Retail, Place Branding and Destination Identity through Shopping Tourism

Cemile Ece and Efnan Ezenel

Introduction

Shopping tourism has become a pervasive phenomenon worldwide. People travel not only to meet their basic needs, but also to experience luxury and unique shopping experiences. Luxury retail is a sector that offers high-end products and services and is closely connected to shopping tourism (Martínez-Ruiz *et al.*, 2010). Luxury brands are featured in unique stores and boutiques around the world to offer customers an exceptional shopping experience.

Luxury retail, the selling of high-end goods and services to affluent consumers, has become an integral part of the tourism industry. For many tourists, shopping in luxury stores is an important element of their travel experience, as they seek unique and exclusive products that they might not be able to find in their home country, or because luxury retailers comprise part of the ambience of tourism landscapes. This trend has spawned a form of tourism known as 'luxury shopping tourism', where travelers plan their trips for the sole (or primary) purpose of shopping in luxury stores (Brochado *et al.*, 2019).

The rise of luxury shopping tourism can be attributed to several factors, including the growing number of affluent consumers worldwide, the increase in international travel and the rise of social media, which has made luxury shopping more visible and desirable (Correia *et al.*, 2020). In addition, luxury brands have recognized the potential of tourism as a source of revenue and have invested heavily in creating flagship stores and immersive shopping experiences that appeal to both locals and tourists.

Likewise, destination marketing, branding and image are important concepts in luxury retail tourism. Each of these concepts plays a critical role in attracting travelers to a destination and creating a positive perception of the location and the products and services offered there (Thirumaran & Raghav, 2017). Retail tourism destinations offer a unique and valuable experience for tourists seeking a diverse shopping experience and cultural immersion. Successful retail tourism destinations offer unique and authentic products and experiences, are easily accessible, are effectively marketed and promoted, collaborate with local organizations and provide excellent customer service.

This chapter examines the evolution of luxury retail in the tourism industry, its impact on destination marketing and branding and presents examples of success in the field. Furthermore, the chapter offers strategic insights for many destinations contributing to the tourism industry.

Luxury Retail in the Tourism Industry

Luxury retail refers to the sale of high-end, premium and often exclusive products, brands or services to affluent and sophisticated consumers. Luxury retail encompasses a wide range of products and sectors, including fashion, jewelry, watches, automobiles, accessories, cosmetics, fragrances, housewares and more (Amatulli & Guido, 2012). Luxury retail has a global presence, with flagship stores in major cities around the world. These stores are designed to provide a comprehensive brand experience and attract affluent customers (Pantano et al., 2018). In addition, luxury brands can have a strong online presence to serve a global customer base. Luxury retail is a highly competitive industry where brands are constantly trying to differentiate themselves, maintain their exclusivity and appeal to the changing tastes of luxury consumers (Jebarajakirthy et al., 2020).

Luxury retail is characterized by its commitment to quality, exclusivity, brand prestige and creating an exceptional shopping experience for affluent consumers. It is a dynamic and evolving industry that is influenced by the state of the economy, fashion trends and changes in consumer behavior (Chevalier & Gutsatz, 2012). Luxury retailers emphasize providing an exceptional shopping experience, including personalized customer service, a luxurious and inviting store environment, and attention to the preferences and needs of individual customers. Shoppers often expect a high level of service, from personal shopping advice to customer service (Cervellon & Coudriet, 2013).

One of the most important characteristics of luxury retail is exclusivity. Luxury brands intentionally limit the availability of their products to convey a sense of scarcity and prestige. Limited editions, customized or individualized options, and unique one-of-a-kind items contribute to this exclusivity (Dion & Arnould, 2011). Luxury products tend to be significantly more expensive than their non-luxury counterparts (Pino et al., 2017). These premium prices are justified by factors such as the use of expensive materials, labor-intensive production and the prestige associated with the brand. Luxury brands invest heavily in building and maintaining their brand image, heritage and reputation. Luxury retail in the tourism industry refers to the sale of high-end, exclusive and premium products or experiences to tourists and travelers who seek upscale and indulgent shopping and leisure opportunities during their trips. It is a subset of luxury retail that specifically caters to tourists and is often closely intertwined with the broader tourism sector. Luxury retail in the tourism sector not only contributes to the revenue of luxury brands but also enhances the overall tourism experience for high-end travelers who are seeking a combination of luxury, culture and leisure during their journeys (Park et al., 2010). It is an important niche within both the luxury retail and tourism sectors.

Luxury retail can provide a significant economic boost to tourist destinations. Affluent tourists are often willing to spend more money on high-end products and experiences, which can contribute significantly to local economies through sales tax revenue, job creation and increased business for luxury retailers and related industries (Iloranta, 2022). Luxury retail also attracts high-end tourists who are willing to spend more money on accommodations, restaurants, entertainment and other experiences. These tourists also often stay longer and spend more, benefitting the destination in ways many other consumers do not. Luxury retail can attract tourists to certain areas or districts of a destination, increasing the number of visitors to other parts of a locale and spreading their expenditures to areas outside the normative tourist zones. This can help promote less visited parts of a destination. In addition, locating luxury stores or boutiques can enhance a destination's brand image and reputation. They can position the destination as upscale and sophisticated, attracting a more affluent and discerning group of travelers.

Influence of Luxury Retail on Destination Marketing

Destination marketing is the strategic planning, promotion and management of a particular place, such as a city, region, or destination, with the goal of attracting visitors, overnight tourists, or other travelers and encouraging them to choose the destination for their leisure or business activities (Pike, 2012). The goal of destination marketing is to improve the visibility, reputation and competitiveness of the destination in the tourism and hospitality industry. Destination marketing is an important component of the tourism industry because it not only spurs economic growth, it also supports sustainable development, cultural preservation, community well-being and positive global interactions (Soteriades, 2012). It is a multifaceted endeavor that involves strategic planning, promotion and management to ensure a destination's attractiveness while benefiting both tourists and the local community (du Rand & Heath, 2006).

The relationship between destination management and luxury tourism is complex and interdependent. Effective destination management and luxury shopping tourism can reinforce and complement each other. As noted, luxury shopping tourism often appeals to high-end travelers seeking a premium shopping experience. A well-managed destination can attract the high-end market by offering a diverse range of luxury retail options and ensuring a smooth, enjoyable visit (Thirumaran & Raghav, 2017). Luxury retail outlets contribute to the overall image and reputation of a destination. Destination management organisations (DMOs) can strategically position their destination as a centre for luxury shopping to attract affluent visitors and enhance the prestige of the destination. Luxury retail tourism can significantly boost a destination's economy by encouraging tourists to spend more money on high-end products and services (Hung *et al.*, 2021). Destination management focuses on creating a positive and memorable visitor experience. High-end shopping is an essential part of this experience for luxury travelers. DMOs can ensure that luxury retail complements other aspects of the destination's offering, such as fine dining, cultural attractions and leisure activities. Effective destination management should include sustainability and responsible tourism practises. Luxury retail can align with these practises by promoting sustainable and ethical products, contributing to the long-term viability and reputation of a destination (Genç, 2019).

The image of a destination is the overall perception and impression that a destination leaves on its target audience. This image is formed based on the preconceptions, knowledge and experiences that tourists have about a destination (Hosany *et al.*, 2007). Factors such as environmental, cultural, social and economic aspects, as well as marketing activities, influence the formation of a destination's image (Beerli & Martin, 2004). Destination image is a fundamental part of branding a destination, which is about highlighting the unique qualities of a destination to differentiate it from others and define its identity (Kladou *et al.*, 2017). Destination branding is done to enhance the tourism potential of a destination and make it attractive to the target audience. Destination branding is the process of creating a unique and recognisable identity for a destination. This involves developing a compelling story, visual identity and messages that set the destination apart from the competition and communicate its core values, culture and attractions to potential visitors (Pike, 2005).

In the context of luxury shopping tourism, destination branding revolves around the luxury shopping experience. A destination can be defined by the presence of shops, boutiques or malls that house luxury brands and offer high-end products, prestigious brands and unique shopping experiences. Destination branding in this context aims to highlight the luxury shopping experience to attract the attention of tourists (Camilleri, 2018). To achieve destination branding, marketing strategies are used to highlight a destination's potential in luxury retail tourism.

These strategies may include efforts to attract luxury retail brands, support local businesses and organise shopping events and festivals. In addition, promotional activities, advertising campaigns and digital marketing are used to increase awareness and reputation in luxury retail.

Luxury retail plays a central role in shaping destination marketing strategies in the tourism industry. This influence stems from the unique appeal and economic impact associated with luxury shopping. Luxury retail facilities such as high-end boutiques, flagship stores of well-known brands and upscale shopping districts greatly enhance the overall appeal of a destination and are a major draw for tourists (Chi *et al.*, 2020). Luxury retail destinations naturally attract wealthy tourists with higher disposable incomes. This market typically spends more on lodging, dining and entertainment, causing destination marketers to realize the importance of attracting high-spending tourists and the critical role of luxury retail as a powerful magnet to achieve this goal (Cervellon & Coudriet, 2013).

Through time, exposure, marketing and word-of-mouth, luxury retail becomes an integral part of a destination's identity, especially when the image of a place is significantly influenced by its association with prestigious brands and high-end shopping experiences (Park *et al.*, 2010). Destination marketing often uses this identity to position the place as a luxurious and sophisticated destination that can attract a specific segment of tourists seeking opulent experiences. Many destinations are also known for their local artisans and craftsmen who create exquisite products, and destination marketing frequently highlights these artisans and their creations, supporting local businesses and preserving traditional craftsmanship while offering unique shopping experiences.

High-end shopping, luxury boutiques and upscale retail districts often create an aura of exclusivity and sophistication (Veríssimo & Loureiro, 2013). When a destination is associated with such luxury shopping experiences, it tends to increase the overall perception of the place. Tourists view such destinations as more glamorous, sophisticated and upscale. The affluent visitors attracted to these places are often regarded as sophisticated and having desire for quality, which can enhance the destination's image (Lee & Kim, 2020).

The marketing strategy behind a destination's image involves creating a compelling, authentic, and differentiated identity, promoting that identity through multiple channels, engaging with the target audience, and continually refining the approach based on feedback and changing market dynamics (Mei, 2022). The goal is to create a positive, appealing and memorable image that attracts visitors and supports the overall growth and development of the destination's tourism industry (Yozcu, 2017). Post-industrial development and rebranding of places through retail tourism are a strategic response to economic and cultural change (Anttirokio, 2015). By leveraging the appeal of high-value shopping experiences and cultural revitalization, destinations can redefine their identity, attract tourists, promote economic growth, and position themselves competitively in the changing world of tourism. Today, many destination marketing organizations are using shopping tourism as a core component of their strategies. Collaboration with retailers, special shopping events and marketing campaigns focused on luxury shopping are used to attract attention and reposition the destination (see Chapter 5). Post-industrial development means a shift away from traditional manufacturing and industrial sectors toward a service-based and knowledge economy (Baksi, 2022). Many destinations that have experienced deindustrialization or economic decline are trying to reinvent themselves. Shopping tourism is one strategy to rejuvenate these areas by leveraging their unique characteristics and providing shopping experiences that appeal to tourists.

Global Urban to Local Boutique Tourism Destinations

Traditionally, tourists have been drawn to major urban centers with their large shopping districts and global retail chains, many of which serve a luxury market (Martínez-Ruiz et al., 2010; Zaidan, 2019). However, there is a growing trend where tourists want to explore smaller, more local areas, often within cities, towns or even rural areas, to discover unique and culturally rich shopping experiences. This shift in tourists' preferences reflects a desire for a closer connection to the places they visit. While major urban centers and global retail chains offer convenience and familiarity, as well as hallmark brands and luxury goods, tourists are increasingly drawn to the charm of smaller, local areas. These areas lure travelers with the promise of authentic encounters and the opportunity to discover hidden treasures (Popescu & Olteanu, 2014).

In these non-mainstream luxury retail destinations, visitors can leave the beaten path and immerse themselves in a world where every shop tells a story. Whether it is a centuries-old family bakery famous for its secret pastry recipes, an art gallery showcasing the works of local talent or a bustling open-air market filled with handmade crafts, each business holds the potential for discovery and a boutique experiences (Karanth & Karanth, 2012). Choosing to support local businesses in these destinations often means directly contributing to the well-being of the community and preserving their unique way of life (García-Milon & Juaneda-Ayensa, 2024).

These local shopping environments represent a compelling change from traditional shopping experiences in luxury urban malls. They embody a particular facet of experiential tourism characterized by the pursuit of authenticity and cultural immersion through consumption. In contrast to the homogenized and standardized merchandise ubiquitous in urban malls, these places offer an exceptional and genuine shopping experience (Brenner & Aguilar, 2002). In these local retail destinations, visitors have the opportunity to discover unique goods, many of which are one-of-a-kind, made with great attention to detail and having a deep connection to the local cultural milieu. Whether it is handicrafts, unique culinary delicacies or hand-woven textiles, each product carries its own story interwoven with the cultural fabric of the destination.

A key element that sets these destinations apart is access to the people behind the crafts. Artisans, boutique owners and local craftspeople are not only purveyors of goods, but also keepers of traditional knowledge and culture. Through dialog with these custodians of craft, visitors can delve deeper into the heritage and traditions of the place (Poelina & Nordensvard, 2018). These interactions go beyond commercial transactions, evolving into cultural exchanges that enrich the visitor's understanding of the destination.

Moreover, shopping at these off-the-beaten-path retail areas goes beyond the mere act of commerce; it becomes a channel for social engagement. Visitors make meaningful connections with the local community, foster cross-cultural dialogs and gain deep insights into the social intricacies of the destination (Lopes *et al.*, 2023). This dynamic exchange transforms the shopping experience into a holistic exploration of the destination's identity. The emergence of what Getz (1993) termed 'tourist shopping villages' is, in many cases, a manifestation of the 'luxuryfication' of rural shopping areas that go through a process of marginal and rural retail, to becoming high-end shopping destinations for urbanites who want to escape the harried life of the cities (Albert, 2019; Timothy, 2024).

Local shopping tourism destinations serve as the epitome of authentic and distinctive consumer experiences, and even many luxury consumers are beginning to seek such experiences. They embody the convergence of heritage preservation, economic uplift and cultural exchange. These destinations have become exemplars of experiential tourism, attracting visitors who seek a deeper connection to the places they explore through the act of shopping (Istenič & Bajec, 2021).

Examples of Luxury Shopping Tourism

Interest in luxury tourism has increased significantly in the tourism industry in recent years. Instead of visiting traditional tourist attractions, many travelers now choose to travel to discover unique luxury shopping experiences (Brochado *et al.*, 2019). This phenomenon shows that luxury is perceived as a lifestyle and an experience. Many destinations now attract tourists with their luxury brands, shopping malls and upscale lifestyles, with Paris, Milan, New York, Dubai, Tokyo and Doha now being counted among the world's leading luxury retail destinations, but there are many others.

Monaco, a key center of luxury retail tourism and often referred to as the 'playground of the rich and famous', is a prime example of luxury retail tourism. Located on the Côte d'Azur, the small principality is known for its opulence, glamor and extravagant lifestyle. At the heart of

Monaco's appeal is its thriving luxury retail scene, which caters to discerning travelers seeking the pinnacle of high-end shopping experiences (d'Hauteserre, 2005; Timothy, 2017). Monaco's reputation as a premier luxury destination is underpinned by its commitment to providing visitors with unparalleled shopping experiences that reflect the timeless charm of the historic principality. Whether one is in search of haute couture, horological masterpieces, bespoke creations or culinary delights, Monaco embodies the epitome of luxury shopping tourism (Kauppinen-Räisänen et al., 2020; Timothy, 2021).

Dubai, a metropolis renowned for its lavishness, magnificence and cutting-edge architectural designs, is a global center for tourism focused on extravagant retail. Situated in the core of the United Arab Emirates, Dubai has become synonymous with exclusive shopping encounters that captivate discerning globetrotters from all corners of the globe (Zaidan, 2016, 2019). In the realm of retail, few counterparts can measure up to The Dubai Mall, offering an unparalleled haven for luxury enthusiasts and claiming to be the world's largest mall. This retail destination hosts an extensive range of esteemed luxury brands, boasting an impressive collection of haute couture fashion. Dubai's allure as a luxury shopping haven extends further to encompass opulent hotels and resorts, showcasing in-house boutiques and personalized shopping services. Visitors are invited to indulge in a seamless amalgamation of retail therapy and sumptuous accommodation (Mehta et al., 2014).

Renowned for its opulence, Hollywood stands as an esteemed hub for extravagant tourism. The fusion of its affluent retail sector with the glitz and glamour of the entertainment industry allows tourists to immerse themselves in the lavishness and sophistication associated with the realm of cinema and notable personalities. Hollywood opens its doors to a realm of opulence and haute couture, from legendary shopping streets to upscale boutiques and exclusive gatherings, rendering it an ultimate magnet for those craving upscale retail encounters (Reynolds, 2012).

Regarded frequently as the 'Fashion Capital of the World', New York City is as a global hub for luxury retail tourism. With its vibrant and diverse neighborhoods, the city offers a unique shopping experience, attracting fashion enthusiasts, trendsetters and elite travelers from all corners of the world (Park et al., 2010). Fifth Avenue, renowned as one of the planet's most esteemed shopping streets, features an impressive lineup of luxury flagship stores. In addition, the charm of Soho and the Meatpacking District lies in their cobblestone streets and trendy boutiques, known for offering designer fashion, stylish footwear and luxury beauty brands. New York City's luxury retail tourism scene presents a dynamic tapestry of iconic shopping streets, upscale boutiques, world-renowned department stores and culturally enriching experiences. It embraces diversity, innovation and style, positioning the city as a premier destination for those seeking the epitome of high-end shopping and fashion exploration (Donzé & Fujioka, 2018).

A hub of luxury retail tourism, Paris, is also famously known as the 'Fashion Capital of the World', just like New York City. France's capital magnetizes elite travelers, trendsetters and fashion enthusiasts owing to its rich history, elegance and significant impact on the fashion industry. In Paris, numerous high-end retailers provide exclusive shopping experiences, allowing customers to indulge in personalized services such as private showings, personal shoppers and bespoke options. With its amalgamation of art, culture and luxury, Paris blends historical buildings housing luxury boutiques alongside cultural institutions, creating an integration of fashion and artistic ambiance (Dion & Borraz, 2015; Rabbiosi, 2015).

Renowned for its modernity and multiculturalism, Singapore has solidified its position as a hub for upscale retail tourism in Asia (Brochado et al., 2019). The city-state showcases a fusion of high-end shopping districts, opulent boutiques and a lively culinary scene. As a result, it magnetizes discerning travelers and fashion enthusiasts from all corners of the globe (Henderson et al., 2011). Boasting an assortment of shopping districts, streets adorned with chic boutiques, and a steadfast commitment to innovation and elegance, Singapore delivers an unparalleled amalgamation of deluxe shopping experiences catered to individual tastes and preferences (Hung et al., 2021).

Hong Kong's status as a global shopping and lifestyle destination is indisputable, as reflected in its flourishing luxury retail tourism sector, which draws millions of consumers from Mainland China and further afield (Hung et al., 2020). The city offers a diverse selection of shopping districts, ranging from bustling urban centers to artsy neighborhoods, ensuring that visitors can immerse themselves in a realm of luxury, culture and fashion, all while savoring the unique blend of Eastern and Western influences that define Hong Kong (Correia et al., 2018). Moreover, Hong Kong's penchant for luxury shopping extends to the realm of gourmet dining, as high-end shopping centers and districts house Michelin-starred restaurants and stylish cafes.

Renowned as an ever-evolving form of travel, luxury retail tourism provides an extraordinary encounter that goes beyond mere shopping. By skillfully merging luxury retail and tourism, the examples noted above present travelers with indelible experiences. These profound encounters serve as a testament to the fact that luxury surpasses being a mere commodity or service, instead becoming a way of life and a conduit for cultural eminence. For explorers seeking the epitome of gratification, luxury retail tourism bestows them with the chance not only to relish in the act of shopping but also to delve into an exploration of diverse cultures and lifestyles found across the globe.

Conclusion and Implications

In conclusion, the symbiotic relationship between luxury retail and the tourism industry has become increasingly pronounced. Tourists no longer embark on journeys solely to explore destinations; many now actively seek immersive shopping experiences as integral parts of their travels. This transformation underscores the pivotal role that destinations can play in elevating the allure of luxury retail, subsequently exerting a profound influence on the branding and image-crafting endeavors of these locales. In this regard, the strategic positioning of a destination through unique luxury retail offerings becomes paramount, where the development of a distinct identity is instrumental (Correia et al., 2019). It is abundantly clear that the processes of branding and image construction wield substantial influence over tourists' decisions when they are in the midst of selecting their travel destinations.

Emerging as a crucial element in destination branding and attracting a larger influx of tourists, luxury retail tourism plays a vital role in the ever-changing landscape. Consequently, this symbiotic relationship not only strengthens the economic growth of a region but also amplifies its competitive advantage in the global tourism industry. Moreover, the influential nature of luxury retail extends its reach to enhance the overall attractiveness of a destination, thereby impacting tourists' preferences and decision-making. With an increasing awareness of this transformative potential, destinations are progressively devising creative marketing strategies to foster closer partnerships with the luxury retail sector (Mishra & Kukreti, 2022).

The considerable potential of luxury retail to shape and redefine the tourism industry cannot be overstated. Destinations can embrace a strategic approach to harness this potential fully, ensuring that it not only leads to the creation of indelible and enchanting experiences for tourists but also positions destinations to seize the manifold opportunities for robust economic growth.

References

Albert, M.L. (2019) Clicks and mortar: The modernization of boutique retail to aid rural revitalization in Mississippi. Unpublished thesis, University of Mississippi.

Amatulli, C. and Guido, G. (2012) Externalised vs. internalised consumption of luxury goods: Propositions and implications for luxury retail marketing. *The International Review of Retail, Distribution and Consumer Research* 22 (2), 189–207.

Anttiroiko, A.V. (2015) City branding as a response to global intercity competition. *Growth and Change* 46 (2), 233–252.

Baksi, A.K. (2022) Branding luxury travel. In A.S. Kotur and S.K. Dixit (eds) *The Emerald Handbook of Luxury Management for Hospitality and Tourism* (pp. 243–270). Emerald.

Beerli, A. and Martin, J.D. (2004) Factors influencing destination image. *Annals of Tourism Research* 31 (3), 657–681.

Brenner, L. and Aguilar, A.G. (2002) Luxury tourism and regional economic development in Mexico. *The Professional Geographer* 54 (4), 500–520.

Brochado, A., Oliveira, C., Rita, P. and Oliveira, F. (2019) Shopping centres beyond purchasing of luxury goods: A tourism perspective. *Annals of Leisure Research* 22 (4), 484–505.

Camilleri, M.A. (ed.) (2018) *Tourism Planning and Destination Marketing*. Emerald.

Cervellon, M.C. and Coudriet, R. (2013) Brand social power in luxury retail: Manifestations of brand dominance over clients in the store. *International Journal of Retail and Distribution Management* 41 (11/12), 869–884.

Chevalier, M. and Gutsatz, M. (2012) *Luxury Retail Management: How the World's Top Brands Provide Quality Product and Service Support*. Wiley.

Chi, H.K., Huang, K.C. and Nguyen, H.M. (2020) Elements of destination brand equity and destination familiarity regarding travel intention. *Journal of Retailing and Consumer Services* 52, 101728.

Correia, A., Kozak, M. and Kim, S. (2018) Luxury shopping orientations of mainland Chinese tourists in Hong Kong: Their shopping destination. *Tourism Economics* 24 (1), 92–108.

Correia, A., Kozak, M. and Del Chiappa, G. (2020) Examining the meaning of luxury in tourism: A mixed-method approach. *Current Issues in Tourism* 23 (8), 952–970.

Correia, A., Kozak, M. and Kim, S. (2019) Investigation of luxury values in shopping tourism using a fuzzy-set approach. *Journal of Travel Research* 58 (1), 77–91.

d'Hauteserre, A.-M. (2005) Tourism, development and sustainability in Monaco: Comparing discourses and practices. *Tourism Geographies* 7 (3), 290–312.

Dion, D. and Arnould, E. (2011) Retail luxury strategy: Assembling charisma through art and magic. *Journal of Retailing* 87 (4), 502–520.

Donzé, P.Y. and Fujioka, R. (eds) (2018) *Global Luxury: Organizational Change and Emerging Markets since the 1970s*. Springer.

Dion, D. and Borraz, S. (2015) Managing heritage brands: A study of the sacralization of heritage stores in the luxury industry. *Journal of Retailing and Consumer Services* 22, 77–84.

du Rand, G.E. and Heath, E. (2006) Towards a framework for food tourism as an element of destination marketing. *Current Issues in Tourism* 9 (3), 206–234.

García-Milon, A. and Juaneda-Ayensa, E. (2024) Local tourist shopping: How purchases in a destination can contribute to sustainability. In T. Walker, E. Demir, G. Machnik-Kekesi and V. Kelly (eds) *Sustainable Tourism: Frameworks, Practices, and Innovative Solutions* (pp. 129–150). Springer.

Genç, R. (2019) Luxury shopping as a new opportunity for tourism market development. *Journal of Tourism and Hospitality Education* 9, 1–8.

Getz, D. (1993) Tourist shopping villages: Development and planning strategies. *Tourism Management* 14 (1), 15–26.

Henderson, J.C., Chee, L., Mun, C.N. and Lee, C. (2011) Shopping, tourism and retailing in Singapore. *Managing Leisure* 16 (1), 36–48.

Hosany, S., Ekinci, Y. and Uysal, M. (2007) Destination image and destination personality. *International Journal of Culture, Tourism and Hospitality Research* 1 (1), 62–81.

Hung, K., Qiu Zhang, H., Guillet, B.D. and Wang, L. (2020) China watching: Luxury consumption and its implications. *Journal of Travel and Tourism Marketing* 37 (5), 577–592.

Hung, K., Ren, L. and Qiu, H. (2021) Luxury shopping abroad: What do Chinese tourists look for? *Tourism Management* 82, 104182.

Iloranta, R. (2022) Luxury tourism–a review of the literature. *European Journal of Tourism Research* 30, 3007–3007.

Istenič, S.P. and Bajec, J.F. (2021) Luxury food tour: Perspectives and dilemmas on the' luxurification' of local culture in tourism product. *Acta geographica Slovenica* 61 (1), 169–184.

Jebarajakirthy, C., Yadav, R. and Shankar, A. (2020) Insights for luxury retailers to reach customers globally. *Marketing Intelligence and Planning* 38 (7), 797–811.

Karanth, K.U. and Karanth, K.K. (2012) A tiger in the drawing room: Can luxury tourism benefit wildlife? *Economic and Political Weekly* 47, 38–43.

Kladou, S., Kavaratzis, M., Rigopoulou, I. and Salonika, E. (2017) The role of brand elements in destination branding. *Journal of Destination Marketing and Management* 6 (4), 426–435.

Kauppinen-Räisänen, H., Mühlbacher, H. and Taishoff, M. (2020) Exploring consumers' subjective shopping experiences in directly operated luxury brand stores. *Journal of Retailing and Consumer Services* 57, 102251.

Lee, S. and Kim, D.Y. (2020) Brand tourism effect in the luxury hotel industry. *Journal of Product and Brand Management* 30 (1), 90–103.

Lopes, J.M., Gomes, S., Durão, M. and Pacheco, R. (2023) The Holy Grail of luxury tourism: A holistic bibliometric overview. *Journal of Quality Assurance in Hospitality and Tourism* 24 (6), 885–908.

Martínez-Ruiz, M.P., Martínez-Caraballo, N. and Amatulli, C. (2010) Tourist destinations and luxury commerce: Business opportunities. *Journal of Place Management and Development* 3 (3), 205–220.

Mehta, S., Jain, A. and Jawale, R. (2014) Impact of tourism on retail shopping in Dubai. *International Journal of Trade, Economics and Finance* 5 (6), 530–535.

Mei, X.Y. (2022) Marketing luxury experiences in an emerging luxury destination of Norway. In A.S. Kotur and S.K. Dixit (eds) *The Emerald Handbook of Luxury Management for Hospitality and Tourism* (pp. 145–161). Emerald.

Mishra, A. and Kukreti, M. (2022) Challenges and prospects for Oman in the making of luxury tourism destination. In A.S. Kotur and S.K. Dixit (eds) *The Emerald Handbook of Luxury Management for Hospitality and Tourism* (pp. 101–121). Emerald.

Pantano, E., Passavanti, R., Priporas, C.V. and Verteramo, S. (2018) To what extent luxury retailing can be smart? *Journal of Retailing and Consumer Services* 43, 94–100.

Park, K.S., Reisinger, Y. and Noh, E.H. (2010) Luxury shopping in tourism. *International Journal of Tourism Research* 12 (2), 164–178.

Pike, S. (2005) Tourism destination branding complexity. *Journal of Product and Brand Management* 14 (4), 258–259.

Pike, S. (2012) *Destination Marketing*. Routledge.

Pino, G., Guido, G. and Nataraajan, R. (2017) Iconic art infusion in luxury retail strategies: Unveiling the potential. *Journal of Global Scholars of Marketing Science* 27 (2), 136–147.

Poelina, A. and Nordensvard, J. (2018) Sustainable luxury tourism, indigenous communities and governance. In M.A. Gardetti and S.S. Muthu (eds) *Sustainable Luxury, Entrepreneurship, and Innovation* (pp. 147–166). Springer.

Popescu, I.V. and Olteanu, V. (2014) Luxury tourism: Characteristics and trends of the behavior of purchase. *Sea: Practical Application of Science* 2 (2), 319–324.

Rabbiosi, C. (2015) Renewing a historical legacy: Tourism, leisure shopping and urban branding in Paris. *Cities* 42, 195–203.

Reynolds, M. (2012) A glamorous gentrification: Public art and urban redevelopment in Hollywood, California. *Journal of Urban Design* 17 (1), 101–115.

Soteriades, M. (2012) Tourism destination marketing: Approaches improving effectiveness and efficiency. *Journal of Hospitality and Tourism Technology* 3 (2), 107–120.

Thirumaran, K. and Raghav, M. (2017) Luxury tourism, developing destinations: Research review and trajectories. *Asian Journal of Tourism Research* 2 (2), 137–158.

Timothy, D.J. (2017) Monaco. In L.L. Lowry (ed.) *The SAGE International Encyclopedia of Travel and Tourism* (pp. 838–840). Sage.

Timothy, D.J. (2021) *Tourism in European Microstates and Dependencies: Geopolitics, Scale and Resource Limitations*. CABI.

Timothy, D.J. (2024) Tourism, shopping and connotations of place. In C.M. Hall (ed.) *The Wiley Blackwell Companion to Tourism* (2nd edn, pp. 501–514). Wiley.

Veríssimo, M. and Loureiro, S.M.C. (2013) Experience marketing and the luxury travel industry. In J.A.C. Santos, F.P. Ribeiro and P. Águas (eds) *Proceedings, Tourism and Management Studies Algarve* 2012 (pp. 296–302). University of Algarve.

Yozcu, O.K. (2017) Competitiveness of Istanbul as a tourism destination for luxury market. *Journal of Tourismology* 3 (2), 2–13.

Zaidan, E. (2016) Tourism shopping and new urban entertainment: A case study of Dubai. *Journal of Vacation Marketing* 22 (1), 29–41.

Zaidan, E. (2019) Shopping tourism and hyper-development in the Middle East and North Africa. In D.J. Timothy (ed.) *Routledge Handbook on Tourism in the Middle East and North Africa* (pp. 365–377). Routledge.

Part 2

Economic, Urban and Spatial Perspectives

6 Tax-Free and Duty-Free Shopping: Benefits to Communities and Tourists

Azila Azmi, Azrul Abdullah and
Mohammad Fadhili Yahaya

Introduction

Shopping activities underscore much of the appeal of tourist destinations, both for purposive shoppers and casual consumers. Although souvenirs and travel supplies (e.g. suntan lotion, sunglasses and toothpaste) are the most common items tourists buy, opportunities to buy higher-end merchandise present themselves frequently in tourist destinations and in transit to the destination. Many people who might otherwise desire to purchase fine merchandise while on holiday are deterred by high prices, particularly regarding highly taxed and otherwise expensive consumer products. To assuage these consumer concerns and to stimulate increased retail sales, most countries have enacted laws and policies that enable non-resident consumers to purchase certain highly-taxed merchandise at tax-free, or lower-taxed, rates as long as the items are not to be consumed within the country where they are purchased (Streng, 1992).

Since the 1960s, duty-free and other forms of tax-free shopping have become widespread and are an important part of the tourist experience in the destination and the spaces in transit. Various programs that eliminate or reduce consumer taxes for tourists (i.e. non-residents), such as value added taxes (VAT), goods and services taxes (GST), alcohol taxes and import duties have been enacted and are now commonplace at departure points and a few other locations throughout the world. This chapter examines the important role of tax-free shopping as a stimulator of retail commerce and an appealing part of the tourism retailscapes of places. It first examines general tax-free retail and its implications, followed by duty-free programs, providing several international examples. It then turns to looking at how duty-free (import taxes) and other tax-free (consumer taxes) shopping can be a stimulator of tourism growth and a motivation for people's choices of a holiday destination.

Tourism, Taxes and Shopping

Taxes are an inevitable part of life. They are enacted by governments to raise revenue for social programs, administrative costs, environmental protection, heritage conservation, health care, education and many other causes. Many different taxes are levied on a place's inhabitants, including excises on income, property values, corporate earnings, inheritances, capital gains, retail merchandise (sales or consumer taxes), automobile usage, alcohol and tobacco, lottery winnings and gambling, services, fuel and gasoline, and many other ordinary elements of everyday consumption. Tourism-specific taxes include lodging taxes, departure fees and certain air ticket fees, sales tax on entrance tickets and taxes on restaurant meals. In addition to tourism taxes, tourists also pay sales/consumer taxes on items they buy in the destination or along the way. Yet, there is a legal concept that most countries ascribe to, which suggests that non-residents should not have to pay consumer taxes on goods that will not be consumed in the destination countries they visit. This is described in greater detail later.

Tourism is one of many countries' most profitable income generators. Tourists are valued not only by tour operators, transportation and lodging providers, as well as government agencies, but also by retailers (Lubis *et al.*, 2020). Hence, it is little wonder that tourists are often given special treatment in some countries because they are seen as an economic blessing for their positive fiscal impacts on private tourism firms, retail businesses and the nation at large. Numerous efforts have been made and incentives given by both government and private companies to boost the number of visitors in a region. The private sector focuses mainly on tourist services, which may include upgrading facilities and creating signature or unique service products to appeal to potential visitors. Governments try to instill tourist-friendly physical and policy environments to accommodate better services, promote positive visitor experiences and develop loyalty with intentions to revisit (Taecharungroj & Tansitpong, 2017).

Nevertheless, between private enterprises and public agencies, the more significant contributor in attracting tourists traditionally has been the government, for it is governments that typically provide and facilitate foreign arrivals and guide market growth. Security, stability and attractive visitor-friendly policies can encourage tourists to choose a country as a holiday destination (Budiarta & Pemayun, 2019). An advantageous retail environment is one such policy framework a country can establish to provide satisfying experiences and induce increased expenditures.

Although destination shopping is often geared towards tourists, retail transactions are obviously not limited to foreign tourists alone. Local shoppers and domestic visitors are also enticed by shopping opportunities. Non-tourist consumers often leave their immediate communities to shop for economic and leisure purposes in other cities or towns, and many even cross a national border to outshop in a neighboring country (Borzooie *et al.*, 2021; Makkonen, 2016, 2023; Timothy, 1995, 2005; Timothy & Butler, 1995) (see Chapter 7). This practice has been well studied and is referred to in the retail, tourism and geography literature as cross-border shopping or outshopping (Fullerton & Walke, 2019; Szytniewski *et al.*, 2018). There is a vast literature on cross-border shopping from both touristic and utilitarian perspectives. This phenomenon itself is not the focus of this chapter, although the notions of tax-free and duty-free shopping are intrinsically connected to the existence of political boundaries (Hussin *et al.*, 2012; Weaver, 2017), for it is at points of departure or arrival that the benefit and performance of tax-free commerce are eventualized. Reduced taxes or tax-free shopping incentives are one of the reasons shopping has flourished in border communities, but it does not take place only in borderlands alone. Tax-free programs are one of the government incentives described above that are introduced to stimulate expenditures and provide unique retail experiences.

In most localities, the majority of goods and services are taxed. As previously noted, the most common of these in a shopping context are sales tax (e.g. GST and VAT) and alcohol and tobacco taxes. In addition, some destinations that have experienced significant overtourism and ecological degradation through mass tourism (or see themselves going in that direction) have implemented tourism-specific environmental taxes (usually added to lodging nights or cruise ship passenger arrivals) to help fund the restoration of natural environments and enforce environmental laws already in place (Palmer & Riera, 2003; Usman & Alola, 2023). Thus, just like in ordinary life, taxes are an inevitable part of tourism.

VAT refunds and tax-free shopping

VATs are a type of sales tax paid to the government each time a purchase takes place along the supply chain. Conversely, standard sales tax is paid only by the consumer at the point of purchase (Ilter & Manahan, 2021). In the countries of Europe, VAT taxes are typically levied nationwide, whereas in the United States, sales taxes are layered and imposed by state, county and city governments. In either case, they are both retail transaction levies imposed by governments.

Tax incentives provided by governments in collaboration with private enterprises have been shown to boost tourists' retail purchases (Wang & Stewart, 2022). Tax-free shopping (TFS) is a common measure established by national or regional governments to stimulate retail sales. TFS schemes allow non-resident travelers to receive a refund on the taxes (sales tax and VAT) they paid when buying merchandise. For the most part, this means foreign travelers only pay the cost of the product or service without having to pay for the same government-imposed levies that residents must pay. As a result, travelers' purchasing power increases, allowing them to buy more merchandise. This incentive has been offered mainly as a tonic to boost the economy of countries by attracting more tourists but more importantly to get them to spend more in the destination.

Although some countries allow shoppers to pay tax-free prices at the point of sale with proof of non-residency and imminent departure from the country (e.g. Japan and Australia), most countries require tourists to seek tax refunds at departure stations, such as airports and cruise ship ports. A prominent retail landscape feature throughout the world is the VAT refund booth at international airports. Some countries, such as Japan and Australia, have established tax-free shops in popular urban tourist zones rather than only at points of departure.

The Japanese have long been regarded as some of the most avid shopper-tourists, and much of their holiday time is spent in retail venues (Kim *et al.*, 2011; Reisinger & Turner, 2002). Japan, likewise, has now become a significant shopping destination for other nationalities, and the country offers tax breaks and other incentives to tourists to encourage more retail activity. For example, famous for its sake liquor, tourists in Japan can buy the famous Japanese spirits at a sake brewery free of sales and alcohol taxes, as long as certain minimum criteria are met, such as the total amount of the sale (excluding the taxes) amounting to at least 5,000 yen (Japan National Tourism Organization, 2024). Japan's tax exemption program for tourists is one of the more frequently cited successes. Tax-free retail in Japan applies to almost all general merchandise, including appliances, electronics, shoes, cosmetics, medicines, tobacco products, food products and alcohol (Japan National Tourism Organization, 2024).

Of the current 193 sovereign states in the world today, at least 170 levy consumer taxes, and many of these allow non-residents to seek VAT and GST reimbursements when leaving the country. In Europe, all 27 European Union countries and all four European Free Trade Association countries (Iceland, Liechtenstein, Norway and Switzerland) offer consumer tax (VAT) refunds for tourists upon departure, although most countries return only a portion of the total VAT spent, and residents of Sweden, Finland and Denmark are unable to claim the tax exemption when returning from Norway (Tax Foundation, 2024). The system is complex, and each country has its own rules, regulations, and minimum purchase requirements to qualify for the refund. With the formation of the European Union, duty-free sales between members states were abolished. For travelers to destinations outside the EU, however, duty-free sales remain. Some duty-free shops are permitted to sell goods to intra-EU travelers, but they are required to charge full-duty prices and local VATs.

VAT rates in Europe vary by country. Of the EU countries, Hungary has the highest VAT (27%), while Luxembourg has the lowest (16%). Because they are not members of the EU, Switzerland and Liechtenstein have very low VATs (8.1%), compared with other European states. Andorra, also not an EU member, has the lowest VAT in Europe – 4.5%

and offers significant duty-free opportunities, which attracts many consumers from nearby Spain and France who cross the border to buy a wide variety of products (Timothy, 2021).

Australia's tax-free program is called the Tourist Refund Scheme. This initiative lets foreign tourists claim a refund on the GST and wine tax spent during the most recent 60 days before leaving the country. Refund claims can be made at international airports and international ship ports (Jin et al., 2020; Wang & Stewart, 2022).

In large federated countries with strong decentralized authority, some individual states or provinces even offer sales tax exemptions for domestic non-residents. For example, the US state of Washington has enacted a sales tax exemption program for out-of-state shopper-visitors who purchase big-ticket items. Many non-residents qualify for sales tax refunds from large purchases of vehicles, watercraft and large vessels, farm machinery and trailers/caravans. For many years, this has stimulated a significant flow of high-end shopping by residents of Washington's neighboring states (and Canada), and as far away as Hawaii and Alaska (Washington Department of Revenue, 2024).

Other US states, including New Hampshire, Alaska, Delaware, Montana and Oregon have no state sales tax, which reduces merchandise costs by as much as 5–7% in some cases. When purchasing large items, such as mobile phones, electronics, appliances, auto parts and recreational vehicles, the missing sales tax can save consumers hundreds of dollars in every transaction. These no-tax states use this advantage to promote themselves as 'tax-free' domestic destinations, and merchants actively advertise in neighboring regions. In New Hampshire, several unique shopping centers thrive along main access routes immediately inside the state's borders, attracting significant outshopping from neighboring Maine, Vermont and Massachusetts (Cammenga, 2020). According to one retail analysis, 'The absence of a statewide sales tax in New Hampshire is a boon for consumers and a unique selling point for businesses. This tax advantage draws visitors from neighboring states, as they flock to New Hampshire for the tax-free shopping experience' (Levy Team, 2024, n.p.).

Duty-free shopping

The first duty-free shop was opened by Brendan O'Regan in 1947 at Ireland's Shannon Airport to provide less-expensive retail opportunities to incoming and outgoing transatlantic air passengers who transited in Ireland. Within the next two decades, the phenomenon had spread to many other airport locations throughout the world, and the trend was set for the future of duty-free shopping (Pinder & Roberts, 2022). Today there are thousands of duty-free shops all over the world, with most of them being part of large multinational conglomerates, such as the DFS (formerly Duty-Free Shoppers) group (based in Hong Kong), which operates over 400 shops in dozens of locations, and Dufry (based in Switzerland), which operates in more than 65 countries at airports, ports and on cruise ships. Thousands of other shops at border crossings are operated by companies that have proliferated almost every corner of the globe (Pinder & Roberts, 2022). In 2022, global expenditures at DFS shops alone amounted to more than 3.7 billion US dollars (Statista, 2024b). In total, duty-free sales, together with sales in other shops located in duty-free areas totaled over 64 billion US dollars (Statista, 2024a). In 2019, before the pandemic, 45% of all duty-free sales occurred in the Asia Pacific region (Duty Free World Council, 2019), and today it remains the most profitable region for duty-free sales (Statista, 2024a). Since its early years, duty-free has come to fill the transit spaces of airports, ports and train stations, and the transit times of passengers worldwide (Hobson, 2000; Timothy, 2005; Weaver, 2017).

Duties are fundamentally import taxes and tariffs. The basic idea behind duty-free shopping is that the merchandise offered never enters the consumer market of the country where it is purchased; the items are destined to be carried out of the country and consumed abroad (Timothy, 2005). Thus, buyers are not required to pay import taxes in the country where they purchase their merchandise. However, this does not exempt individuals from paying duties when they enter another country. In fact, the country of arrival will usually charge import taxes on any items valued above the personal import exemption. In the United States, for example, the personal import allowance is generally $800 USD per person with certain limitations, such as the country visited and the amount of time spent abroad. Duty-free simply means that buyers do not have to pay import taxes in more than one country. This is why duty-free shopping is almost always located at exit gateways, such as airport departure areas, border crossings, harbors and train stations in international departure areas, as well as on cruise ships, flights and boats that navigate international waters and airspace (Doong et al., 2012; Hu & Zhang, 2025). Because most duty-free products have not entered the taxable marketplace, generally they are also free of local GSTs and VATs.

Unlike ordinary tax-free transactions, duty-free provides a much bigger incentive. Generally, duty-free also encompasses general tax-free, again because the merchandise never enters the domestic market. Unlike general tax-free incentives, which require individuals to apply for a refund, duty-free is applied during the retail encounter, with prices already reflecting an absence of import levies. Duty-free vendors are authorized to import items to sell to consumers who are in the process of leaving a country. In theory, this means that the cost of merchandise should be lower than at ordinary stores where residents also shop. However, price manipulation still occurs in the duty-free space, often meaning that the price differentials are negligible. Most duty-free products are highly taxed imported items, such as perfumes, jewelry, tobacco, alcohol and sweets.

According to a study by Peter and Anandkumar (2011), the most commonly sold items at Dubai's airport duty-free stores are perfume (14.5%), gold (11.4%), electronics (7.8%), confectionary (7.5%), cosmetics (6.1%) and sundry other items (52.7%), including souvenirs, clothing and so on.

A handful of countries (e.g. Australia, Chile, Malaysia, Turkey, China and the United Arab Emirates) allow incoming duty-free sales. Inbound retail points provide opportunities for travelers to purchase untaxed products immediately before entering the customs area. This, then, makes their goods subject to duties and local taxes on anything above their individual duty-free allowance (Christiansen & Smith, 2008). This gives travelers an opportunity to purchase items they might have missed at their original departure point and eases the burden of having to lug bags of items onto the airplane.

Tax-Free Retail and Shopping Tourism

The TFS global phenomenon has grown in popularity over the past few decades. Tax refunds for foreign tourists have become an established practice within global tourism (Bush & Storey, 2016; Defrasne, 2020; Timothy, 2018). In the past decade, TFS has grown by double digits globally (Statista, 2024a). From a policy and legal standpoint, TFS programs aim to ensure tax neutrality in tourist spending on exported goods. In addition to these policy considerations, countries that implement TFS enjoy a variety of other advantages (Bush & Storey, 2016). For instance, a TFS system encourages merchants to issue invoices and report proper sales to tax authorities, thus reducing the size of a country's shadow economy. Accordingly, the costs of implementing a TFS scheme are more than offset by the economic benefits. Specifically, a TFS system encourages foreign tourists' spending and stimulates retail sales (CEBR, 2020) directly in retail and in other areas of tourism commerce, such as lodging, dining, sightseeing and transportation.

Volumes of research have shown that international tourists positively impact the local economy (see Jansen-Verbeke, 1991; Liberato *et al.*, 2020). They are more inclined to stay longer and spend more money than domestic tourists on souvenirs and products that may not be available or affordable in their home countries. Several international studies (e.g. Aasness & Nygård, 2014; Einav *et al.*, 2014; Yamagishi & Nagai, 2024) describe how shopping is an essential part of people's destination decision-making and is an important component of a destination's tourism product mix.

As previously noted, offering tax-free retail to foreign tourists is an increasingly common strategy to grow tourism and retain visitation levels. Duty-free shops are extremely profitable for cities and regions with international airports, border crossings and ship ports (Azmi *et al.*, 2017; Lukić, 2012), and many such places use their tax-exempt status as a marketing tool to promote retail tourism. In one of the US examples mentioned previously, 'at one time, Delaware actually used its highway welcome sign to remind motorists that Delaware is the "Home of Tax-Free Shopping"' (quoted in Cammenga, 2020: 4). Tax-free shopping increases tourists' propensity to buy retail goods (Dimanche, 2003). Duty-free shops enhance the destination (Azmi *et al.*, 2015, 2017; Hussin *et al.*, 2012) and can help create more enriching retail experiences among visitors, even those who had not intended to browse in a duty-free store.

Some studies have found that offering tax-free shopping programs for international tourists can spark more interest in visiting a destination and stimulate shopping tourism more generally (e.g. Aasness & Nygård, 2014; Baker *et al.*, 2021; Dimanche, 2003). Tourists who have no forethought to purchase certain goods might be encouraged to do so when an economic incentive is provided, especially when the savings value is greater than they would ordinarily get in their home countries. The application of TFS is believed to boost international visitor arrivals and spending because the visitors can claim sales tax refunds (Banús, 2020). This strategy not only financially benefits shopper-tourists, but also fosters a strategic competitive advantage for the destination.

Tax-free shopping can be a major factor that tourists consider when deciding where to travel. In some cases, the higher the taxes, the less attractive a destination will be for foreign visitors (Kalaiselvan, 2022), causing them to choose a different holiday destination. By the same token, lower taxes or a tax-free fiscal environment can be a significant draw for foreign tourists and out-of-town consumers.

In most Asian countries, TFS adds significant appeal for regional and domestic tourists but far less so among Westerners, as they are more attracted to the charm of Asia's natural and cultural heritage (Aliman *et al.*, 2014). Among Chinese tourists, the presence of duty-free shops and tax refunds is an extremely important incentive for choosing a destination (Xu & McGehee, 2012).

There are many reasons tourists choose a holiday destination, but shopping is clearly an important factor, even if it is a secondary activity (Timothy, 2005). For shopping tourists – those whose primary purpose for traveling is to shop – retail opportunities are a major motivator, and competitive prices, including tax rates, may help in choosing a destination. For shopping tourists, high taxes or the implementation of new taxes can be a significant deterrent, particularly in luxury shopping contexts where big-ticket items may be heavily taxed. By the same token, national tax refund programs can also wield a significant influence in choosing a destination (Michalkó *et al.*, 2014; Wang & Stewart, 2022).

Because the sales tax authority in the United States is at the state and local level rather than at the national level, there are no nationwide tax refund schemes. Louisiana is the only state in the United States to offer a statewide tax refund for international visitors. Foreign tourists with a valid passport and a roundtrip international transportation ticket valid within 90 days are eligible for a refund of sales tax under the Louisiana Tax-Free Shopping program (Dimanche, 2003). In Louisiana, TFS increases tourists' propensity to buy retail merchandise. Meanwhile, in Canada and the UK, TFS has increased the number of inbound tourists, grown the average expenditures per tourist, and increased overall employment in tourism (CEBR, 2017, 2020).

According to King *et al.* (2005: 103), duty-free shops have come to play a pivotal role in the sales and operations of inbound tours to Australia. 'In some cases, duty free shops…have evolved into de facto [inbound tour operators], arranging many elements of the itinerary'. Often, duty-free or tax-free shops pay a fee to tour companies to ensure exclusive rights as retail stopovers on package tours, and they may even provide tour guides and motorcoaches for tour groups. Tour guides in this scenario usually receive a commission from the tax-free shop (King *et al.*, 2005).

To promote TFS, the island of Hainan, China – an officially designated duty-free island – increased its tax-free per person exemption from 30,000 RMB to 100,000 RMB in 2020. In 2021, duty-free sales in the province comprised approximately half of the island's tourism revenue, and three more duty-free shops were opened that year. The world's largest duty-free complex was opened in the north of the island near Haikou in 2022, adding nearly 300,000 square meters of additional tax-free shopping space to the island's retail landscape (Xinhua, 2022). Similarly, Langkawi Island, Malaysia, was declared a duty-free island on January 1, 1987. All goods on the island are exempted from tax payments except petroleum-derived items. To be tax-exempt, tourists need to stay on the island for at least 48 hours (Royal Malaysia Customs Department, 2021). It is one of the most famous island destinations in the country, largely owing to its duty-free status. Domestic Malaysian tourists are especially appreciative of Langkawi's special tax status and flock there for shopping tourism purposes (LADA, 2017).

Challenges and Opportunities for Tax-Free Shopping

TFS programs can offer financial benefits to host countries at large through direct spending and multipliers and also affect the expenditures of international tourists and the sales receipts at retail establishments on a business-to-business level (Baker *et al.*, 2021). Promoting TFS refunds is not easy. The process is tedious and requires a great deal of public-private sector collaboration, political advocacy, as well as significant legislation to bypass or waive traditional tax laws. Yet, shopping tourism appears to be worth the effort. TFS schemes are subject to regulations that vary by country, including minimum spending in a single shop or on a single item, as well as restrictions on the types of products that can be claimed and the amount of time one must spend in the country of purchase. Likewise, refunds can only be claimed on exported goods, so devising ways of distinguishing between what will be consumed in-country versus the country visitors are going to, can be a challenge.

For tourists, the refund process is cumbersome. Refund-eligible receipts must be obtained, followed by long queues at the airport validation desk, and sometimes substantial wait times for reimbursements to be received. New technology could, however, help simplify the process. Mobile phone apps and improved transmission of sales data are being introduced to help streamline the refunding process.

Likewise, buying tax-free goods abroad does not mean travelers are exempted from paying applicable taxes on their purchases when they arrive at home. There is often a level of ignorance about how duty-free and tax-free systems work, which can create considerable misunderstandings and result in dissatisfied customers. Similarly, that prices are not always cheaper in duty-free shops compared to downtown stores owing to price manipulations, high airport rental fees and other factors, can cause confusion and ill-will towards the place where merchandise was purchased.

From the destination's perspective, such programs are expensive to operate, and although few countries return the entire VAT amounts to tourists, the returned money is funds that did not remain in the destination economy. Nonetheless, the system helps merchants and retailers keep better records and provide more honest accounting to the government and provides a major impetus for tourism growth and development, which then has spinoff effects in other areas of the economy.

There is no conclusive evidence to show that TFS programs work perfectly as a strategy to promote shopping tourism. Yet, several studies have highlighted that tax-free incentives in some countries do influence tourist expenditures and thereby benefit local economies through various multipliers (see Dimanche, 2003; Einav *et al.*, 2014; Liberato *et al.*, 2020). Likewise, even in the face of tax-free retailing, government revenue is generated through shopping tourism as tourists spend more on other taxed goods, eat in restaurants and stay overnight in registered accommodations.

Conclusion

Tourists love to spend, and many countries have devised ways of increasing tourists' expenditures. One such means is the enactment of duty-free and general tax-free schemes that aim to reduce tourists' costs of buying local or imported items. This option is unavailable to most destination residents unless, in some cases, they too are departing their home country, for the absence of levies indicates that the merchandise will be consumed abroad. While not always a less expensive option, in theory at least, such programs should save non-resident visitors considerable amounts of money on highly-taxed merchandise. For some people, duty-free shopping is a salient part of the culture of traveling, a 'must-do' activity, whether or not it is actually cost-effective. Others are drawn into duty-free shopping through browsing behavior during transit down-time.

Tax-free shopping opportunities play into some people's choice of holidays, and many people plan their vacation destinations based on the importance of shopping as a motivation for travel. Governments have unique opportunities to legislate lower retail prices through tax incentives to stimulate tourism growth that benefits not only the retail sector, but also the transportation, accommodations, food services and attractions sectors as well through the broader ripple effects of tourism. Many shopping destinations use their duty-free and general tax-free designations as key marketing tools, and locations that host international departure points are in an exclusive position to benefit from this unique form of shopping tourism.

In the present and future, it is likely that the world's debordering processes (e.g. supranational clusters, such as the European Union) and the elimination of many border institutions will diminish some incentives for duty-free and tax-free shopping as many trade agreements aim to eliminate or severely reduce tax and price differences between member states. On the contrary, in parts of the world currently experiencing rebordering, such as between the EU and its external neighbors, or the US and Mexico, and many others, the erection of new barriers and tighter trade regulations could see a resurgence of tax- and duty-free shopping activities worldwide. Also, of particular concern for countries with regard to tax-free shopping, is the growing prevalence of online retail, much of which goes untaxed and unreported, regardless of how governments try to regulate taxation on online sales and imports. These variables in today's complex environment of geopolitics and internet-based buying will no doubt raise additional questions about taxes and tax-free sales that governments and international organizations must address going forward, and future research is needed in this arena.

References

Aasness, J. and Nygård, O.E. (2014) Revenue functions and Dupuit curves for indirect taxes with cross-border shopping. *International Tax and Public Finance* 21 (2), 272–297.

Aliman, N.K., Hashim, S.M., Wahid, S.D.M. and Harudin, S. (2014) The effects of destination image on trip behavior: Evidence from Langkawi Island, Malaysia. *European Journal of Business and Social Sciences* 3 (3), 279–291.

Azmi, A., Hamid, I.A., Ahmad, J.A. and Ramli, R.A. (2017) Tourism supply chain perspectives on border shopping development at Padang Besar, Malaysia. In A. Saufi, I.R. Andilolo, N. Othman and A.A. Lew (eds) *Balancing Development and Sustainability in Tourism Destinations: Proceedings of the Tourism Outlook Conference* 2015 (pp. 291–300). Springer.

Azmi, A., Sulaiman, S., Asri, D. and Razali, M.A. (2015) Shopping tourism and trading activities at the border town of Malaysia-Thailand: A case study in Padang Besar. *International Academic Research Journal of Social Science* 1 (2), 83–88.

Baker, S.R., Johnson, S. and Kueng, L. (2021) Shopping for lower sales tax rates. *American Economic Journal: Macroeconomics* 13 (3), 209–250.

Banús, S.C.H. (2020) Comparative analysis of VAT refunds systems to foreign tourists in Argentina, Colombia, Ecuador, and Uruguay: The case of the tax-free shopping. Paper presented at the 4th International Scientific Conference EMAN, 2020.

Borzooie, P., Lak, A. and Timothy, D.J. (2021) Designing urban customs and border marketplaces: A model and case study from Lotfabad, Iran. *Journal of Borderlands Studies* 36 (3), 469–486.

Budiarta, I.P. and Pemayun, I.D.G.A. (2019) Policy model in increasing tourist visits in Alas Kedaton tourism object. In Proceedings of the International Conference on Applied Science and Technology, 2019 – Social Sciences Track (iCASTSS 2019). Online: http://dx.doi.org/10.2991/icastss-19.2019.78

Bush, H. and Storey, D. (2016) *Economic Impact of Duty Free and Travel Retail in Europe*. Paris: Duty Free World Council.

Cammenga, J. (2020) State and local sales tax rates, midyear 2020. *Fiscal Fact*, 716, 1–7.

CEBR (2017) *Understanding the Value of Tax-free Shopping to Destination Economies. Study No 1: The United Kingdom*. Centre for Economics and Business Research.

CEBR (2020) *Understanding the Potential of Tax-free Shopping in Destination Economies: Canada Study*. Centre for Economics and Business Research.

Christiansen, V. and Smith, S. (2008) Optimal commodity taxation with duty-free shopping. *International Tax and Public Finance* 15, 274–296.

Defrasne, C. (2020) Who benefits from tax free shopping? Customers behavioral responses to countries VAT refund strategies. Unpublished master's thesis, Paris School of Economics.

Dimanche, F. (2003) The Louisiana tax free shopping program for international visitors: A case study. *Journal of Travel Research* 41 (3), 311–314.

Doong, H.S., Wang, H.C. and Law, R. (2012) An examination of the determinants of in-flight duty-free shopping: Hedonic and utilitarian motivations. *International Journal of Tourism Research* 14 (3), 303–306.

Duty Free World Council (2019) *Economic Impact of Duty Free and Travel Retail in Asia Pacific*. DFWC.

Einav, L., Knoepfle, D., Levin, J. and Sundaresan, N. (2014) Consumer behavior in online shopping is affected by sales tax. Online: https://eprints.lse.ac.uk/58453/

Fullerton, T.M. and Walke, A.G. (2019) Cross-border shopping and employment patterns in the southwestern United States. *Journal of International Commerce, Economics and Policy* 10 (03), 1950015.

Hobson, J.P. (2000) Tourist shopping in transit: The case of BAA plc. *Journal of Vacation Marketing* 6 (2), 170–183.

Hu, T. and Zhang, Y. (2025) Exploring tourist shopping from the perspective of duty-free shopping: An analysis of online reviews. *Journal of Vacation Marketing* 31 (1), 52–66.

Hussin, F., Abdullah, N., Maamor, S. and Abdullah, H. (2012) Border economy: Issues and problems faced by traders in the Rantau Panjang Duty-free Zone. *Journal of Sociological Research* 3 (2), 46–56.

Ilter, C. and Manahan, M. (2021) The application of sales tax in the US and how it differs from value added tax. *Journal of Taxation of Investments* 38 (2), 61–65.

Jansen-Verbeke, M. (1991) Leisure shopping: A major concept for the tourism industry? *Tourism Management* 12 (1), 9–14.

Japan National Tourism Organization (2024) Frequently asked questions: Shopping. Online: https://faq.japan-travel.jnto.go.jp/en/faq/articles/102054

Jin, H., Moscardo, G. and Murphy, L. (2020) Exploring Chinese outbound tourist shopping: A social practice framework. *Journal of Travel Research* 59 (1), 156–172.

Kalaiselvan, A. (2022) Evaluation of the impact of goods and service tax implementation on the entry rate of foreign tourists. *Journal of Global Economy, Business and Finance* 4 (11), 79–82.

Kim, S.S., Timothy, D.J. and Hwang, J. (2011) Understanding Japanese tourists' shopping preferences using the Decision Tree Analysis method. *Tourism Management* 32 (3), 544–554.

King, B., Prideaux, B., Hobson, P. and Hobson, L.D. (2005) Modelling the structure of the Korean package tour industry in Australia. In Conference Proceedings, *ANZMAC 2005: Broadening the Boundaries* (pp. 103–108). Crawley: University of Western Australia Press.

Lembaga Pembangunan Langkawi (LADA) (2017) Laman Web Rasmi LADA. Online: http://www.lada.gov.my/v2/

Levy Team (2024) Fiscal freedom: Analyzing New Hampshire's exemption from general sales tax. Online: https://www.levy.company/startups/new-hampshire-sales-tax#:~:text=The%20absence%20of%20a%20statewide,the%20tax%2Dfree%20shopping%20experience.

Liberato, D., Liberato, P. and Silva, M. (2020) Shopping tourism: Comparative analysis of the cities of Oporto and Lisbon as shopping destinations. In V. Katsoni and T. Spyriadis (eds) *Cultural and Tourism Innovation in the Digital Era* (pp. 365–379). Springer.

Lubis, H., Rohmatillah, N. and Rahmatina, D. (2020) Strategy of tourism village development based on local wisdom. *Jurnal Ilmu Sosial dan Humaniora* 9 (2), 320–329.

Lukić, R. (2012) The effects of cross-border tourist purchases on retail performance. *The European Journal of Applied Economics* 9 (2), 21–26.

Makkonen, T. (2016) Cross-border shopping and tourism destination marketing: The case of Southern Jutland, Denmark. *Scandinavian Journal of Hospitality and Tourism* 16 (1), 36–50.

Makkonen, T. (2023) Outshopping abroad: Cross-border shopping tourism and the competitive advantage of borders. In D.J. Timothy and A. Gelbman (eds) *Routledge Handbook of Borders and Tourism* (pp. 269–280). Routledge.

Michalkó, G., Rátz, T., Hinek, M. and Tömöri, M. (2014) Shopping tourism in Hungary during the period of the economic crisis. *Tourism Economics* 20 (6), 1319–1336.

Palmer, T. and Riera, A. (2003) Tourism and environmental taxes. With special reference to the 'Balearic ecotax'. *Tourism Management* 24 (6), 665–674.

Peter, S. and Anandkumar, V. (2011) A study on the sources of competitive advantage of Dubai as a shopping tourism destination. In *Proceedings of the International Business Research Conference*. Online: http://www.wbiconpro.com/510-Sangeeta.pdf.

Pinder, D. and Roberts, J. (2022) Airport luxury retail. In P-Y. Donzé, V. Pouilliard and J. Roberts (eds) *The Oxford Handbook of Luxury Business* (pp. 379–401). Oxford University Press.

Reisinger, Y. and Turner, L.W. (2002) The determination of shopping satisfaction of Japanese tourists visiting Hawaii and the Gold Coast compared. *Journal of Travel Research* 41 (2), 167–176.

Royal Malaysia Customs Department (2021) *Panduan Am Cukai Pelancongan 2017 (Versi 2)*. Online: https://www.myttx.customs.gov.my/wp-content/uploads/2022/09/Panduan-Am-Cukai-Pelancongan-V2-03082021.pdf

Statista (2024a) Duty free and travel retail sales worldwide in 2010 to 2022. Online: https://www.statista.com/statistics/477856/global-travel-retail-sales/#:~:text=World%3A%20duty%20free%20and%20travel%20retail%20sales%202010%2D2022andtext=In%202022%2C%20global%20duty%20free,the%20previous%20year's%2052%20billion.

Statista (2024b) Turnover of DFS Group worldwide from 2013 to 2022. Online: https://www.statista.com/statistics/565818/global-turnover-of-dfs/

Streng, W.P. (1992) 'Treaty shopping': Tax treaty 'limitation of benefits' issues. *Houston Journal of International Law* 15 (1), 789.

Szytniewski, B.B., Spierings, B. and van der Velde, M. (2018) Socio-cultural proximity, daily life and shopping tourism in the Dutch–German border region. In J. Diaz Soria Inmaculada and J. Diaz Soria Inmaculada (eds) *Proximity and Intraregional Aspects of Tourism* (pp. 60–74). Routledge.

Taecharungroj, V. and Tansitpong, P. (2017) Attractions, attitude, and activities: Tourism attributes that drive destination loyalty for international tourists visiting Thailand. *International Journal of Tourism Policy* 7 (2), 129–150.

Tax Foundation (2024) VAT rates in Europe, 2023. Online: https://taxfoundation.org/data/all/eu/value-added-tax-2023-vat-rates-europe/#:~:text=More%20than%20170%20countries%20worldwide,VAT%20is%20a%20consumption%20tax

Timothy, D.J. (1995) Political boundaries and tourism: Borders as tourist attractions. *Tourism Management* 16 (7), 525–532.

Timothy, D.J. (2005) *Shopping Tourism, Retailing and Leisure*. Channel View Publications.

Timothy, D.J. (2018) Shopping tourism. In S. Agarwal, G. Busby and R. Huang (eds) *Special Interest Tourism: Concepts, Contexts and Cases* (pp. 134–144). CABI.

Timothy, D.J. (2021) *Tourism in European Microstates and Dependencies: Geopolitics, Scale and Resource Limitations*. CABI.

Timothy, D.J. and Butler, R.W. (1995) Cross-border shopping: A North American perspective. *Annals of Tourism Research* 22 (1), 16–34.

Usman, O. and Alola, A.A. (2023) How do environmental taxes influence the effect of tourism on environmental performance? Evidence from EU countries. *Current Issues in Tourism* 26 (24), 4034-4051.

Wang, T. and Stewart, M. (2022) The law and policy of VAT tourist tax refund schemes: A comparative analysis. *World Tax Journal* 14 (2), 285–330.

Washington Department of Revenue (2024) Sales tax exemption for non-residents. Online: https://dor.wa.gov/taxes-rates/retail-sales-tax/sales-tax-exemption-nonresidents#:~:text=All%20nonresidents%20may%20be%20exempt,are%20interstate%20or%20foreign%20sales)

Weaver, A. (2017) Complementary contrasts in a mobile world: 'In-betweenness' and the selling of duty-free products. *International Journal of Tourism Research* 19 (1), 80–88.

Xinhua (2022) China's Hainan sees soaring duty-free sales in 2021. Online: https://www.globaltimes.cn/page/202201/1243985.shtml

Xu, Y. and McGehee, N.G. (2012) Shopping behavior of Chinese tourists visiting the United States: Letting the shoppers do the talking. *Tourism Management* 33 (2), 427–430.

Yamagishi, D. and Nagai, H. (2024) Development of a tax-free shopping environment in Japan: An analysis of its representations in a financial newspaper. *Tourism Planning and Development* 21 (5), 530–549.

7 Border Crossings, Retail and Shopping Tourism

Dallen J. Timothy and Gülsel Çiftci

Introduction

As each contribution in this volume has made clear, one of the most pervasive activities undertaken by tourists is shopping. Almost everyone shops while on holiday, and retail activities have come to dominate many leisure journeys. Souvenirs, including handicrafts and food items, have long been regarded as the 'material culture' of tourism (Hitchcock & Teague, 2019), and even in ancient days, souvenirs were a significant consumer item brought home from great distances by explorers, traders, pilgrims and warriors (Birch, 1998; Evans, 1999; Ron & Timothy, 2019). Shopping in tourism generally can be seen broadly from two perspectives. The first is shopping as the primary motive for undertaking a journey. The second is shopping as a secondary tourist activity. The first type is most obvious in the growing trend of luxury shopping, where people's primary motive for travel is to purchase luxury items with brand recognition. Many affluent consumers travel for the sole purpose of buying big-ticket items such as clothing, shoes and handbags, accessories, cars, jewelry, furs and other such luxury merchandise. However, even less affluent consumers buy while they travel and may be motivated to travel by opportunities to purchase utilitarian, household goods.

Despite the explosive growth of shopping tourism since the Second World War, the increasing affluence of the mid-20th century, and the retail opportunities that have come to accompany this trend, there is a type of retail mobility that has taken place quietly without the flare of typical shopping tourism, sometimes of dubious legality, and usually with utilitarian motives being more salient than leisure motives. However, it expresses leisure elements for certain. This phenomenon is typically known in the retail literature as international outshopping, and in the social sciences as cross-border shopping.

This chapter describes the phenomenon of cross-border shopping and how it manifests in borderland regions throughout the world. First, the chapter discusses cross-border shopping, exploring both utilitarian and leisure perspectives. It then delves into the concept of border retail spaces and planning in great detail. Finally, it examines contemporary trends observed in cross-border shopping.

Cross-Border Shopping: Utilitarian and Leisure Perspectives

For centuries, people have left their home environments to acquire or consume merchandise elsewhere. However, since the evolution of the modern state in the late Middle Ages and early modern period, and the Westphalian concept of state sovereignty, national borders have evolved from nebulous zones of control (or lack thereof) into the bounded states that we know today. With the development of national borders, particularly since the 1600s, economies, histories, polities and societies have evolved differently on opposite sides of political borders, and human mobilities have been hampered. From a politico-economic perspective, protectionist measures have been taken at national borders to safeguard populations, natural resources and domestic industries. One key protectionist measure enacted by states has been tariffs, duties, quotas and other trade barriers. This has led to differential development on opposite sides of the same political divide over the last few centuries, often creating considerable contrasts in social and economic development between adjacent countries and their inhabitants.

These contrasts have had a significant bearing on the types of merchandise produced and the prices demanded for different products. Likewise, national population sizes and consumer markets have determined economies of scale, with sparsely populated countries having an insufficient ability to manufacture or import merchandise in large enough quantities to reduce the per-item cost. Such contrasts have stimulated the growth of cross-border shopping where consumers cross state borders to buy in other jurisdictions.

While cross-border retail activity has garnered attention from economists, geographers and retail scholars for decades, only recently has it begun to be understood more holistically from a tourism perspective (e.g. Azmi *et al.*, 2015, 2017; Di Matteo & Di Matteo, 1996; Makkonen, 2016, 2023; Stoffelen & Timothy, 2023; Timothy, 1999; Timothy & Butler, 1995). Although the official statistical definition of a *tourist* is someone who travels away from home temporarily and stays away at least one night, the meaning of *tourism* is far more encompassing. It entails the totality of all services that facilitate travel in one's home environment (pre-travel), in transit and in the destination, as well as the experiences, opportunities and outcomes that derive from travel. Thus, while day-trippers may not fit the statistical definition of tourists, they certainly do in fact participate in the broader phenomenon of tourism, regardless of whether they have leisure motives or cross state frontiers strictly for economic and utilitarian purposes and return home the same day (Timothy *et al.*, 2022).

Cross-border shopping refers to people traveling to foreign but adjacent areas to buy goods and services for themselves or others. This phenomenon has also come to encompass petty trade, where individuals cross borders to buy merchandise that they then take home to sell at a higher price (Michalkó et al., 2014; Michalkó & Timothy, 2001; Szytniewski et al., 2020; Timothy et al., 2022). Most cross-border shopping entails day trips, or same-day returns, motivated overwhelmingly by utilitarian purposes. Even though practical shopping and petty-trade do not appear to be touristic activities in the traditional sense, they are effectively part of the tourism system, particularly when consumers spend additional time abroad, dine out, engage in other activities and otherwise contribute to the neighboring economy (Timothy, 2024).

Several conditions typically need to exist for cross-border shopping to emerge and gain success (Leimgruber, 1988; Studzińska et al., 2018; Timothy, 2005). First, there must be enough of a contrast between two sides of a border in terms of prices, tax rates, product quality, merchandise selection and service quality to stimulate an interest in crossing. The very existence of differences or otherness attracts people across political boundaries to purchase items that might not be available at home or which are far less expensive abroad.

Second, exchange rates between currencies need to be advantageous enough to warrant traveling abroad for retail purposes. When a country's currency is strong, its residents are more inclined to shop in a neighboring jurisdiction where the currency is weaker. In the US–Canada context, there are considerable fluctuations in north-to-south or south-to-north shopping depending on the strength of the US dollar or the Canadian dollar (Timothy & Butler, 1995). This, together with the issue of product costs and taxes more generally, can be a major stimulant for transfrontier commerce.

Third, potential consumers must be aware of what is available to buy on the other side. This knowledge usually develops through word-of-mouth, personal experience, social media, advertising and family traditions. Many borderland families have a long tradition of crossing a boundary to shop, as a means of saving money, spending quality time together, visiting relatives on the other side, or simply a pleasurable getaway from home (Castaño et al., 2010). Many families in northern Finland have undertaken shopping excursions to Sweden for generations to buy household items and groceries, making a family outing not just one of economic necessity but also one of family togetherness (Timothy & Więckowski, 2023).

Fourth, consumers have to be willing and able to cross in terms of personal mobility, especially in light of cultural, political and economic differences, which can act as significant barriers to cross-boundary travel. Borders are often perceived to be psychological barriers when they divide different cultures, languages and political and economic systems. Thus, geographically near but culturally distant destinations on the other side of a border can create a significant perceived or psychological distance and level of discomfort that may keep some would-be consumers from shopping abroad (Timothy, 2001). Likewise, the powers of the state that pervade border environments (e.g. security, border controls and fiscal administration) can sometimes make border crossings unwelcoming and intimidating, adding a functional distance between localities that might in actuality be relatively close (Timothy, 1995, 2001; Więckowski & Timothy, 2021).

Finally, the international frontier has to be open enough that consumers can cross with as little effort as possible. Boundaries that are open to local traffic usually see the greatest growth in cross-border shopping. For example, Bulgarian nationals may cross the Turkish border using their national ID cards instead of a passport. This process is restricted to the Edirne and Kırklareli provinces. Arriving at the border checkpoint, Bulgarian nationals are required to present their national ID cards to the Turkish border officials for verification. Once the officials have confirmed the authenticity of the documents and the legitimacy of the consumer's intensions, the traveler is allowed to cross the Turkish border. This option, which is both fast and convenient, eliminates the need for obtaining a passport, resulting in notable cost reduction and time savings, which effectively encourage transfrontier commercial activity. Likewise, border crossings that require the least number of documents and those that have the easiest import regulations tend to be the busiest for transfrontier retail.

Based on their research about Canadians shopping in the United States, Timothy and Butler (1995) proposed a spatial model that reflects some of these concepts and people's motives for participating in transfrontier shopping, the types of merchandise they buy, and the distance they have to travel to the border (Figure 7.1). In most cases (and this has been confirmed by research in other settings (e.g. Bygvrå, 2009, 2019; Spierings & van der Velde, 2013)), there is a significant distance–decay in effect. This means that the further consumers live from the border (as represented by the bottom line of the box and diagram), the less often they will cross but the higher the value of the merchandise they buy will be (represented by the width of the base of the triangle). Traveling great distances to a neighboring country may be advantageous if the price and tax differentials are significant enough to merit such a journey (Leick *et al.*, 2021). For example, traveling two hours to save $350 on a new lawnmower or refrigerator may be a worthwhile endeavor, even if the buyers must pay import duties. Such a trip may also include some ludic activities and elements of leisure, such as dining out, going to the cinema, or visiting a park or museum. Likewise, holiday times and special events tend to stimulate additional cross-border shopping (Klamár & Kozoň, 2022), the value of which will create more positive and enjoyable experiences (Sharma *et al.*,

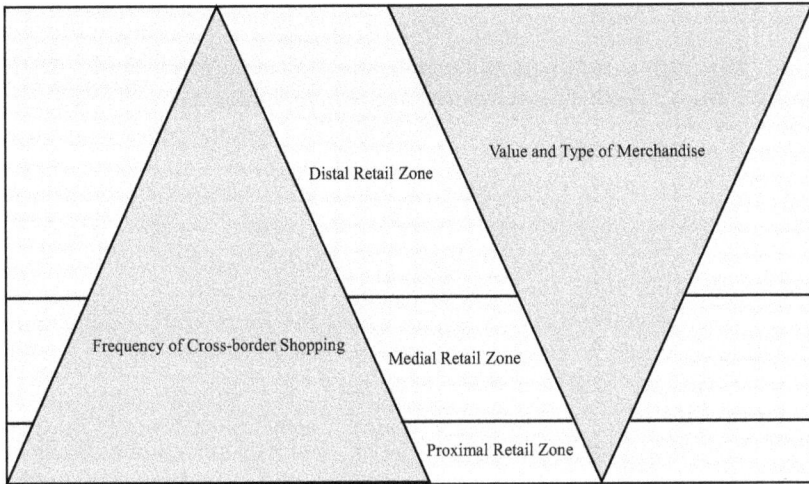

Figure 7.1 Distance–decay model of cross-border shopping (after Timothy & Butler, 1995)

2018). Conversely, the closer consumers live to the border, the more often they cross and the more they tend to buy daily necessities, such as food, petrol, school supplies and clothing (represented by the tip of the upside-down triangle). The destination abroad may lose its 'exoticness' when consumers' daily lives and ordinary action spaces begin to extend across the border (Ramsey et al., 2019; Spierings & van der Velde, 2013; Szytniewski et al., 2017). Their more frequent trips abroad are less likely to be pleasure-oriented – more purely utilitarian – although many people find pleasure in acquiring bargains and discounts (Anderson et al., 2014; Cox et al., 2005; Timothy, 2005), and the proximal shoppers might also be just as inclined to pursue leisure activities while abroad.

There is significant agreement that cross-border shopping is becoming an increasingly entertaining and recreational activity, now far beyond the purely utilitarian motives families and individuals used to act on simply to save money (Bar-Kołelis & Wendt, 2018; Bar-Kołelis & Wiskulski, 2012; Timothy, 2005; Timothy & Butler, 1995; Yuan et al., 2013). Some scholars have also pointed to the notion that cross-border shopping is now often regarded as an immersive cultural experience that can help consumers experience a foreign culture, a historic city environment or simply enjoy the 'exotic otherness' of being abroad (Bar-Kołelis & Wiskulski, 2012; Ramsey et al., 2019; Spierings & van der Velde, 2013). Russian and Ukrainian shoppers in Poland have long appreciated the distinct experience of shopping in the EU, which added a significant dimension of pleasure to their retail experiences (Bar-Kołelis & Wendt, 2018; Smętkowski et al., 2017).

European examples

Cross-border shopping takes place in nearly all borderlands, where transfrontier mobility is permitted and where contrasts are sufficient to stimulate demand (Timothy, 2014, 2018). The majority of studies have focused on borders in North America and Europe, although the phenomenon is also strong in Latin America, Asia and Africa and is beginning to receive growing research attention in those regions (Azmi et al., 2017; Borzooie, 2021; Dilla et al., 2022; Jin et al., 2019; Kuncharin & Mohamed, 2013; Lau et al., 2005; Risal et al., 2022; Rogerson, 2018).

One of the goals of the European Union (EU) is to stabilize and standardize economic conditions between member states. This includes taxation (see Chapter 6). The value-added tax (VAT) in Europe is a consumer tax that is levied on services and products throughout the entire production chain, including at the point of sale. Most European countries have fairly standardized VATs, although even within the EU there are substantial differences. For example, Hungary has the highest VAT rate (27%), and Luxembourg's (16%) is the lowest. This does not always determine final consumer prices, however. Until quite recently, prices for many household products were lower in Hungary than in neighboring Slovakia, Romania and Austria, causing surges of day-trippers to visit supermarkets and other retailers in Hungary (Michalkó et al., 2014; Michalkó & Timothy, 2001; Tömöri, 2010, 2011). In the past few years, however, prices have risen dramatically in Hungary compared to its neighbors owing to rising inflation and unfavorable exchange rates between the euro and Hungary's forint, resulting in many Hungarians now outshopping in the same countries whose citizens used to shop in Hungary (Mohi, 2023).

Despite the European single market's efforts to equalize prices and taxes, each member state still has some control over its internal taxes and merchandise costs. This means that cross-border shopping continues to be a common phenomenon along many intra-EU borders. For instance, French consumers regularly shop in Spain for food products, wine and ordinary household goods because of price differences. Belgian motorists regularly fill their tanks in Luxembourg, where petrol can be as much as 30 cents per liter cheaper. Dutch consumers regularly cross into Belgium for beer, gasoline, tobacco and chocolate (Timothy, 2024). Swiss consumers shop in Germany for an assortment of items, including cosmetics, clothing, groceries, pharmaceuticals, toys, furniture, alcohol, fuel and many other goods (Ramsey et al., 2019). Such retail patterns exist throughout Europe and in many corners of the globe.

Although the transfrontier asymmetries in tax rates have traditionally been a major stimulator of transborder commerce, in Europe today there is not always a direct correlation between VAT rates and cross-border shopping. In fact, in many cases, the retail trends seem to contradict common sense. Although currently Germany's VAT is 19% and Poland's is 23%, there remains a strong flow of German consumers to Poland who shop for ordinary household items, tobacco, food, home décor and toys, owing to Poland's lower prices on certain items (Więckowski, 2010; Więckowski & Timothy, 2021). This pattern also reveals that the debordering processes of the European Union and its affiliated Schengen Agreement have created transfrontier daily living spaces among local borderlanders for whom the border and the border bazaars no longer serve a novelty purpose. According to Szytniewski and Spierings (2018), however, for some consumers, especially those who live away from the border, there is an element of leisure and intrigue associated with German consumers crossing into Poland and appreciating the uniqueness of the border market and spending a little time abroad.

Conversely, being outside the European Union, Switzerland and Liechtenstein have low VATs (8.1%), compared to their neighbors, but this does not translate into incoming shopping from France, Italy, Germany or Austria, because prices tend to be more expensive in Switzerland overall (with the exception of some electronics) despite the low VAT. This is largely because of the country's small economy of scale, lack of competition and high import restrictions and taxes on agricultural and industrial products to protect its farming sector (Kluser, 2023).

Also, not being formal members of the European Union, some of Europe's microstates are able to use their lower taxes or duty-free status to appeal to shopper-tourists. Andorra has the lowest VAT in Europe (4.5%) which, unlike in Switzerland, does in fact reach the Andorran consumer. In addition to its low VAT, Andorra is a duty-free country, which attracts high numbers of shoppers from France (VAT 20%) and Spain (VAT 21%) who visit by private car or chartered buses to purchase duty-free products such as jewelry, perfume, cosmetics, alcohol, tobacco, clothing and shoes (Timothy, 2021). Likewise, Italy's VAT is 22%, whereas the VAT in San Marino is 17%. Although this asymmetry does not provide the deep savings here as it does in Andorra, it does create some tax advantages for Italian consumers and foreign businesses headquartered in the tiny microstate.

Even the smallest country in the world, the Vatican City, experiences its own unique sort of shopping 'tourism'. In addition to a handful of cafes and restaurants, there are several retail opportunities inside the Vatican. Many residents of Rome mail packages and letters, including bulk Christmas cards, from the Vatican's post offices. The price is essentially the same as Poste Italiane's rates, but Poste Vaticane is regarded as more efficient and trustworthy than Italy's postal services (Timothy, 2021). Some Romans have permission to shop in the country's supermarket, fill prescriptions at its pharmacy, and fill up their tanks with petrol at its only gas station. The upper level of the historic Vatican railway station has also been converted into a duty-free shop, where qualified residents of Rome can buy high-end goods at tax-free prices. There are no customs excises entering the microstate, and Italy imposes no taxes on goods entering from the Vatican. Likewise, the microstate levies no sales, VAT, or fuel taxes, thus making grocery items and petrol significantly less expensive than in surrounding Italy (Timothy, 2021).

Border Retail Spaces and Planning

As alluded to previously, most traditional cross-border shopping in most places developed organically as economic and political conditions encouraged and facilitated its development. Today, however, given that most localities desire tourism growth, shopping in border communities now plays a greater role in borderland economic development and is treated as a legitimate form of tourism that requires planning and marketing (Timothy, 2023). One manifestation of this is the intentionality of developing retail spaces in border areas.

A unique border shopping phenomenon is the 'free-trade zone' – an area established near international borders or ports, where customs laws and levies are waived. These special economic zones are in areas with easy access to foreign trade and, like duty-free shopping generally, any merchandise bought in these areas is meant to be exported and therefore is not taxed in the country of purchase. By law, such areas are granted either extraterritoriality or items that are not subject to import taxes because they never enter the consumer marketplace. Although there are dozens of free-trade zones throughout the world relatively few are devoted especially to tourism and retail. One example of the latter is the Corozal Free Zone (CFZ) in Belize. Although CFZ caters mainly to large import companies and other business, many of its individual retailers cater to the shopping whims of Mexican shoppers. This free-trade zone is located within the 700-m territory of Belize between the Mexican border and Belize's customs area. This unique geographical position enables Mexican shoppers to purchase many imported items from all over the world at duty-free prices (Timothy, 2024).

The geographic planning principle of clustering is an important part of the planned cross-border shopping phenomenon. In several places, border markets have been intentionally built as clearinghouses for goods for sale to foreign consumers. In some cases, informal marketplaces that developed adjacent to national borders have now been formalized and are an important part of the national economy. In other localities, border markets were built purposively to be markets that draw people across the frontier. One example of this is the tiny Vietnamese market that hugs the Chinese border near Detian waterfall in Cao Bằng province, Vietnam. The market was established specifically to cater to Chinese tourists visiting the famous waterfall. There, vendors sell Vietnamese food items, sweets, clothing, shoes and sundry other merchandise that is far less expensive (and more exotic) than in China. In Cedynia, Poland, the

Polenmarkt Hohenwutzen was built adjacent to the German border with the sole purpose of attracting Germans for imitation brand names, groceries, hot meals, household goods, alcohol and inexpensive haircuts. In the city of Lotfabad, Iran, a 'border marketplace' was built intentionally to be a focal point of Turkmenistani shoppers who cross from the country's capital and largest city, Ashgabat, to shop in Lotfabad near the international border (Borzooie et al., 2021). Similarly, Indonesia has an ongoing development project to build commercial clusters at its land borders with Papua New Guinea, East Timor and Malaysia that invite people from those countries to walk a few meters into Indonesia, shop, dine, seek entertainment and return home the same day (Risal et al., 2022).

A mix of organic and purposive development can be seen at some border crossings where shops cluster along a particular road, traffic route, or pedestrian corridor near border crossings. This has been a common pattern in Mexico's northern border towns for decades, as advantages south of the line have drawn American day visitors and overnight tourists to shops, restaurants, bars and other retail establishments (Arreola & Curtis, 1993). Today, few souvenir shops remain, and the retail landscapes of most of Mexico's northern border towns now resemble clusters of pharmacies, dental offices and medical clinics (Timothy, 2020). Similarly, many of the streets adjacent to border crossing points between PR China and Macau and Hong Kong are lined with baby formula shops, pharmacies and other small retailers catering to the needs of Mainland Chinese consumers.

Other localities have seen the rapid construction of shopping centers and mega malls near busy border crossings that cater both to local consumers and shoppers from the other side (Bar-Kołelis, 2013; Sullivan et al., 2012). The Nordby Shopping Center in Strömstad, Sweden, was built very close to the main border crossing near the population center of Oslo, Norway. It opened in 1983 to critical acclaim and is now considered Sweden's largest shopping center. It provides a much less expensive option for Norwegians to buy household products, clothing, sporting goods, toys, pet supplies and pharmaceuticals in its 123 shops. Sweets are a particularly lucrative product at Norby and elsewhere along Sweden's border with Norway, as sugared candy and soft drinks are highly taxed in Norway but not in Sweden.

Current Trends in Cross-Border Shopping

Several unique trends can be identified in the realm of cross-border shopping that differentiate it from other forms of retail tourism. First, many borderland destinations have resorted to marketing their retail offerings on the other side of the border. For the most part, cross-border shopping has emerged organically where a competitive advantage on one side drew consumers across to buy. Deliberate promotional efforts were not traditionally something done in this context, but today retail localities are beginning to realize the need to market and brand themselves as transfrontier shopping destinations owing to increased competition from other places and online retail, and with reduced tax advantages, particularly within the EU and in other supranational trading blocs.

Second, cross-border shopping is now being seen more broadly than ever before. Whereas traditional outshopping has dealt primarily with no-frills consumerism, its leisure elements are now more in focus, with destination communities providing other ancillary sorts of attractions to keep people in town longer and increase their expenditures. Even in utilitarian-motivated contexts, the ludic or hedonic side of transfrontier buying is becoming an increasingly important focus among shoppers and destination planners (Kim & Sullivan, 2003; Leick et al., 2021; Szytniewski & Spierings, 2018). Likewise, cross-border shopping, even that of a purely utilitarian nature, can stimulate other forms of tourism and have positive impacts on other sectors, including longer stays and increased expenditures on recreational activities (Leick et al., 2021; Malkowski et al., 2020).

Thirdly, as part of this broadening understanding of cross-border shopping, the purchase of second homes/seasonal homes and even healthcare is now being realized as part of the transfrontier movement of retail capital (i.e. shopping). Borderland medical tourism is a prime example of this growing phenomenon and is an increasingly salient form of border tourism in many of the world's border regions (Cuevas Contreras & Zizaldra Hernandez, 2023; Oberle & Arreola, 2004). Likewise, vacation home ownership in border areas, which are often characterized by pristine natural environments, is becoming a progressively commonplace retail phenomenon, particularly in the case of west Europeans purchasing second homes in Eastern Europe (Hannonen et al., 2015; Opačić & Koderman, 2018).

Fourth, as previously noted, economic development agencies and tourism planners in national borderlands must now consider retail tourism in their broader planning efforts and policymaking. This includes not only marketing and providing deeper experiences but also considerations of crisis management and securitization of border areas for foreign shoppers. Shopping is frequently seen as an ideal form of tourism in the peripheral and economically marginal borderlands, which may otherwise have few development options (Spierings & van der Velde, 2013; Timothy, 2001). Such is the case in the Polish borderlands it shares with all its neighbors, but in particular its borders with Ukraine, Russia and Belarus (Malkowski, 2019; Malkowski et al., 2020; Smętkowski et al., 2017).

Fifth, for mass sales, online retailing appears now to be overtaking the marketplace, often replacing the need for physically traveling across a border. International commerce leaders, including eBay, Alibaba, Temu and Amazon, provide digital cross-border shopping that is safer, less expensive and the goods returnable, even if the online experience does take away the corporeal experience of being abroad.

Finally, even in places where tax and import legislation and supranational treaties aim to reduce border differences, the phenomenon of cross-border shopping continues to flourish, both because it is a tradition and because there are still sufficient price differences to merit a trip abroad in many cases (e.g. Germans buying in Poland).

Benefits of Cross-Border Shopping

Cross-border shopping has several benefits that positively impact both sides of the border. Whereas the implications of transfrontier shopping are obvious for merchants and communities on the retail side of the border, the customer-generating side also benefits considerably. As the phenomenon grows in border communities, consumers from other parts of the origin country are drawn to the borderlands, where they often spend considerable funds on their own side on items such as lodging, food and petrol before or after their international retail excursion. As Makkonen (2016) rightfully notes, in many places, cross-border shopping has become a significant tourist attraction that benefits the economies on both sides of an international boundary.

Specifically, one of the most obvious advantages is the generation of revenue for businesses on both sides. When consumers purchase goods from businesses in another country, the host country's businesses benefit from increased sales, while the consumers' home country may benefit from import taxes levied on the goods purchased abroad. In addition, the increased demand for goods in border regions often leads to the creation of new jobs in retail and related industries on both sides. Likewise, cross-border shopping can lead to lower prices on both sides of the border as businesses compete to attract customers or offer special deals to keep them home. This competition stimulated by international outshopping may also result in businesses selling better quality merchandise and services. In turn, this benefits consumers who can enjoy better value for money.

Cultural exchange is another potential benefit as consumers interact with retailers and people from different countries. As consumers explore different places, products and services, they may develop a greater understanding and appreciation of cultures, traditions and products different from their own (Castaño et al., 2010; Kuncharin & Mohamed, 2013; van der Velde & Spierings, 2010). This can help foster cultural awareness and create more harmonious relationships between neighboring countries, particularly at the local level, just as other forms of local-international tourism have done (Gelbman & Schweitzer, 2023; Leandro, 2021; Park et al., 2022; Prokkola, 2023).

In summary, cross-border shopping fosters economic cooperation, trade, cultural exchanges between neighboring countries, benefiting both sides economically and socially.

Some of this seems counterintuitive, especially when the source country campaigns to play into the sense of duty to the homeland or nationalism to try to get citizens to stay home to shop (Timothy & Butler, 1995). In one of the biggest waves of international outshopping in modern times, Canadians by the millions spent billions of dollars in US border communities throughout the early 1990s owing to the strength of their dollar against the US currency. American merchants offered many incentives such as lower prices, special deals, free delivery across the border, and accepting Canadian currency at par (Timothy, 1999; Timothy & Butler, 1995). In an effort to quell the leakage of Canadian cash to the United States, the Canadian government and individual provincial governments initiated promotional campaigns to discourage residents from traveling southward to shop. Playing on patriotic notions of national support and attempts to guilt Canadians into staying home, billboards, advertisements and posters in Canadian shops beseeched Canadians to spend their money in the homeland to benefit its economy rather than that of its neighbor to the south.

Conclusion

Cross-border shopping is an interesting intersection of economic, social and cultural dynamics within borderland regions worldwide. The act of crossing political borders for retail purposes has evolved from a practical need to a multifaceted activity encompassing leisure, cultural exchange and even healthcare and real estate investments. Cross-border shopping reflects the complex interplay of factors such as price differentials, currency exchange rates, product availability, cultural influences, spatial and psychological distances and ease of crossing borders.

Traditional models of outshopping have primarily focused on practical needs and cost savings, but contemporary trends highlight the growing importance of leisure elements and experiential aspects in shaping consumer behavior. The intentional planning of cross-border shopping destinations underscores the recognition of retail tourism as a strategic component of economic development in border regions.

However, cross-border shopping also has its challenges and complexities, such as taxation policies, import regulations and fluctuations in currency values, which continue to influence consumer behavior and shape patterns of cross-border commerce. Moreover, the rise of online retailing presents both opportunities and threats to traditional cross-border shopping models. Despite these challenges, cross-border shopping remains a dynamic and resilient phenomenon, with benefits extending to both sides of the border. The cross-border flow of retail capital contributes to the vitality and interconnectedness of borderland economies and societies. Whether through increased consumer spending, job creation, or cultural exchanges, cross-border shopping enriches the experiences of both residents and visitors alike. By embracing innovation, collaboration, and strategic planning, border communities can harness the potential of retail tourism to foster sustainable development.

In future studies of cross-border shopping, researchers should explore several avenues to deepen our understanding of this phenomenon and its implications. First, studying the effects of digital platforms and online marketplaces on cross-border shopping habits can provide valuable insights into transfrontier consumer behavior. It is essential to examine how e-commerce and online retail are transforming conventional models of cross-border shopping and explore the factors that influence customers' choices to shop online instead of physically crossing borders.

Consumer behavior is an important topic worth exploring. Conducting in-depth studies on consumer motivations and decision-making processes in cross-border shopping can provide valuable insights. It is also important to understand the interplay between utilitarian motives, such as price differentials and product availability, and hedonic motives, such as leisure experiences and cultural immersion, in shaping cross-border shopping behaviors.

Cross-border destination marketing and branding can be used to develop new strategies to promote borderland regions as transfrontier shopping destinations. Investigating how destination branding and promotional campaigns impact consumer perceptions and intentions to engage in cross-border shopping activities is crucial.

Retail planning is an important aspect when exploring the development of borderland regions. It involves assessing the impact of clustering strategies, free-trade zones and the development of border markets on local economies, tourism growth and community well-being. By studying such initiatives, we can gain insights into how they contribute to the overall growth and success of a region, including what are often marginalized areas of the state (Timothy, 2023).

To streamline border processes and enhance efficiency, it is crucial to explore how governments and stakeholders can effectively balance economic development goals with considerations regarding taxation, customs regulations, security measures and the facilitation of cross-border mobility. It is also crucial to explore the potential for cross-border collaboration and regional integration to tackle challenges and seize opportunities related to cross-border shopping. Policymakers need to assess how international treaties, bilateral agreements and cross-border partnerships can facilitate retail trade and promote economic cooperation across borders.

Ensuring a sustainable cross-border shopping environment requires vigilant monitoring and analysis of evolving retail dynamics, such as changes in consumer preferences, market trends and geopolitical factors. It is imperative to investigate emerging patterns of cross-border shopping in response to evolving economic, social and technological landscapes to enable this manifestation of shopping tourism to continue thriving and benefitting the people who travel to shop and the communities and merchants in the destinations they choose to visit.

References

Anderson, K.C., Knight, D.K., Pookulangara, S. and Josiam, B. (2014) Influence of hedonic and utilitarian motivations on retailer loyalty and purchase intention: A Facebook perspective. *Journal of Retailing and Consumer Services* 21 (5), 773–779.

Arreola, D.D. and Curtis, J.R. (1993) *The Mexican Border Cities: Landscape Anatomy and Place Personality.* University of Arizona Press.

Azmi, A., Hamid, I.A., Ahmad, J.A. and Ramli, R.A. (2017) Tourism supply chain perspectives on border shopping development at Padang Besar, Malaysia. In *Balancing Development and Sustainability in Tourism Destinations: Proceedings of the Tourism Outlook Conference* 2015 (pp. 291–300). Springer.

Azmi, A., Sulaiman, S., Asri, D. and Razali, M.A. (2015) Shopping tourism and trading activities at the border town of Malaysia-Thailand: A case study in Padang Besar. *International Academic Research Journal of Social Science* 1 (2), 83–88.

Bar-Kołelis, D. (2013) New retail locations and cross-border shopping in Poland. *Journal of Geography, Politics and Society* 3 (1), 19–31.

Bar-Kołelis, D. and Wendt, J.A. (2018) Comparison of cross-border shopping tourism activities at the Polish and Romanian external borders of European Union. *Geographia Polonica* 91 (1), 113–125.

Bar-Kołelis, D. and Wiskulski, T. (2012) Cross-border shopping at Polish borders: Tri-city and the Russian tourists. *GeoJournal of Tourism and Geosites* 9 (5), 43–51.

Birch, D.J. (1998) *Pilgrimage to Rome in the Middle Ages: Continuity and Change.* Boydell Press.

Borzooie, P., Lak, A. and Timothy, D.J. (2021) Designing urban customs and border marketplaces: A model and case study from Lotfabad, Iran. *Journal of Borderlands Studies* 36 (3), 469–486.

Bygvrå, S. (2009) Distance and cross-border shopping for alcohol: Evidence from Danes' cross-border shopping 1986–2003. *Nordic Studies on Alcohol and Drugs* 26 (2), 141–163.

Bygvrå, S. (2019) Cross-border shopping: Just like domestic shopping? A comparative study. *GeoJournal* 84 (2), 497–518.

Castaño, R., Perez, M.E. and Quintanilla, C. (2010) Cross-border shopping: Family narratives. *Qualitative Market Research: An International Journal* 13 (1), 45–57.

Cox, A.D., Cox, D. and Anderson, R.D. (2005) Reassessing the pleasures of store shopping. *Journal of Business Research* 58 (3), 250–259.

Cuevas Contreras, T. and Zizaldra Hernandez, I. (2023) Borders and healthcare: Medical mobility, globalization and borderland tourism. In D.J. Timothy and A. Gelbman (eds) *Routledge Handbook of Borders and Tourism* (pp. 281–295). Routledge.

Dilla, H., Cabezas, M.F. and Figueroa, M.T. (2022) Notes for a discussion on Latin American cross-border regions. *Journal of Borderlands Studies* 37 (3), 435–451.

Di Matteo, L. and Di Matteo, R. (1996) An analysis of Canadian cross-border travel. *Annals of Tourism Research* 23 (1), 103–122.

Evans, G. (1999) *Souvenirs: From Roman Times to the Present Day.* National Museums of Scotland.

Gelbman, A. and Schweitzer, R. (2023) Tourism in protected areas and transboundary parks for peace. In D.J. Timothy and A. Gelbman (eds) *Routledge Handbook of Borders and Tourism* (pp. 366–378). Routledge.

Hannonen, O., Tuulentie, S. and Pitkänen, K. (2015) Borders and second home tourism: Norwegian and Russian second home owners in Finnish border areas. *Journal of Borderlands Studies* 30 (1), 53–67.

Hitchcock, M. and Teague, K. (2019) *Souvenirs: The Material Culture of Tourism.* Routledge.

Jin, H., Moscardo, G. and Murphy, L. (2019) Exploring Chinese outbound tourist shopping: A social practice framework. *Journal of Travel Research* 59 (1), 156–172.

Kim, E.Y. and Sullivan, P. (2003) Cross-border tourism and shopping: Consumer segmentation. *E-Review of Tourism Research* 1 (1), 14–20.

Klamár, R. and Kozoň, J. (2022) Cross-border shopping tourism—case study to compare two regions of north-eastern Slovakia. *Folia Geographica* 64 (2), 46–85.

Kluser, F. (2023) *Cross-Border Shopping: Evidence from Swiss Household Consumption.* Universität Bern, Departement Volkswirtschaft.

Kuncharin, W. and Mohamed, B. (2013) Cross-border shopping motivation, behaviours and ethnocentrism of Malaysians in Hatyai, Thailand. *International Journal of Social, Human Science and Engineering* 7 (4), 291–301.

Lau, H.F., Sin, L.Y.M. and Chan, K.K.C. (2005) Chinese cross-border shopping: An empirical study. *Journal of Hospitality and Tourism Research* 29 (1), 110–133.

Leandro, F.J. (2021) Tourism as an instrument of cultural de-bordering and economic connectivity: The case of Macau. In J. Tavares da Silva, Z. Breda and F. Carbone (eds) *Role and Impact of Tourism in Peacebuilding and Conflict Transformation* (pp. 148–163). IGI Global.

Leick, B., Schewe, T. and Kivedal, B.K. (2021) Tourism development and border asymmetries: An exploratory analysis of market-driven cross-border shopping tourism. *Tourism Planning and Development* 18 (6), 673–698.

Leimgruber, W. (1988) Border trade: The boundary as an incentive and an obstacle to shopping trips. *Nordia* 22 (1), 53–60.

Makkonen, T. (2016) Cross-border shopping and tourism destination marketing: The case of Southern Jutland, Denmark. *Scandinavian Journal of Hospitality and Tourism* 16 (1), 36–50.

Makkonen, T. (2023) Outshopping abroad: Cross-border shopping tourism and the competitive advantage of borders. In D.J. Timothy and A. Gelbman (eds) *Routledge Handbook of Borders and Tourism* (pp. 269–280). Routledge.

Malkowski, A. (2019) Shopping tourism as a factor in the development of peripheral areas on the example of the Polish-Ukrainian borderland. *Scientific Papers of Silesian University of Technology*, 139, 301–312.

Malkowski, A., Mickiewicz, B. and Malkowska, A. (2020) Shopping tourism as a factor in the development of peripheral areas: The case of the Polish-German borderland. *European Research Studies Journal* 23 (3), 238–248.

Michalkó, G., Rátz, T., Hinek, M. and Tömöri, M. (2014) Shopping tourism in Hungary during the period of the economic crisis. *Tourism Economics* 20 (6), 1319–1336.

Michalkó, G. and Timothy, D.J. (2001) Cross-border shopping in Hungary: Causes and effects. *Visions in Leisure and Business* 20 (1), 4–22.

Mohi, P. (2023) More and more Hungarians are shopping in Slovakia: Is it worth it? Online: https://dailynewshungary.com/more-and-more-hungarians-are-shopping-in-slovakia-is-it-really-worth-it/

Oberle, A.P. and Arreola, D.D. (2004) Mexican medical border towns: A case study of Algodones, Baja California. *Journal of Borderlands Studies* 19 (2), 27–44.

Opačić, V.T. and Koderman, M. (2018) From socialist Yugoslavia to the European Union: Second home development in Croatia and Slovenia. In C.M. Hall and D.K. Müller (eds) *The Routledge Handbook of Second Home Tourism and Mobilities* (pp. 167–178). Routledge.

Park, J., Tse, S., Mi, S.D. and Song, H. (2022) A model for cross-border tourism governance in the greater Bay area. *Journal of China Tourism Research* 18 (6), 1259–1283.

Prokkola, E-K. (2023) Cross-border tourism initiatives in the European Union. In D.J. Timothy and A. Gelbman (eds) *Routledge Handbook of Borders and Tourism* (pp. 353–365). Routledge.

Ramsey, D., Thimm, T. and Hehn, L. (2019) Cross-border shopping tourism: A Switzerland-Germany case study. *European Journal of Tourism, Hospitality and Recreation* 9 (1), 3–17.

Risal, S., Irawanto, I., Hergianasari, P., Rahma, A.A. and Pramudiana, I.D. (2022) Border development model on the human development aspects for sustainable community. *Journal Borneo Administrator* 18 (2), 125–138.

Rogerson, C.M. (2018) Informal sector city tourism: Cross-border shoppers in Johannesburg. *GeoJournal of Tourism and Geosites* 22 (2), 381–392.

Ron, A.S. and Timothy, D.J. (2019) *Contemporary Christian Travel: Pilgrimage, Practice and Place*. Channel View Publications.

Sharma, P., Chen, I.S. and Luk, S.T. (2018) Tourist shoppers' evaluation of retail service: A study of cross-border versus international outshoppers. *Journal of Hospitality and Tourism Research* 42 (3), 392–419.

Smętkowski, M., Németh, S. and Eskelinen, H. (2017) Cross-border shopping at the EU's Eastern edge: The cases of Finnish-Russian and Polish-Ukrainian border regions. *Europa Regional* 24 (1/2), 50–64.

Spierings, B. and van der Velde, M. (2013) Cross-border differences and unfamiliarity: Shopping mobility in the Dutch-German Rhine-Waal Euroregion. *European Planning Studies* 21 (1), 5–23.

Stoffelen, A. and Timothy, D.J. (2023) Bordering, ordering and othering through tourism: The tourism geographies of borders. *Tourism Geographies* 25 (8), 1974–1992.

Studzińska, D., Sivkoz, A. and Domaniewski, S. (2018) Russian cross-border shopping tourists in the Finnish and Polish borderlands. *Norsk Geografisk Tidsskrift* 72 (2), 115–126.

Sullivan, P., Bonn, M.A., Bhardwaj, V. and DuPont, A. (2012) Mexican national cross-border shopping: Exploration of retail tourism. *Journal of Retailing and Consumer Services* 19 (6), 596–604.

Szytniewski, B.B. and Spierings, B. (2018) Place image formation and cross-border shopping: German shoppers in the Polish bazaar in Słubice. *Tijdschrift voor Economische en Sociale Geografie* 109 (2), 295–308.

Szytniewski, B.B., Spierings, B. and van der Velde, M. (2017) Socio-cultural proximity, daily life and shopping tourism in the Dutch-German border region. *Tourism Geographies* 19 (1), 63–77.

Szytniewski, B.B., Spierings, B. and van der Velde, M. (2020) Stretching the border: Shopping, petty trade and everyday life experiences in the Polish–Ukrainian borderland. *International Journal of Urban and Regional Research* 44 (3), 469–483.

Timothy, D.J. (1995) Political boundaries and tourism: Borders as tourist attractions. *Tourism Management* 16 (7), 525–532.

Timothy, D.J. (1999) Cross-border shopping: Tourism in the Canada-United States borderlands. *Visions in Leisure and Business* 17 (4), 4–18.

Timothy, D.J. (2001) *Tourism and Political Boundaries*. Routledge.

Timothy, D.J. (2005) *Shopping Tourism, Retailing and Leisure*. Channel View Publications.

Timothy, D.J. (2014) Trends in tourism, shopping, and retailing. In A.A. Lew, C.M. Hall and A.M. Williams (eds) *The Wiley Blackwell Companion to Tourism* (pp. 378–388). Wiley Blackwell.

Timothy, D.J. (2018) Shopping tourism. In S. Agarwal, G. Busby and R. Huang (eds) *Special Interest Tourism: Concepts, Contexts and Cases* (pp. 134–144). CABI.

Timothy, D.J. (2020) Borderscapes and tourismscapes: The place of postcards in Mexican border town tourism. *Geographia Polonica* 93 (4), 553–568.

Timothy, D.J. (2021) *Tourism in European Microstates and Dependencies: Geopolitics, Scale and Resource Limitations*. CABI.

Timothy, D.J. (2023) National peripheries, borderlands and tourism planning. In P.F. Xie (ed.) *Handbook on Tourism Planning* (pp. 132–145). Edward Elgar.

Timothy, D.J. (2024) Tourism, shopping and connotations of place. In C.M. Hall (ed.) *The Wiley Blackwell Companion to Tourism* (2nd edn, pp. 501–514). Wiley.

Timothy, D.J. and Butler, R.W. (1995) Cross-border shopping: A North American perspective. *Annals of Tourism Research* 22 (1), 16–34.

Timothy, D.J., Michalkó, G. and Irimiás, A. (2022) Unconventional tourist mobility: A geography-oriented theoretical framework. *Sustainability* 14 (11), 6494.

Timothy, D.J. and Więckowski, M. (2023) Borders, heritage, and memory. In D.J. Timothy and A. Gelbman (eds) *Routledge Handbook of Borders and Tourism* (pp. 219–240). Routledge.

Tömöri, M. (2010) Investigating shopping tourism along the borders of Hungary: A theoretical perspective. *GeoJournal of Tourism and Geosites* 6 (2), 202–210.

Tömöri, M. (2011) The role of the 'DebOra' cross-border Eurometropolis in the Hungarian-Romanian CBC relations: A case study of shopping tourism in Debrecen and Oradea. *Eurolimes* 11, 170–179.

van der Velde, M. and Spierings, B. (2010) Consumer mobility and the communication of difference: Reflecting on cross-border shopping practices and experiences in the Dutch-German borderland. *Journal of Borderlands Studies* 25 (3–4), 191–205.

Więckowski, M. (2010) Tourism development in the borderlands of Poland. *Geographia Polonica* 83 (2), 67–81.

Więckowski, M. and Timothy, D.J. (2021) Tourism and an evolving international boundary: Bordering, debordering and rebordering on Usedom Island, Poland-Germany. *Journal of Destination Marketing and Management* 22, 100647.

Yuan, J., Fowler, D.C., Goh, B.K. and Lauderdale, M.K. (2013) Mexican cross-border shoppers' motivations to the USA. *International Journal of Culture, Tourism and Hospitality Research* 7 (4), 394–410.

8 Shopping Tourism and Retail-led Urban Regeneration in Historic Cities

Azadeh Lak and Pantea Hakimian

Introduction

Since the 1990s, urban decline has plagued many historic city centers and traditional retail areas. In response, retail-based regeneration has emerged as a crucial urban planning strategy to improve the quality of urban life and revitalize deteriorated retail areas using various development agents and employing bottom-up initiatives such as community-based institutions and organizations (Claxton & Siora, 2008). Urban decline in historic cities came about through a combination of forces, one of which is the closure of traditional commercial areas as new peri-urban and suburban shopping centers were built and with the increasing prevalence of convenient online shopping. This has had a chain reaction within urban areas where local economies have been devastated and the cultural and heritage values of many cities have been severely diminished (Kim & Jang, 2017). Accordingly, many governments have allocated sizable funds to regenerating and promoting these derelict commercial areas as a way of turning them into tourist destinations by linking them to the broader heritagescapes of the city and refocusing their retail role (Bourlessas *et al.*, 2022; Kim & Jang, 2017; Loda *et al.*, 2020; Mermet, 2017). Several retail-led regeneration strategies, such as town center management, business improvement districts and town management organizations, have highlighted the role of retail as an engine of economic growth in deprived areas (Dixon, 2005).

Shopping is one of the most influential drivers of city tourism and urban regeneration, and has expanded to the point that a type of urban tourism, leisure shopping, is now widespread and significantly influences the image of cities in the competitive global marketplace (Zaidan, 2019). In their studies, Pearce (1999) and Rabbiosi (2015) demonstrate that the historical characteristics of Paris as a capital of style and fashion, and an otherwise commercial metropolis, are fundamental features of the successful promotion of tourism in the city. Retail as a social incubator (Oldenburg, 1989) and part of cities' intangible cultural heritage (Zukin, 2012) is a foundational truism in many shopping tourism studies. Accordingly, urban environments as spaces of consumption have come to resemble 'malls without walls' (Featherstone, 2007; Gold & Ward, 1996; Gotham, 2002).

To link urban renewal with tourism, Swarbrooke (2000) introduced five regeneration strategies through tourism development, a few of which overlap significantly: business tourism-led strategies, event tourism-led strategies, attraction tourism-led strategies, retail tourism-led strategies and cultural tourism-led strategies. Likewise, Jansen-Verbeke (1986) emphasizes the historical core of the city as a 'leisure product', categorizing the components of urban tourism into three categories: primary elements (e.g. attractions, activities and entertainment complexes), secondary elements (e.g. facilities and support services, such as accommodations, restaurants and shopping facilities) and conditional elements (e.g. tourism-related infrastructure such as transportation, education and information).

Most retail and urban development studies have been conducted in the Global North (Dawson *et al.*, 2020), with only a few having been done in the Global South (e.g. Faria & Whitesell, 2021; Santad, 2013). A few studies have been conducted on traditional markets (bazaars and souqs) in the Middle East, focusing primarily on architecture, restoration, aesthetic designs, guilds of traditional merchants, how they were formed throughout history and their modern-day role in urban tourism (Boussaa, 2015; Gutberlet, 2024; Orbaşli, 2019; Zaidan, 2019). Other studies have focused on the rise and decline of other retail spaces in city centers and suburban areas in the developed and developing worlds (Escudero-Gómez, 2024; Guimarães, 2019; Susser & Schneider, 2003), with many politicians and industry leaders concerned about how to arrest the decline of shopping in the urban milieu.

For tourism development and urban planners, successful urban regeneration with an emphasis on shopping tourism depends on understanding how to provide the most suitable services and products to tourists and knowing their purchase patterns and priorities and the factors influencing them. This necessity should cause companies, organizations and tourism destinations to understand consumer behavior, adjust their marketing strategies, engage in spatial restoration interventions and provide the necessary infrastructure to support tourism.

Thus, identifying and describing the factors affecting shopping tourism in traditional markets and taking advantage of the valuable socioeconomic potential of such areas, can contribute to the development of urban tourism, including shopping, in historic cities. Accordingly, it is crucial to learn which variables have the most significant impact on regenerating the historical fabric of cities with a focus on the development of shopping tourism. Thus, in this chapter we first present a theoretical framework of influential factors to answer this question. Then, urban regeneration experts in urban administration who are engaged in the policymaking of urban regeneration in the historical cities of Iran were surveyed to gain empirical insight. Finally, the urban regeneration success model focused on shopping tourism was developed with Smart_PLS and is presented in this chapter.

Tourism, Shopping Tourism and Retail-Led Urban Regeneration

Shopping tourism is defined by tourists who shop outside their usual environment, which is a determining factor in their decision to travel (Choi et al., 2016b). Shopping is an inevitable part of tourism, and the crossover between shopping and tourism creates unique urban retail landscapes in tourist destinations and, according to Zukin (2018), the landscape of consumption is a landscape of power in which various actors, including the public, policymakers, tourism planners, tourists, shopkeepers, residents, developers and real estate agents are active (Zukin & Maguire, 2004). In the context of urban shopping tourism, space and the presence of people in their urban environments is important in tourism planning and urban design (Gottdiener, 1995). For example, shopping centers do not exist without consumer culture and sufficient demand, and at the same time, they are essential public action spaces in late-capitalist societies (Rabbiosi, 2011).

Rabbiosi (2011) suggests that shopping tourism destinations may be classified as destinations with particular goods that are widely known (unique handicrafts, outlets); destinations that offer a variety of goods with higher quality, higher volume, or lower prices; places that are centers of creativity and design (fashion areas); and destination malls such as shopping centers and large-scale retail markets and outlet stores. Shopping is one of the three most essential tourism activities, along with dining and sightseeing (McKercher, 2020; Timothy, 2005). The importance of shopping in the tourism economy even influences the urban brand of many countries, with some cities and/or countries becoming well-known shopping destinations both through intentional branding efforts and natural organic processes. However, unfortunately, shopping's role in urban branding has not been adequately addressed in tourism and urban planning literature (Choi et al., 2016a; Yüksel, 2013).

Recent approaches to shopping tourism have sought to combine shopping and retail activities with the components of urban entertainment, leading to the notion of 'retailtainment' (Stephenson et al., 2010), which was earlier called 'shoppertainment' in the contexts of specific shopping centers in the 1990s (Butler, 1991; Timothy, 2005). Shoppertainment and retailtainment have been successfully adopted in luxury tourism centers such as Dubai, integrating shopping centers with leisure and entertainment complexes and activities such as sightseeing, cultural experiences, gourmet dining and entertainment (Zaidan, 2016).

Retail-led urban regeneration has always been seen both positively and negatively (Kim & Jang, 2017; Ruming *et al.*, 2018). On the one hand, it has played an essential role in the rapid economic recovery of cities. A case study of Liverpool showed that at the turn of the 21st century, several factors, including the dominance of a new administrative and political structure in the city council, the establishment of a clear vision for the city, and a retail-led catalyst project contributed to the rapid economic recovery of that British city (Parker & Garnell, 2006). On the other hand, despite the advantages of shopping-based development, there is always a chance of commercial and retail stagnation in historic cities. Lowe's (2005a) study shows the opposite – that the development of retail and shopping centers did not cause stagnation in Southampton, England, but instead brought about a significant synergy with the city's enhanced retailscape, the retail development being an essential element in the city's regeneration successes.

Similar successes can be found throughout the world in central city revitalization programs and waterfront development efforts. Famous waterfront developments in cities such as Sydney, Baltimore, Toronto, Liverpool and Singapore have had salient impacts on urban revitalization and socioeconomic development, and most of the urban regeneration in these localities has been retail based, catering both to local consumers and out-of-town tourists and creating engaging clusters of retailtainment (Chang & Huang, 2011; Edwards *et al.*, 2008).

Similarly, in certain cities, retail and shopping have been at the forefront of mass urban development. Some of the best-known examples include Dubai, Doha and Singapore, but there are many others. Although criticized widely for its unsustainable growth and negative ecological impacts, the hyperdevelopment of Dubai and Abu Dhabi has led to those cities' (and indeed the entire UAE) being branded as a world-class luxury shopping destination where mass tourism is now characterized by rapid urbanization, pervasive entertainment complexes, high standards of living and an abundance of high-end retail intermixed with various amusements, festivals, sporting events and extravagant second-home/vacation ownership among the world's elite (Alhosani & Zaidan, 2014; Zaidan, 2019).

Despite this trend towards mass urbanization and hyperdevelopment, traditional markets (e.g. souqs, bazaars and farmers markets) have for centuries been a focal point of community life, socioeconomic functions and the physical fabric of historic cities. These were traditionally spaces of social interaction and networking, economic trade and consumption, social identity formation and places of community pride. Today, these same marketplaces provide opportunities for tourists (and locals) to buy a wide range of goods and local cultural products. Spending time in local markets can create positive emotions and feelings in visitors and provide deeper cultural experience and a better understanding of places and peoples; possessing a historic marketplace can act as a competitive advantage over other destinations (Dixon, 2005; Henderson, 2000; Hsieh & Chang, 2006; Kikuchi & Ryan, 2007). As such, many of these historic settings throughout the world continue to be important social venues for local residents, and significant tourist attractions for outside visitors. In traditional markets, visitors can get a taste of local culinary delights, mix and mingle with destination inhabitants, and become culturally immersed, which so many tourists crave today. Nonetheless, with modernization and rapid urbanization since the 1970s, but particularly in the 2000s, traditional markets have diminished in their socioeconomic roles and as tourist attractions, although many cities have tried to maintain their marketplaces as salient features of the urban landscape and utilize them as heritage tourist attractions (Boussaa, 2015; Gutberlet, 2024).

From a tourist destination perspective, tourists' commitment to shopping depends on many intrinsic and extrinsic factors, such as gender, age, socioeconomic status, family life, currency exchange rates, lifestyle habits, the purpose of travel, mode of transportation, accommodation types, social interaction, attitude towards destination cultures and the nationality of the buyer (Henderson *et al.*, 2011; Jin *et al.*, 2021; Turner & Reisinger, 2001), which may determine spending patterns and gift-buying behaviors (Barutçu *et al.*, 2011; Park, 2000; Timothy, 2005; Tosun *et al.*, 2007). Other studies have highlighted the merchandise and venue attributes that help simulate tourist shopping in the destination, such as product features and prices, service and business performance, brand recognition, store layout and environment and adequate ancillary services (Soleimani & Mohammadnejad, 2015; Turner & Reisinger, 2001).

Destination-based factors that affect the growth of shopping tourism include price differentials (e.g. low-cost destinations, low prices at the destination), destination characteristics (e.g. places known for particular goods, destination popularity, shopping routes and zones, festivals and events, the presence of shopping centers and local hospitality) and unique products (e.g. souvenirs and handicrafts, and special tax breaks for tourists) (Asadi *et al.*, 2020; Timothy, 2005). Similarly, other place-specific geographic and demographic features are important in facilitating and encouraging the development of shopping tourism. These include the clustering of retail establishments and complementary businesses, adequate pedestrian space and appropriate signage, co-mingling of local residents and tourists, maintaining environmental quality, efforts to create a sense of place, planning events and entertainment, access to free WiFi and solid marketing efforts (McKercher, 2020; Timothy, 2005).

Many research studies have shown that all of these personal, retail and destination characteristics help create satisfying shopping tourism

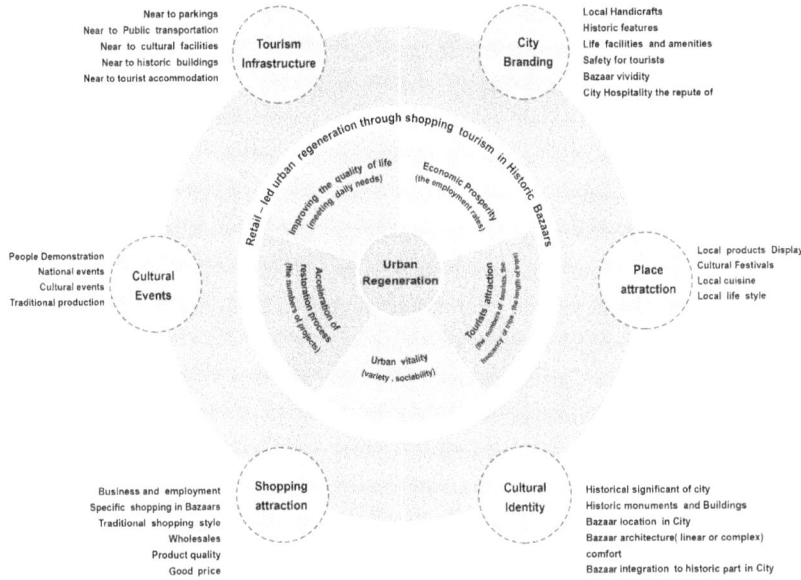

Figure 8.1 The conceptual framework of the study

experiences and can help maintain a thriving retailscape and perhaps renew one that is on the verge of decline (Soleimani & Mohammadnejad, 2015). Pourzakarya and Fadaei Nezhad Bahramjerdi (2019) contend that shopping areas – also doubling as areas of cultural development and innovation – have the best chance of success with innovative urban planning and the re-creation of a culture-based urban sense of place. Although these principles pertain to all urban shopping contexts, they are perhaps most effectual and visible in historic city centers and traditional marketplaces. Based upon the knowledge gleaned from the literature and the importance of culture in maintaining an urban sense of place that relates to traditional shopping venues, for the study presented here, the development process was divided into four categories: cultural setting, local cultural and creative industries (CCIs), cultural and creative tourism and cultural infrastructure that aligns with the needs of urban regeneration. Figure 8.1 displays the relationship between regeneration success and shopping tourism development factors in a conceptual model.

Empirical Case Study: Historic Cities of Iran

Materials and methods

The authors are from Iran and have access to urban settings in the country for analysis. Figure 8.2 illustrates examples of traditional bazaars and markets in some if Iran's most historic cities. Our study aimed to

Figure 8.2 Examples of traditional Iranian bazaars

determine how shopping tourism factors and shopping attractions in traditional markets affect urban regeneration using data from municipal employees with academic degrees in urban planning and design. The participants were selected using purposive and snowball sampling. This study used a cross-sectional online survey with a self-administered questionnaire. Following previous studies in the literature, the questionnaire was developed based on the criteria of success of urban regeneration and the factors affecting commercial tourism, public policy and shopping-oriented regeneration. The questionnaire was revised and validated based on the feedback from five members of the same survey group. The target population was selected from municipal employees in several historic cities in Iran (see Table 8.1). The questionnaires were distributed by email, WhatsApp and other electronic social networks.

Table 8.1 Participants' demographic data

Variables	Categories	Frequency	Percentage
Gender	Female	23	23
	Male	77	77
Age	30–40 years	34	34
	41–50 years	46	46
	50–60 years	20	20
Education	MS in Urban Planning Design	84	84
	PhD in Urban Planning & Design	16	16
Municipality	Tehran	39	39
	Esfahan	12	12
	Shiraz	11	11
	Kerman	8	8
	Qazvin	8	8
	Yazd	7	7
	Kashan	7	7
	Tabriz	8	8
Total		100	100

Of 150 municipal staff of the eight historic cities in Iran (Tehran, Tabriz, Isfahan, Shiraz, Kerman, Kashan, Yazd and Qazvin) agreeing to complete the questionnaire, 100 responses from 23 female and 77 male staff were received. Most participants were 41–50 years old (45%) and mainly from Tehran's municipal government (51%). Table 8.1 shows the participants' demographic characteristics.

As noted at the outset, this study aims to identify and describe the factors affecting shopping tourism in traditional markets in Iran's historic cities and examine the potential of these areas for retail-based tourism development and urban renewal. Besides demographic data, the survey aimed to assess various dimensions of urban tourism (e.g. city branding, historical-spatial attractiveness of the city, the place's cultural identity, the attractiveness of local shopping, the display of unique events and activities and the extent of urban tourism infrastructure), criteria for success of the urban regeneration process, including economic prosperity (e.g. increased employment), attracting and retaining tourists (e.g. numbers of tourists, frequency of travel and length of stay), the vitality of the urban environment (diversity and attendance), accelerating the regeneration process of the historical fabric (number of projects) and improving quality of life and meeting daily needs.

The sample size was estimated using structural equation model analysis. Pai *et al.* (2021) argue that structural equation model analysis needs the N:q ratio of 5 to 1, or 5 participants for each estimated parameter in our framework. Hence, a minimum of 125 participants needed to be recruited for our study.

Analysis

Participants' responses were recorded in an Excel sheet and analyzed with Smart PLS1 (version 3.3.0). We also performed descriptive statistics on the demographic data with IBM SPSS (version 21). Structural equation modeling (SEM) was used to evaluate the interaction of various factors such as city branding, events, place-based attractions, shopping attractions, cultural identity and tourist infrastructure and amenities that affected the success of urban regeneration in traditional bazaars in old cities. The data were analyzed using SEM with the help of Smart PLS software. Then, the findings of the SEM–PLS analysis and the *t*-statistics were presented and discussed.

Results

Measurement model

The measurement model must be evaluated and improved before calculating the structural model. The outer loadings of all the associated survey questions were computed and statistically examined to ascertain if they were significant. The indicators with statistically insignificant outer loadings and poor correlations with the underlying constructs, were left out of the study. When modifying the model, the AIC (Akaike information criterion), BIC (Bayesian information criterion) and HQC (Hannan–Quinn information criterion) values were evaluated as Smart-PLS (version 3.3) outputs. Hair *et al.* (2019) suggest that outer loadings larger than 0.708 are desirable since they may account for more than 50% of the variations in an indicator. Almost all the outer loadings of the selected items were significant ($p < 0.708$) (Hair *et al.*, 2019).

Moreover, the construct reliability and validity were assessed by calculating the usual internal consistency measures. With 2000 samples, a bootstrapping approach was used to estimate the composite reliability measures for the dimensions, as shown in Table 8.2. The internal consistency of the constructs was confirmed by the composite reliability values ranging from 0.6 to 0.9 for most of the constructs (Hair *et al.*, 2019). The constructs with *p*-values less than 0.05 were considered to be significant.

The convergent validity was assessed by calculating the average variance values (AVE), as displayed in Table 8.3. The AVE values were almost 0.5 and were thus significant (Hair *et al.*, 2019). Hence, convergent validity was established.

Table 8.2 Composite reliability measures of constructs, bootstrapping with 2000 samples

Variables	Original sample (O)	Sample mean (M)	Standard deviation (STDEV)	t-statistics (|O/STDEV|)	p-values
City branding	0.814	0.861	0.128	6.978	0.000
Events	0.494	0.328	0.138	3.869	0.005
Place attractions	0.452	0.313	0.176	3.005	0.046
Cultural identity	0.437	0.574	0.114	5.605	0.000
Shopping attraction	0.698	0.696	0.043	10.827	0.000
Tourist infrastructures and amenities	0.524	0.268	0.163	6.967	0.048
Urban regeneration	0.541	0.542	0.074	7.434	0.000

Table 8.3 Average values of constructs, bootstrapping with 2000 samples

Variables	Original sample (O)	Sample mean (M)	Standard deviation (STDEV)	t-statistics (|O/STDEV|)	p-values
City branding	0.613	0.564	0.014	10.410	0.000
Events	0.439	0.229	0.037	3.632	0.000
Place attractions	0.479	0.368	0.035	11.495	0.000
Cultural identity	0.497	0.373	0.036	7.978	0.010
Shopping attraction	0.498	0.522	0.028	17.356	0.000
Tourist infrastructures and amenities	0.588	0.396	0.045	10.216	0.020
Urban regeneration	0.468	0.481	0.037	14.182	0.000

The Heterotrait and Monotrait (HTMT) Heterotrait and Monotrait ratios in Table 8.4 were calculated to evaluate the discriminant validity of the constructs. As can be seen, nearly every ratio is smaller than 0.85 (although there are a few HTMT ratios above 1.0), overall demonstrating the distinctness of the structures (Hair *et al.*, 2019).

Before executing the structural equation model, the collinearity of the constructs was evaluated. The variance inflation factor (VIF) values in Table 8.5 are less than 3, indicating that the constructs were not multi-collinear. Figure 8.3 shows the route coefficients and external loadings for

Table 8.4 Heterotrait and Monotrait (HTMT) ratios of constructs.

Variables	Events	Place Attractions	Cultural Identity Health	Shopping Attraction	Tourist Infrastructures and Amenities
City branding	0.249				
Events	0.284	1.032			
Place attractions	0.219	0.544	1.029		
Cultural identity	0.138	0.531	0.703	0.476	
Shopping Attraction	0.258	0.580	0.602	0.642	1.117
Tourist Infrastructures and Amenities	0.267	0.668	0.491	0.764	0.349

Table 8.5 Variance inflation factor (VIF) values of the constructs

Variables	Urban regeneration
City branding	1.027
Events	1.180
Place attractions	1.088
Retail attraction	1.045
Cultural identity	1.128
Tourism infrastructure and amenities	1.050

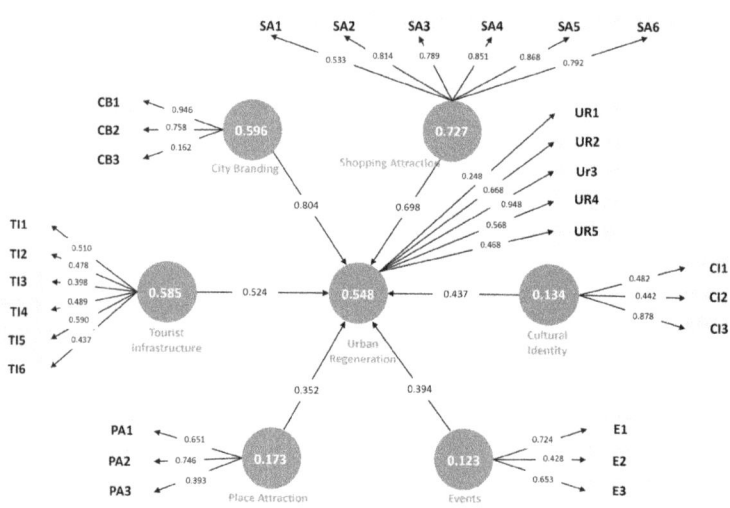

Figure 8.3 Path coefficients and outer loadings for the direct relationship between the successful regeneration and shopping tourism development

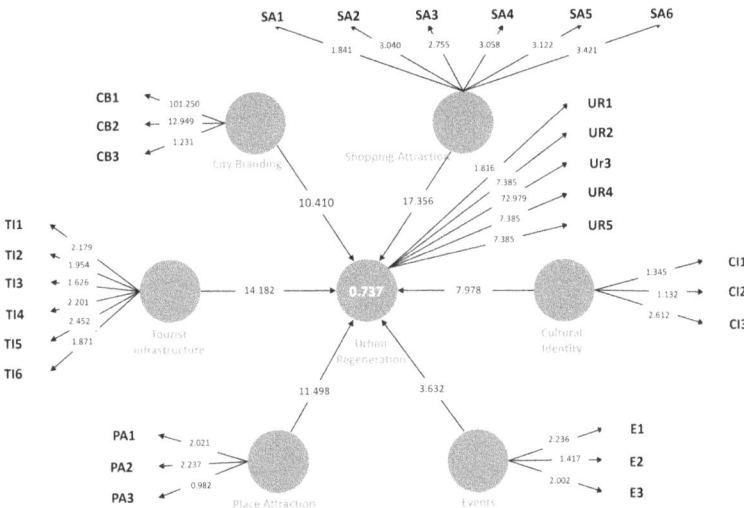

Figure 8.4 *t*-statistics for the inner and outer models

the direct association between the influencing elements and urban regeneration, including economic success, visitor length of stay, vibrancy, promotion of quality of life and facilitation of urban renewal initiatives. The percentage of variations in the dependent construct explained by the independent constructs is shown by the R^2 value as a measure of the explanatory power of the structural model. The strength of the link is moderate to relatively strong ($R^2 = 0.628$) (Hair *et al.*, 2019).

The estimated PLS-SEM and associated *t*-statistics for indicators and constructs in each model are shown in Figure 8.3. The significance of the route coefficients and indicators is examined using the *t*-statistics (Figure 8.4). These *t*-statistics matching *p*-values are less than 0.1. (0.05). As a result, the related effects are significant at the 10% (5%) significance level.

Table 8.6 shows the total effect values for each construct. As can be seen, the attractiveness of shopping and products has a strong positive effect on the success of urban regeneration in traditional markets. The city brand and the availability of necessary tourism amenities and infrastructure (e.g. shopping) are practical factors contributing to the realization of urban regeneration. Other factors, such as local attractiveness, cultural identity and various religious and national events, can also be influential in the realization of urban regeneration.

Discussion

Many historic cities, especially in developing countries, have faced physical and socioeconomic declines of city centers and traditional markets as historical retail areas (González & Waley, 2013; Venkov, 2022).

Table 8.6 Path coefficients, *t*-statistics and *p*-values of the structural model

Variables	Original Sample (O)	Sample Mean (M)	Standard Deviation (STDEV)	t-statistics (|O/STDEV|)	p-values
City branding -> U.R	0.613	0.593	0.049	3.970	0.086
Events -> U.R	0.439	0.115	0.059	1.950	0.059
Place attractions -> U.R	0.479	0.175	0.071	1.998	0.054
Cultural identity -> U.R	0.497	0.125	0.067	1.990	0.047
Retail attraction -> U.R	0.498	0.706	0.063	8.554	0.000
Tourist infrastructures -> U.R	0.588	0.558	0.080	3.288	0.022

Note: t-test statistics > 1.96 or p-value < 0.05 is considered statistically significant (Pai et al., 2021).

Thus, urban policymakers are looking for policies to promote historic centers, especially traditional marketplaces, and to support small and independent retailers in these urban retailscapes. In addition to urban decline in the central areas of historic cities, lifestyle changes caused by increased income levels and changes in consumption patterns have caused residents to use supermarkets and online shopping, reducing the prosperity of and functionality of many traditional markets (Dutch-Brown *et al.*, 2021). In addition, the rapid growth of e-commerce during the COVID-19 pandemic, along with a decrease in urban tourism, has intensified the decline of the shopping life of historic city centers and traditional markets (Van Eck *et al.*, 2020). Thus, there is a need for approaches to promoting shopping tourism and attracting tourists to regenerate traditional marketplaces and other urban retail spaces.

The empirical study presented here sought to identify factors affecting the development of shopping tourism for the successful regeneration of traditional markets and their broader urban settings. The results indicate that shopping tourism might be a successful tool for urban regeneration only when the essential elements of tourism development are promoted. In addition, the prosperity of businesses can be enhanced by marketing activities to raise funds to increase income, improve the city's social, economic and environmental conditions as a result of urban branding, and regenerate the city based on its greater retail and heritage potential (Rabbiosi, 2015). This study indicates that the success of tourism depends on the realization of economic prosperity, attracting and retaining tourists, increasing the vitality of the urban environment, accelerating the process of regenerating the urban historic fabric, and improving the quality of life of residents. Based on our analysis, two major groups of factors may be helpful in ensuring successful regeneration based on shopping tourism:

- The first group is comprises physical features and destination-specific elements that help create an appealing ambience. The most strategic elements of shopping tourism development are the brand and regional reputation in attracting tourists (e.g. the attractiveness of the place and the hospitability of its residents). This is followed by destination qualities that truly enhance the immersive experience overall, including the diversity and attractiveness of shopping along with business diversity, effective marketing, provision of employment, the occurrence of special cultural events and festivals, the presence of unique and rare religious ceremonies, enthralling intangible heritage (e.g. language, customs, ethnic or cultural identities and lifestyles) and the physical features of the urban landscape. These include, among others, historic buildings and monuments – especially those located in or near traditional markets, and a wide range of retail options. Our findings align with those of other studies that have looked at the factors of success in retail-led urban regeneration (e.g. Iliev, 2021; Rabbiosi, 2011).
- The second group of influential variables include infrastructure and amenities that support tourism. This entails typical hospitality and transportation services, but also includes wellness amenities, such as recreation areas, medical centers, sanitary facilities and open green spaces.

Conclusion

Shopping-based tourism has been shown to be a critical tool in helping to restore market towns and other retail districts and revive retail activities. Although the focus of this study was historic marketplaces in historic towns, many of its precepts can also be extrapolated for urban retailscapes in general, whether historic center retail zones or newer suburban shopping clusters. Entrepreneurship and business incentives are necessary to improve the physical quality of bazaars and commercial centers in historic cities (Lowe, 2005b; Rabbiosi, 2015). Shopping tourism in traditional markets and city centres is dependent on the retail sales of local products to tourists who may be in town for other purposes, such as spending leisure time, taking care of business, participating in religious practices, visiting attractions or attending special events. However, in some cases, shopping malls and retail zones may be a catalyst for visiting, an 'anchor attraction' that brings tourists to the city to shop (Butler, 1991; Lorch & Smith, 1993; Thanasi-Boçe *et al.*, 2021; Timothy, 2005). Several of the urban malls in Dubai are characteristic of this. Thus, the improvement of prosperity in retail districts depends on the tourist-consumers who are attracted to the city specifically to shop or by visitors who are in town for other purposes. The urban brand plays a significant role in attracting visitors and local consumers who want to visit attractive marketplaces and bustling spaces of retail otherness (e.g. malls and shopping centers).

For this reason, the market is slowly and spontaneously witnessing an increase in offering suitable products for shopping tourism. Thus, considering that a large part of the physical appeal of a city's historic fabric comprises markets and related commercial establishments (Gutberlet, 2024; Pourjafar et al., 2014), there is considerable potential for increased prosperity through growth in income generation and residents' entrepreneurial efforts. In addition, special events and festivals can incentivize locals to shop and draw in consumers from further afield to visit local markets and other urban shopping complexes, especially those in declining city centers. As noted earlier in the context of waterfront development, this can help stimulate less productive parts of the city to become increasingly productive, renewed with more vitality, and increase resident employment, outside investments, and stimulate tourist expenditures. Thus, spaces that have been marginalized over the past half century can once again become zones of thriving commerce and tourism-oriented retail (Borucka et al., 2022; Jones, 2017). Meanwhile, urban managers must create suitable platforms for implementing innovative and collaborative projects focusing on physical renovation, market expansion and fostering a notable sense of place.

One of the limitations of the study presented in this chapter was the small sample size and the lack of knowledgeable experts on urban regeneration and shopping tourism programs. While such a small sample size creates certain limitations with SEMs and other analytical tools, the study does in fact produce some useful findings that can help other historic communities with traditional marketplaces and expanding modern retailscapes. To further the knowledge base on shopping-induced urban development, qualitative studies on tourist shopping in traditional marketplaces and in other retail zones of historic cities are needed. In doing so, exploring residents' and tourists' lived experiences in traditional markets and modern retail zones can help identify other factors contributing to the success of urban regeneration through shopping tourism.

The experience of reviving traditional markets and retail areas in various parts of the world supports the findings from Iran (Kim & Jang, 2017; McKercher, 2020), indicating that a lack of prosperity in local commercial areas can undermine local and national economies and deteriorate the cultures and histories connected to traditional commercial areas, as maintaining traditions and conserving urban heritage may be of secondary or tertiary value when livelihoods and basic survival are at stake (Timothy & Boyd, 2003). To strengthen traditional retail areas, governments can implement certain policies during urban regeneration processes. These policies include developing infrastructure, enhancing merchants' business management skills and implementing location marketing strategies in each area. By connecting retailscapes with historical and cultural sites close to commercial districts in urban regeneration plans, these initiatives can assist in promoting local traditional commercial areas and transforming them into tourist destinations.

References

Alhosani, N. and Zaidan, E. (2014) Shopping tourism and destination development: Dubai as a case study. *The Arab World Geographer* 17 (1), 66–81.

Asadi, M., Ghanbari, A. and Alizadeh Aqdam, M.B. (2020) An evaluation of the role of shopping tourism in the development of urban tourism: Case study of Tanakura Market in Urmia. *Urban Tourism* 7 (3), 1–16.

Barutçu, S., Doğan, H. and Üngüren, E. (2011) Tourists' perception and satisfaction of shopping in Alanya region: A comparative analysis of different nationalities. *Procedia-Social and Behavioral Sciences* 24, 1049–1059.

Borucka, J., Czyż, P., Gasco, G., Mazurkiewicz, W., Nałęcz, D. and Szczepański, M. (2022) Market regeneration in line with sustainable urban development. *Sustainability* 14 (18), 11690.

Bourlessas, P., Cenere, S. and Vanolo, A. (2022) The work of foodification: An analysis of food gentrification in Turin, Italy. *Urban Geography* 43 (9), 1328–1349.

Boussaa, D. (2015) Urban regeneration, sustainability and urban heritage: A case study of Souk Waqif, Doha. *Journal of Urban Regeneration and Renewal* 8 (4), 389–400.

Butler, R.W. (1991) West Edmonton Mall as a tourist attraction. *Canadian Geographer* 35 (3), 287–295.

Chang, T.C. and Huang, S. (2011) Reclaiming the city: Waterfront development in Singapore. *Urban Studies* 48 (10), 2085–2100.

Choi, M.J., Heo, C.Y. and Law, R. (2016a) Developing a typology of Chinese shopping tourists: An application of the Schwartz model of universal human values. *Journal of Travel and Tourism Marketing* 33 (2), 141–161.

Choi, M.J., Heo, C.Y. and Law, R. (2016b) Progress in shopping tourism. *Journal of Travel and Tourism Marketing* 33 (1), 1–24.

Claxton, R. and Siora, G. (2008) *Retail-led Regeneration: Why it Matters to Our Communities*. Prague: DTZ Consulting.

Dawson, J., Findlay, A. and Sparks, L. (2020) *The Retailing Reader*. Routledge.

Dixon, T.J. (2005) The role of retailing in urban regeneration. *Local Economy* 20 (2), 168–182.

Dutch-Brown, N., Grzybowski, L., Romahn, A. and Verboven, F. (2021) Are online markets more integrated than traditional markets? Evidence from consumer electronics. *Journal of International Economics* 131, 103476.

Edwards, D., Griffin, T. and Hallar, B. (2008) Darling Harbour: Looking back and moving forward. In B. Hayllar, T. Griffin and D. Edwards (eds) *City Spaces-Tourist Places: Urban Tourism Precincts* (pp. 275–294). Routledge.

Escudero-Gómez, L.A. (2024) Shopping centers challenging decline: Competitive strategies in three case studies from Madrid's urban area. *Journal of Retailing and Consumer Services* 79, 103826.

Faria, C. and Whitesell, D. (2021) Global retail capital and urban futures: Feminist postcolonial perspectives. *Geography Compass* 15 (1), 12551.

Featherstone, M. (2007) *Consumer Culture and Postmodernism*. Sage.

Gold, J.R. and Ward, S.V. (1996) *Place Promotion: The Use of Publicity and Marketing to Sell Towns and Regions*. Wiley.

González, S. and Waley, P. (2013) Traditional retail markets: The new gentrification frontier? *Antipode* 45 (4), 965–983.

Gotham, K.F. (2002) Marketing Mardi Gras: Commodification, spectacle and the political economy of tourism in New Orleans. *Urban Studies* 39 (10), 1735–1756.

Gottdiener, M. (1995) *Postmodern Semiotics: Material Culture and the Forms of Postmodern Life*. Blackwell.

Guimarães, P.P.C. (2019) Shopping centres in decline: Analysis of demalling in Lisbon. *Cities* 87, 21–29.

Gutberlet, M. (2024) *Overtourism and Cruise Tourism in Emerging Destiantions on the Arabian Peninsula*. Routledge.

Hair, J.F., Risher, J.J., Sarstedt, M., Ringle, C.M. (2019) When to use and how to report the results of PLS-SEM. *European Business Review* 31 (1), 2–24.

Henderson, J.C. (2000) Food hawkers and tourism in Singapore. *International Journal of Hospitality Management* 19 (2), 109–117.

Henderson, J.C., Chee, L., Mun, C.N. and Lee, C. (2011) Shopping, tourism and retailing in Singapore. *Managing Leisure* 16 (1), 36–48.

Hsieh, A-T. and Chang, J. (2006) Shopping and tourist night markets in Taiwan. *Tourism Management* 27 (1), 138–145.

Iliev, D. (2021) Urban regeneration and changes driven by tourism and the 'Skopje 2014' Project. *Transylvanian Review of Administrative Sciences* 17 (62), 94–114.

Jansen-Verbeke, M. (1986) Inner-city tourism: Resources, tourists and promoters. *Annals of Tourism Research* 13 (1), 79–100.

Jin, H., Moscardo, G. and Murphy, L. (2021) Unfolding Chinese tourist shopping practices: An observational study. *Tourism Review* 76 (2), 460–472.

Jones, A.L. (2017) Regenerating urban waterfronts – Creating better futures –From commercial and leisure market places to cultural quarters and innovation districts. *Planning Practice and Research* 32 (3), 333–344.

Kikuchi, A. and Ryan, C. (2007) Street markets as tourist attractions –Victoria Market, Auckland, New Zealand. *International Journal of Tourism Research* 9 (4), 297–300.

Kim, H-R. and Jang, Y. (2017) Lessons from good and bad practices in retail-led urban regeneration projects in the Republic of Korea. *Cities* 61, 36–47.

Loda, M., Bonati, S. and Puttilli, M. (2020) History to eat: The foodification of the historic centre of Florence. *Cities* 103, 102746.

Lorch, B.J. and Smith, M.J. (1993) Pedestrian movement and the downtown enclosed shopping center. *Journal of the American Planning Association* 59 (1), 75–86.

Lowe, M. (2005a) Revitalizing inner city retail? The impact of the West Quay development on Southampton. *International Journal of Retail and Distribution Management* 33 (9), 658–668.

Lowe, M. (2005b) The regional shopping centre in the inner city: A study of retail-led urban regeneration. *Urban Studies* 42 (3), 449–470.

McKercher, B. (2020) Anatomy of successful tourism shopping districts. *International Journal of Tourism Cities* 6 (4), 831–846.

Mermet, A.C. (2017) Global retail capital and the city: Towards an intensification of gentrification. *Urban Geography* 38 (8), 1158–1181.

Oldenburg, R. (1989) *The Great Good Place: Cafés, Coffee Shops, Community Centers, Beauty Parlors, General Stores, Bars, Hangouts, and How They Get You through the Day*. Paragon House Publishers.

Orbaşli, A. (2019) Urban heritage in the Middle East: Heritage, tourism and the shaping of new identities. In D.J. Timothy (ed.) *Routledge Handbook on Tourism in the Middle East and North Africa* (pp. 95–105). Routledge.

Pai, S., Patil, V., Kamath, R., Mahendra, M., Singhal, D.K. and Bhat, V. (2021) Work-life balance amongst dental professionals during the COVID-19 pandemic – A structural equation modelling approach. *Plos One* 16 (8), e0256663.

Park, M.K. (2000) Social and cultural factors influencing tourists' souvenir-purchasing behavior: A comparative study on Japanese 'Omiyage' and Korean 'Sunmul'. *Journal of Travel and Tourism Marketing* 9 (1–2), 81–91.

Parker, C. and Garnell, C. (2006) Regeneration and retail in Liverpool: A new approach. *Journal of Retail and Leisure Property* 5 (4), 292–304.
Pearce, D.G. (1999) Tourism in Paris: Studies at the microscale. *Annals of TourismResearch* 26 (1), 77–97.
Pourjafar, M., Amini, M., Varzaneh, E.H. and Mahdavinejad, M. (2014) Role of bazaars as a unifying factor in traditional cities of Iran: The Isfahan bazaar. *Frontiers of Architectural Research* 3 (1), 10–19.
Pourzakarya, M. and Fadaei Nezhad Bahramjerdi, S. (2019) A study of culture-led regeneration approach in creative city building: Developing an analytical framework for the regeneration of cultural and creative quarter. *The Monthly Scientific Journal of Bagh-e Nazar* 16 (77), 5–14.
Rabbiosi, C. (2011) The invention of shopping tourism: The discursive repositioning of landscape in an Italian retail-led case. *Journal of Tourism and Cultural Change* 9 (2), 70–86.
Rabbiosi, C. (2015) Renewing a historical legacy: Tourism, leisure shopping and urban branding in Paris. *Cities* 42, 195–203.
Ruming, K., Mee, K., McGuirk, P. and Sweeney, J. (2018) Shopping centre-led regeneration. In K. Ruming (ed.) *Urban Regeneration in Australia: Policies, Processes and Projects of Contemporary Urban Change* (pp. 268–294). Routledge.
Santad, C. (2013) Urban Development and the Socio-spatial Transformation of Retail Areas: A Case Study of Provincial Towns in Thailand. Unpublished PhD thesis, University of Warwick.
Soleimani, M. and Mohammadnejad, A. (2015) Measuring tourists' satisfaction of the Mahabad City with an emphasis on shopping tourism. *Journal of Urban Economics and Management* 3 (10), 139–155.
Stephenson, M.L., Russell, K.A. and Edgar, D. (2010) Islamic hospitality in the UAE: Indigenization of products and human capital. *Journal of Islamic Marketing* 1 (1), 9–24.
Susser, I. and Schneider, J. (2003) *Wounded Cities: Destruction and Reconstruction in a Globalized World*. Berg.
Swarbrooke, J. (2000) Tourism, economic development and urban regeneration: A critical evaluation. In M. Robinson (ed.) *Developments in Urban and Rural Tourism* (pp. 269–285). Business Education Publishers.
Thanasi-Boçe, M., Kwiatek, P. and Labadze, L. (2021) The importance of distance and attraction in patronizing a shopping mall. *Journal of Place Management and Development* 14 (2), 222–238.
Timothy, D.J. (2005) *Shopping Tourism, Retailing and Leisure*. Channel View Publications.
Timothy, D.J. and Boyd, S.W. (2003) *Heritage Tourism*. Prentice Hall.
Tosun, C., Temizkan, S.P., Timothy, D.J. and Fyall, A. (2007) Tourist shopping experiences and satisfaction. *International Journal of Tourism Research* 9 (2), 87–102.
Turner, L.W. and Reisinger, Y. (2001) Shopping satisfaction for domestic tourists. *Journal of Retailing and Consumer Services* 8 (1), 15–27.
Van Eck, E., Van Melik, R. and Schapendonk, J. (2020) Marketplaces as public spaces in times of the Covid-19 coronavirus outbreak: First reflections. *Tijdschrift voor Economische en Sociale Geografie* 111 (3), 373–386.
Venkov, N.A. (2022) Marketplace decline heads east: Neoliberal reform, socio-spatial sorting and patterns of decline at Sofia's public markets. In C. Sezer and R. van Melik (eds) *Marketplaces: Movements, Representations, and Practices* (pp. 88–100). Routledge.
Yüksel, F. (2013) The streetscape: Effects on shopping tourists' product/service quality inferences and their approach behaviors. *Journal of Quality Assurance in Hospitality and Tourism* 14 (2), 101–122.
Zaidan, E. (2016) Tourism shopping and new urban entertainment: A case study of Dubai. *Journal of Vacation Marketing* 22 (1), 29–41.

Zaidan, E. (2019) Shopping, tourism and hyper-development in the Middle East and North Africa. In D.J. Timothy (ed.) *Routledge Handbook on Tourism in the Middle East and North Africa* (pp. 365–377). Routledge.

Zukin, S. (2012) The social production of urban cultural heritage: Identity and ecosystem on an Amsterdam shopping street. *City, Culture and Society* 3 (4), 281–291.

Zukin, S. (2018) *Point of Purchase: How Shopping Changed American Culture*. Routledge.

Zukin, S. and Maguire, J.S. (2004) Consumers and consumption. *Annual Review of Sociology* 30 (1), 173–197.

9 Economic Success in Unexpected Places: Spatial Anomalies in Swedish Retailing

Roger Marjavaara, Elin Nilsson and Magnus Strömgren

Introduction

For as long as people have met and gathered in small villages, in larger cities, at religious events or along travel routes, for shorter or longer periods of time, people have engaged in trading goods – the forerunner of today's retailing. Hence, retail is an economic sector with a long history (Stobart & Howard, 2019). Most early retail took place in town centres, establishing the relationship between trade and urban agglomeration (Coleman, 2006). A general rule has been that retail is located where the customers are (May, 1989), owing to the economic potential of proximity to larger consumer markets. This is true not only for retailing, but for almost all economic activities in liberal markets (Scott, 1998), and the general trend is a drainage of investments, companies and people in sparsely populated rural and peripheral localities in favour of urban agglomerations (Lindgren et al., 2017), at least when it comes to physical retailing, which is the focus of this chapter.

The decline of physical retail in rural areas is accentuated by the development of e-commerce, increasing the competition for local retailers, by giving consumers access to a wider selection of products at more favourable prices (Postnord, 2021). Further, there is a trend towards physical concentration in the retail sector which brings the development of generally larger and fewer retail stores, mainly located in proximity to urban centres (Amcoff et al., 2015; Timothy, 2005; Wang & Du, 2021). All in all, this means there is a centripetal force at play, leading to retail establishments gravitating towards population centres.

There are, however, examples of individual retail firms that prosper and even grow in localities that are theoretically unfavourable. These locations are far from population centres and characterised by population decline. Geographical location is perhaps not everything that explains the economic success of a retail firm, but as Brown (1989: 450) argues, 'Location is, nevertheless crucial'. Usual explanations for the success of these retail firms are tourism (McIntyre, 2012) and enhanced individual consumer mobility. Previous research has shown that incentives to travel to other places for engagement in retail is usually related to price differences and/or product availability (Makkonen, 2016; Timothy, 2005). One prominent example is cross-border shopping, where people cross political boundaries to take advantage of retail opportunities (e.g. lower taxes, greater merchandise selection, greater critical mass of retail outlets, cheaper products and the 'foreign' novelty appeal) in a foreign country or neighbouring region (Makkonen, 2023; Timothy & Butler, 1995) (see Chapter 7).

At times, however, consumers travel far to buy products that are available closer to home, even if they are not offered at considerably lower prices than in the home area. This is described in a tourism context by Prideaux (2002), as the need for attractions in peripheries to be of greater significance than in urban cores, to overcome the disadvantageous location. It is therefore interesting to examine retail firms that have a conventionally thought of unfavourable location, and do not offer unique products at considerably lower prices, but still seem to survive, and even prosper. This chapter focuses on these retail firms by discussing their characteristics and competitive advantages that seem to counteract their adverse locations. The chapter will proceed by outlining recent developments in the geography of retailing and what is known about what we term 'The anomalies of retailing'. We also present examples of this phenomenon and discuss possible explanations in the context of Sweden.

The Spatial Concentration of Retailing

For a long time, the clustering tendency of economic activities has been observed (cf. Marshall, 1890). This has accelerated during the 20th century (Moretti, 2013) and is today an inevitable force, affecting all nations of the world. One important reason is the general trends of migration and urbanization (Newbold, 2010) directed towards city centres, resulting in positive benefits for firms located there. In retail, Öner (2014) claims that one of the most notable characteristics is the location sensitiveness of retail compared to many other economic sectors, owing to retail's dependence on access to a large consumer base. This means, roughly, that where many people live, there are many stores. Other important locational factors are strong local purchasing power and increasing population volumes (Howard, 1992; Kirkup & Rafiq, 1994), which prevail in proximity to urban centres. For example, in Sweden, 50 municipalities out of a total of 290 generate two-thirds of the total turnover in retailing (HUI, 2011).

The spatial concentration of retailing was noted at an early stage. Christaller (1966) observed that places are ordered in a hierarchy, where high-order places are comparatively infrequent and located far away from each other. These central places serve their hinterland with a wide selection of goods and services. Low-order places occur more frequently and are more closely located to one another. Low-order places offer a limited variety of goods and services, which are more sensitive to distance decay. As Christaller points out, the number of inhabitants is an important indicator of place centrality and variation in retail supply.

Not only do retail firms tend to locate where people live, they also benefit symbiotically from co-location with other retail firms and therefore tend to cluster. This was noted by Hotelling (1929), who argued that retail firms adjust their geographical market position depending on competitors, with the process leading to firms being co-located or symbiotically clustered. Porter (1998, 2000) argues that the spatial co-location of retail establishments, in similar or related industries, also leads to specific advantages for the individual firms, such as access to specialized knowledge and skilled labour, firms being able to share costs for common services, and other benefits. According to Brown (1989) and Timothy (2005), the clustering of similar retail outlets is a universal trait. Spatial clustering is also beneficial to customers, with imperfect knowledge of the market and heterogenic needs, leading to multipurpose shopping or cross-shopping (Miller et al., 1999). Further, the relative increase in competition of being co-located with competitors is outweighed by the additional demand created by the retail agglomeration (Brown, 1989). Retail firms that can co-locate with other firms tend to grow faster and have greater longevity owing to the advantages provided by a crucial mass of sellers (Sunley, 2000). This is observable in the decreasing number of smaller retail outlets in favour of large shopping centres and retail 'parks', where an increasing share of retail, especially durable goods, takes place (Bergström & Fölster, 2005; Erlandsson, 1995). This phenomenon can be observed in Sweden, which is among the leading countries of Europe in terms of shopping centre development (Birkin et al., 2002), and elsewhere. A retail location is also a long-term investment that is hard to disestablish once it is established, further favouring urban locations with lower economic risks (Ghosh & McLafferty, 1987). Finally, there is also a trend in retail market concentration, where a few big companies account for an increasing share of the market. In 2018, only 14 companies generated half of the turnover in Swedish retailing, while the other half was generated by around 58,000 firms (Svensk Handel, 2018). Hence, there are many factors pointing toward the benefit of co-location in close proximity to urban centres rather than retailers being located far from urban agglomerations, individually, and in small local markets without obvious opportunities for economic growth.

Consumer Behaviour

Retail and the hospitality industry in Sweden are facing a challenging transformation (Ekström & Jönsson, 2022), where consumers' needs, expectations and behaviours are constantly changing. We are facing new shifts in retailing at a time when disruptive changes are happening quickly. Increased digitization and the COVID-19 pandemic have created conditions for a rapidly growing e-commerce-based retail environment. This, in turn, has changed the balance of power between customers and service providers. There are fewer physical stores and in-person employees (Svensk Handel, 2019), but at the same time a new and strong entrepreneurship is emerging during this transformational time, not always based in urban areas. These new firms work digitally and innovatively and can act quickly based on evolving conditions. Companies must now be both dynamic and innovative to keep up with new developments (Svensk Handel, 2019) and enhance the customer experience.

Previous research has shown the importance of understanding the customer experience (Grewal *et al.*, 2009; Kim & So, 2022; Verhoef *et al.*, 2009). Customer expectations for a 'total experience' have also increased with technological development, with consumers today demanding a personal, immediate, engaging and interactive experience (Priporas *et al.*, 2017). In the transformation, and the ongoing development of digital social media, virtual space has become increasingly important (Ballantyne & Nilsson, 2017). The increased use of smartphones and mobile applications has changed the way consumers interact with businesses (Dinsmore *et al.*, 2017; Fang, 2019). Digital services challenge traditional concepts of what constitutes a consumer experience and customer value (Ballantyne & Nilsson, 2017). This opens opportunities for customers and companies. To take advantage of these opportunities, however, it is important to be responsive to changing customer needs; companies need to be adaptive and agile. The importance of physical place is therefore challenged in the new retail settings that have emerged.

Developments in many cities highlight the link between tourism and shopping, especially acknowledging that shopping and dining can in fact be the most important attractions in the city centre (Morgan, 2006; Warnaby *et al.*, 2005; Zaidan, 2019). The individual consumer can experience a visit to a city or a shopping mall as an overall experience without clear differences between the retail experience and visits to cafes, restaurants, museums, theatres and other places that attract visitors (Zaidan, 2016). The identification of attractive combinations of retail and hospitality services is therefore important for service businesses, as well as for urban place-making and marketing. The topic of shopping is a key subject of exploration in tourism research, and consumers like to combine shopping with holidays and vacations (Sullivan *et al.*, 2012; Timothy, 2005).

The identification of such combinations is particularly important for small and medium-sized cities where downtown districts may be the only choice for secondary entertainment beyond the area's primary tourist attraction(s) (Runyan, 2006). Research also suggests that the number and diversity of businesses will determine whether consumers choose to shop (Runyan, 2006). The higher number of shops and the more diversity in retail venues, the greater the likelihood will be that people visiting a city will engage in multi-faceted shopping (Ghosh, 1986) – a type of shopping that blends into the overall leisurescapes of the city and urban sightseeing, favouring urban localities.

Retail Tourism beyond the Urban Agglomeration

Shopping has become one of the most common leisure and tourism activities today (Rabbiosi, 2011; Timothy, 2005), constituting an important part of the tourism experience. Spending time and money in a shopping centre, a street market, a tax-free shop at an airport or border crossing, or an exclusive boutique in the fashionable quarters of a major city is something almost everyone on holiday engages in. Major tourism destinations are key shopping destinations, and vice versa. In Sweden (during 2019) 35% of tourism expenditures, or 108 billion SEK (9.9 billion EUR; 10.42 billion USD), derived from retail, dwarfing expenditures on lodging, restaurants and transportation (Svensk Handel, 2020), and 35,000 employment opportunities were created from tourist expenditures in retail alone. While tourist-generated shopping accounts for only 14% of all retail earnings in Sweden (Svensk Handel, 2020), it is nevertheless disproportionately important for certain localities.

Nonetheless, shopping as a means of attracting tourists is often forgotten or even disregarded. In academic research the nexus between tourism and retail has not yet attracted the attention it deserves, given the scale of the phenomenon. Tourism can be the boost local retailers need to remain fiscally viable, especially when local demand is low and/or deteriorating (Löffler, 2007). By attracting consumers from elsewhere, outshopping can also take a lead role in economic regeneration, especially in places having a hard time attracting other industries (Jansen-Verbeke, 1991). This is key in expanding local economic activities, as well as increasing job opportunities, economic affluence and welfare. However, retail is rarely seen as an export industry, but rather as a service industry for the benefit of local consumers. However, if unique and successful, it serves the same purpose for local economies as traditional exporting manufacturing industries (Williams, 2012). In a sense, without tourism or mobile consumers, retail is a zero-sum game.

Of course, much tourist-derived retail consumption is generated in major cities throughout the world. For example, the Mall of America, the biggest mall in the United States, is said to be the most potent tourist attraction in all of Minnesota (Underhill, 2004). Shopping in high-street boutiques in major urban tourist destinations, such as London, Paris and Rome, generates high expenditures. In Sweden, major urban centres generally have an inflow of consumption from the nearby hinterland as described by Christaller (1966), especially when it comes to durable goods (Handelsfakta, 2022).

Even though the benefits of locating retail firms in cities are evident, not all retail firms follow this trend. Rural and peripheral areas are not totally devoid of customers. People still live there and visit, and their demands are met by local retailers. However, owing to the centripetal trends within retailing and a diminishing local customer base, the preconditions to conduct retail becomes weaker, especially for durable goods, which need large volumes of customers to make a profit (Amcoff *et al.*, 2015). However, poor conditions and demand in rural and peripheral localities can be compensated through temporal mobility (i.e. tourism).

Cross-border shopping is also an important part of shopping tourism in Sweden (see Chapter 7). Places adjacent to the border of Norway, and to some extent Finland, have higher retail indexes owing to tourism consumption and proximity to the border (Handelsfakta, 2022). Further, places with long traditions of tourism also show an inflow of retail consumption. However, in some cases investments in new retail establishments and the success and persistence of existing retail happen in locations that are hard to explain by the attraction of urban agglomerations, proximity to an international border, traditional tourism developments, or other previously noted factors of success such as clustering or peri-urban location. These are the exceptions we call 'the anomalies of retailing', or retail firms that become major shopping attractions despite lacking the obvious preconditions of success. What is it then that makes these firms overcome the distance their customers must travel? Interestingly, there is a lack of theoretical explanation of this phenomenon.

Swedish Retail Tourism Anomalies

In this section we present examples of anomalies in Swedish retailing that can provide support for the concepts presented in this chapter. However, these cases do not represent the general situation in Sweden. To do so, further extensive empirical enquiry is needed. The cases are merely illustrative of the phenomenon as described. See Figure 9.1 for the geographical context of the cases presented.

Gekås Ullared – the largest retail store in Sweden

On the Swedish west coast, around 100 km south of the city of Gothenburg, lies the village of Ullared. With a population of 748 residents

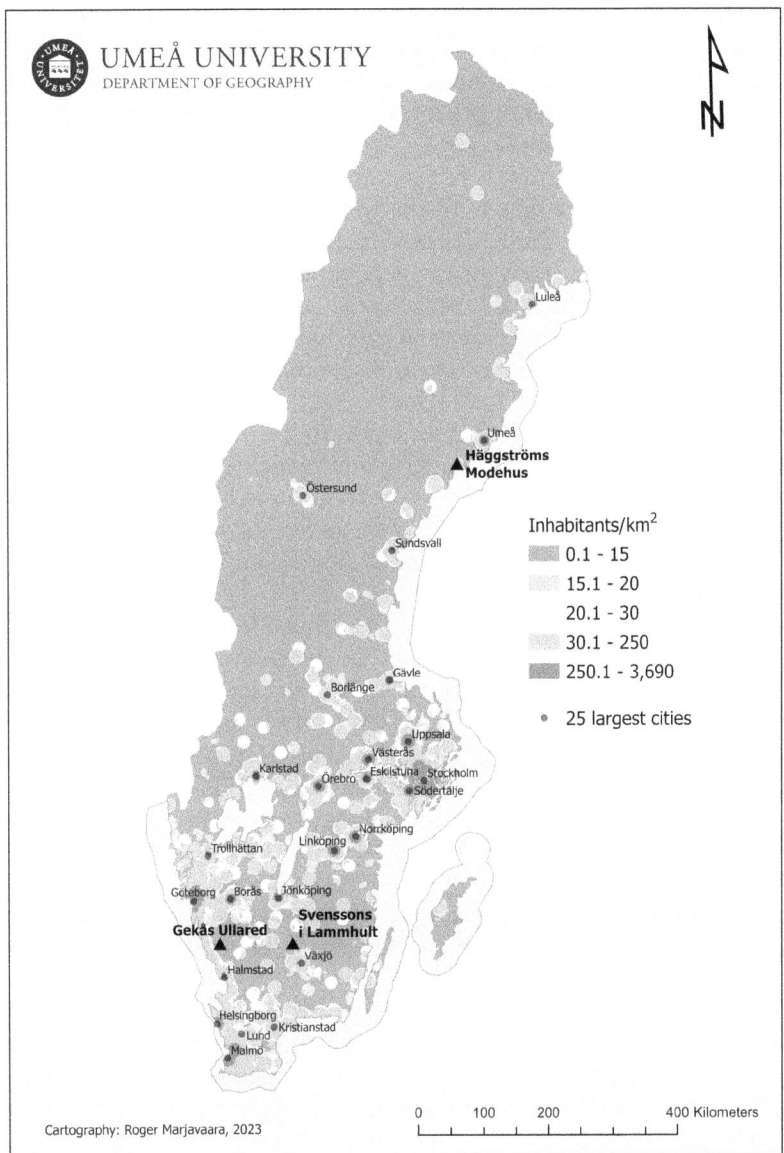

Figure 9.1 Population density in Sweden and the 25 largest cities in 2020, alongside the location of the described cases. Source: Statistics Sweden, 2023

(Statistics Sweden, 2022b), Ullared is by all accounts a rural settlement. Even though it is a small village, Ullared hosts an icon in Swedish retailing: Gekås. According to Näslund and Edsta (2015), Gekås is the largest individual store in Sweden. With 44,500 m² of retail space (Gekås, 2022), it is approximately twice the size of an average IKEA store (Terry-Armstrong,

2012). Gekås was inaugurated in 1963 by the entrepreneur Göran Karlsson, who started selling clothes in the basement of a house in the village. Today, Gekås has grown into a major retail destination. It attracts around 4.9 million visitors a year, generating a total turnover of 3.96 billion SEK in 2021 (513 million EUR; 538 million USD) (Retriever Business, 2022). The establishment has seen a growth in earnings every year from 2012 to 2019 (Retriever Business, 2022), until the pandemic hit with imposed restrictions on shopping. In 2021, Gekås had around 800 employees (Retriever Business, 2022), more than the entire population of the village.

Based on their own data, the average customer is a woman, age 43, living in the southwestern part of Sweden, on a day trip to Ullared, travelling on average 230 km one-way, and spending around 3,300 SEK (302 EUR; 317 USD) on annual basis (Gekås, 2022). This means that the store's catchment area is rather big, incorporating a large part of southern Sweden, as well as parts of Denmark and Norway. During peak season, all the goods in the entire store is emptied three times a week, and the waiting line for entering the store can be up to 1.4 km long (see Figure 9.2) (Gekås, 2022). The standing record number of customers for one day is 29,200 (Gekås, 2022).

The primary incentive for consumers drawn to Gekås is the generally lower prices (Hjelmgren, 2010). Further, Hjelmgren (2010) argues that prices are kept low owing to relatively low costs for the property and low rents, given its rather peripheral location, and economies of scale in terms of purchasing goods. Likewise, the store is privately owned and not listed on the stock exchange, giving the company protection from shareholders' interest in returning short-term profits. Low prices are of course important; however, given the distance people travel to visit Gekås, and the costs associated with getting there, the economic advantage is slim, according to Ekström (2010). Gekås' products are what consumers can normally find closer to home, in the larger population centres surrounding Ullared, such as Borås (pop. 114,091), Halmstad (pop. 104,573), Kungsbacka (pop. 85,301), Jönköping (pop. 143,579) and Sweden's second largest city Gothenburg (pop. 587,549) (Statistics Sweden, 2022a) (see Figure 9.1). The store's ancillary services are mentioned by Hjelmgren (2010) as being part of the total appeal of Gekås. These include restaurants, cafes, a campground and a hotel, as well as a tourist information and a day-care for dogs.

Gekås does not use traditional marketing channels. Rather, word-of-mouth is the most common means of spreading information and reputation. Word is spread between friends, relatives and colleagues about the Gekås total experience (Ekström, 2010). In addition, Gekås has been portrayed in a Swedish reality TV show, which demonstrated everyday life among workers and customers (Eriksson, 2015). This has boosted the store's popularity and increased visitor numbers. Shopping at Gekås has become a total experience, and even waiting in long queues to enter the store can be something consumers discuss with their peers (see Figure 9.2).

Figure 9.2 Customers queuing in line to enter Gekås in Ullared during peak season in July 2021. Photo: Roger Marjavaara, 2021

This experience and the extra services offered at Gekås stresses that the total encounter is important, not just the utilitarian aspects of shopping (Timothy, 2005).

Even though Gekås offers lower prices than many other retailers, the journey to and from Ullared diminishes the economic benefits. Gekås is also located where retail opportunities have generally disappeared, the store offers similar products to those of its competitors in more favourable locations, and Ullared does not provide the amenities of a traditional tourist destination. Furthermore, Gekås has not relocated or opened other outlets closer to major markets, which indicates that it attracts sufficient numbers of customers, even though its real location is less advantageous. Gekås, therefore, is an interesting case and can be considered an anomaly in Swedish retailing.

Svenssons i Lammhult – designer furniture in the countryside

Lammhult is a village in southeastern Sweden with a population of 1,751 (Statistics Sweden, 2022b). It is in Kronoberg County, some 35 km north of the municipality administrative centre of Växjö (pop. 71,282) (see Figure 9.1) (Statistics Sweden, 2022b). The general area has a long history of furniture manufacturing and retail, and this is also the case in Lammhult village itself. Today, there are three furniture stores in town. In addition to Nilssons Möbler and Norrgavel, Lammhult houses

Svenssons i Lammhult (English *Svenssons in Lammhult*) – the largest and oldest of the three companies (Möbelriket, 2022).

The roots of Svenssons i Lammhult can be traced to the early 20th century, when Erik Svensson started a carpentry in the area. The business expanded continuously and was later complemented by the establishment of a furniture store. During most of the 20th century, the company, at the time called Svenssons Möbler, was owned by different generations of the Svensson family. The furniture sold was partly designed and manufactured in-house but also included designer furniture produced elsewhere. In 1991, in connection with a change in ownership, the company was given its current name Svenssons i Lammhult and started to concentrate solely on furniture retail. Following another change in ownership in 2007, the company was bought in 2021 by Nordic Nest, which is owned by the BHG Group (Svenssons, 2022).

In 2021, Svenssons i Lammhult had 64 employees and a turnover of 388 million SEK (35.5 million EUR; 37.3 million USD) – almost twice as much as in 2012 (Retriever Business, 2022). During the 21st century, the company established additional stores in the three main metropolitan areas of Sweden: Stockholm, Gothenburg and Malmö (Svenssons, 2022). Its current product portfolio includes furniture and home décor from over 200 well-known brands. Compared to the market catchment areas of the additional stores opened recently, the customer base and purchasing power at the Lammhult location is unremarkable. Although the company has announced a downsizing in Lammhult related to merchandising, customer service and e-commerce staff, the company still views the original retail location as a central corporate asset that it intends to continue to support (SVT, 2022). Given the pronounced countryside character of the area, the continuous success of the original store in Lammhult appears to be a retail anomaly.

Häggströms Modehus – high fashion in the forest

Häggströms Modehus (English *Häggström's Fashion Store*) is the biggest clothing store in northern Sweden. It is situated in the village of Lögdeå (see Figure 9.1), between the cities of Örnsköldsvik (pop. 55,823) and Umeå (pop. 130,997) (Statistics Sweden, 2022a), with a population of only 571 (Statistics Sweden, 2022b). Häggströms Modehus was inaugurated in 1922 and is a third-generation family-owned company. In 2015, the famous former ice hockey player Peter Forsberg became part owner, through the company Forspro. Earnings in 2021 were 16.4 million SEK (1.5 million EUR; 1.6 million USD) (Retriever Business, 2022). In common with most retail businesses, the pandemic caused a decrease in visitor numbers and spending, but the store is experiencing a recovery. Over the years, Häggströms Modehus has evolved from a tailoring business to a fashion shop. In 1972, the first expansion of the store took place, and since then it has expanded several times. It is now a two-storey fashion store, with 2300 m² of retail space (Häggströms, 2022) and 10 employees (Retriever Business, 2022).

Häggströms Modehus offers thousands of branded items for women, men and children in the store and online. The brands on sale are similar to what can be found in most cities, including Calvin Klein, Tiger of Sweden, Lyle and Scott, Desigual and Bitte Kai Rand, as well as cheaper brands such as Bestsellers and B-young. The concept behind this shop's success is that customers should not have to go to several stores; instead, all their retail needs can be met at Häggströms Modehus. There is also a café in the store, Hildas Café. In several media interviews, the owner emphasizes that he views his staff as family and that they are as dedicated and engaged as he is. The owner, David Häggström, argues:

> The only constant is change. That's a bit how I see our business. So much has happened since grandfather's time – we have had to keep up with our times and adapt our business in step with the changes. But what has been constant since opening is our personal service, which has been a stated ambition since opening in 1922. (VK, 2022, n.p.)

Häggströms Modehus was the first store in Sweden with a membership card, first in the north of Sweden to use TV commercials, and the first to establish its own online store (Edstrand, 2022). Häggströms Modehus satisfies the definition of an anomaly. It is located relatively far from major markets and is not part of any sort of major tourist area. Although Häggströms has a seemingly disadvantageous location compared its competitors, it remains profitable and attractions enough customers to stay in business for more than a century.

Conclusion and Future Studies

In this chapter we have discussed the trend of spatial concentration of retailing, but also the phenomenon of individual retailers that buck this normative trend and survive well on their own, in isolation from other shopping establishments and often far from population centres. Cases from Sweden were described to illustrate these patterns and trends. Research has so far not adequately addressed this phenomenon, and there is a general lack of theoretical constructs to help explain the existence of these anomalies. To develop this concept further and to explore such situations more comparatively, additional empirical inquiries are needed. However, there are some hypothetical explanations behind the persistence and success of these anomalies.

We can conclude that the example cases are all in the original locations where they were inaugurated, meaning that there is likely a path dependency at play. It seems that being located where the retail firm was founded is important and that there is a matter of place loyalty, sense of place and rootedness, not necessarily only profit maximization and economic efficiency. Hypothetically, location can be a way of differentiating retail stores from their competitors. These firms have been in operation for a long time, in the case of Häggströms Modehus 100+ years, and they have invested in the location and have become synonymous with the place as a sort of local 'place brand' or iconic image. In the case of Gekås, it is common for people throughout Sweden to refer to the place, Ullared (the village name), when they actually mean the store Gekås. For Svenssons i Lammhult, not least owing to its current name, the brand of the company is intrinsically linked to its place of origin.

Despite their relative isolation, all of these shopping venues perform relatively well. This relates to the issue of ownership and management. The cases are all privately owned firms that have some degree of autonomy. Gekås is owned by a single individual with a history within the firm. Häggströms Modehus is owned by the third generation of the founding family and co-owned by a private company based in the region. Svenssons i Lammhult was recently bought by a publicly listed company. Hence, there are three levels or types of ownership. It appears that the individual companies have an important say in the future of operations in the original location. This appears to work, as all of the cases continue to do business in their places of origin, have not shut down or relocated as many retail firms have done. Svenssons i Lammhult has opened branches in other localities that are more economically logical (e.g. the store in Stockholm), but the company has not terminated operations in Lammhult, and it remains an integral part of that community. In most cases, rural areas have smaller and decreasing local markets. This, in combination with lower purchasing power, makes Svenssons decision to remain in Lammhult even more remarkable. The area in and around Lammhult has a long history of furniture production and sales, and today there are still several furniture makers in the region. Thus, the outlet has become part of the area's heritage and perceived advantages generally associated with complementary clustering may have played a part in the company's decision to stay. The autonomy in decision-making regarding daily operations and perhaps at higher strategic levels may be of crucial importance. This potentially gives local management the opportunity to finetune the organization in response to customer demand because the decisions can be made at the grassroots level where operational and context-specific knowledge is highest.

Another area of interest is the impact of staffing. In a situation of increased competition from e-commerce and larger retailers with more competitive locations, staffing policy is important. As indicated by Hjelmgren (2010), Gekås in Ullared values its staff and puts a great deal of effort into creating satisfied employees. Likewise, Häggströms Modehus claims to treat its staff as family members. This translates into giving customers individual and custom service experiences, thereby increasing customer satisfaction and loyalty. Thus, low staff turnover and high investments in staff members' well-being may contribute to explaining the successful performance of these retailers.

The anomalies of retailing might also draw economic benefits from their relatively peripheral locations in terms of reducing operational costs. In these marginal places, property prices are generally lower owing to lower competition for space, leading to lower rent, property tax and real estate costs. In addition, wages are generally lower in rural areas, increasing profit margins and allowing retailers to employ more staff, potentially increasing the quality of customers' retail experiences. Staff loyalty might also be driven by a lack of alternative jobs/employers, which would naturally lead to employees being interested in maintaining the well-being of the shopping venue.

It is also of interest to examine this phenomenon from a demand perspective. What are shoppers' motives in engaging in activities in these rather illogical places? We believe it is a matter of escapism, where 'getting away for a while' may be an important motive. Here, the social element of rural shopping might trump more utilitarian reasons for buying products in more rational localities. It might also be a matter of cognitive proximity, where mental closeness to the store is more influential than the physical distance. In other words, if the consumers are sure their needs will be met, they might prefer to travel a longer distance than take their chances with another store located closer. These are all strong possibilities that influence people's decisions to patronize retail establishments that make little geographic sense, but these arguments remain to be proven by more in-depth empirical analysis. This is the agenda for future studies utilizing quantitative and qualitative data on store performance, firm ownership, corporate policies, place features and customer motivations. Many interesting questions remain to be answered in explaining why these shopping anomalies occur – what makes retail succeed where it 'is not supposed to be'.

References

Amcoff, J., Mohall, M., Waxell, A. and Östh, J. (2015) *Detaljhandelns förändrade geografi*. Uppsala Universitet.

Ballantyne, D. and Nilsson, E. (2017) All that is solid melts into air: The servicescape in digital service-space. *Journal of Services Marketing* 31 (3), 226–235.

Bergström, F. and Fölster, S. (2005) *Kampen om köpkraften: Handeln i framtiden*. Forma Publishing.

Birkin, M., Clarke, G. and Clarke, M. (2002) *Retail Geography and Intelligent Network Planning*. Wiley.

Brown, S. (1989) Retail location theory: The legacy of Harold Hotelling. *Journal of Retailing* 65 (4), 450–470.

Christaller, W. (1966) *Central Places in Southern Germany* (C.W. Baskin, trans.). Prentice Hall (original work published 1933).

Coleman, P. (2006) *Shopping Environments: Evolution, Planning and Design*. Architectural Press.

Dinsmore, J.B., Swani, K. and Dugan R.G. (2017) To free or not to free: Trait predictors of mobile app purchasing tendencies. *Psychology and Marketing* 34 (2), 227–244.

Edstrand, M. (2022) Modehus firade 100 år. 16 october 2022. *Örnsköldsviks Allehanda*.

Ekström, K.M. (2010) Familjekonsumtion, konformitet och distinktion. In K.M. Ekström, J. Hagberg, D. Hjelmgren, M. Lind and N. Salomonson (eds) *Fenomenet Ullared: En förstudie* (pp. 19–30). Högskolan i Borås.

Ekström, K.M. and Jönsson, H. (2022) Orchestrating retail in small cities. *Journal of Retailing and Consumer Services* 68 103008.

Eriksson, G. (2015) Ridicule as a strategy for the recontextualization of the working class: A multimodal analysis of class-making on Swedish reality television. *Critical Discourse Studies* 12 (1), 20–38.

Erlandsson, U. (1995) Varuhandeln. In C.G. Alvstam (ed.) *Sveriges Nationalatlas – Industri och service* (pp. 124–129). SNA.

Fang, Y.H. (2019) An app a day keeps a customer connected: Explicating loyalty to brands and branded applications through the lens of affordance and service-dominant logic. *Information and Management* 56 (3), 377–391.

Gekås (2022) *Om oss*. https://www.gekas.se/om-oss/ (accessed 13 December 2022).

Ghosh, A. (1986) The value of a mall and other insights from a revised central place model, *Journal of Retailing* 62 (1), 79–97.

Ghosh, A. and McLafferty, L. (1987) *Location Strategies for Retail and Service Firms*. Lexington Books.

Grewal, D., Levy, M. and Kumar, V. (2009) Customer experience management in retailing: An organizing framework. *Journal of Retailing* 85 (1), 1–14.

Handelsfakta (2022) *Regional handel*. https://handelsfakta.se/handeln-sverige/ (accessed 12 December 2022).

Hjelmgren, D. (2010) Att ge kunderna positiva upplevelser i samband med deras vistelse i Ullared. In K.M. Ekström, J. Hagberg, D. Hjelmgren, M. Lind and N. Salomonson (eds) *Fenomenet Ullared: En förstudie* (pp. 9–18). Högskolan i Borås.

Hotelling, H. (1929) Stability in competition. *The Economic Journal* 39 (153), 41–57.

Howard, E. (1992) Evaluating the success of out-of-town regional shopping centres. *The International Review of Retail, Distribution and Consumer Research* 2 (1), 59–80.

HUI (2011) *Kampen om köpkraften: Handeln i Framtiden*. HUI Research.

Häggströms (2022) *Om oss*. https://haggstromsmodehus.se/content/4-om-oss (accessed 15 December 2022)

Jansen-Verbeke, M. (1991) Leisure shopping: A magic concept for the tourism industry? *Tourism Management* 12 (1), 9–14.

Kim, H. and So, K.K.F. (2022) Two decades of customer experience research in hospitality and tourism: A bibliometric analysis and thematic content analysis. *International Journal of Hospitality Management* 100, 103082.

Kirkup, M. and Rafiq, M. (1994) Managing tenant mix in new shopping centres. *International Journal of Retail and Distribution Management* 22 (6), 29–37.

Lindgren, U., Borggren, J., Karlsson, S., Eriksson, R.H. and Timmermans, B. (2017) Is there an end to the concentration of businesses and people? In C. Keskitalo (ed.) *Globalisation and Change in Forest Ownership and Forest Use* (pp. 139–181). Palgrave Macmillan.

Löffler, G. (2007) The impact of tourism on the local supply structure of goods and services in peripheral areas: The example of northern Sweden. In D.K. Müller and B. Jansson (eds) *Tourism in Peripheries: Perspective from the Far North and South* (pp. 69–84). CABI.

Makkonen, T. (2016) Cross-border shopping and tourism destination marketing: The case of southern Jutland, Denmark. *Scandinavian Journal of Hospitality and Tourism* 16 (1), 36–50.

Makkonen, T. (2023) Outshopping abroad: Cross-border shopping tourism and the competitive advantage of borders. In D.J. Timothy and A. Gelbman (eds) *Routledge Handbook of Borders and Tourism* (pp. 269–280). Routledge.

Marshall, A. (1890) *Principles of Economics*. Macmillan.

May, E.G. (1989) A retail odyssey. *Journal of Retailing* 65 (3), 356–367.

McIntyre, C. (2012) *Tourism and Retail: The Psychogeography of Liminal Consumption.* Routledge.

Miller, C.E., Reardon, J. and McCorkle, D.E. (1999) The effects of competition on retail structure: An examination of intratype, intertype, and intercategory competition. *Journal of Marketing* 63 (4), 107–120.

Moretti, E. (2013) *The New Geography of Jobs.* Mariner Books.

Morgan, M. (2006) Making space for experiences. *Journal of Retail and Leisure Property* 5 (4), 305–313.

Möbelriket (2022) *Möbelbutikerna.* https://www.mobelriket.se/a/c/mobelbutikerna (accessed 15 December 2022).

Näslund, N.E. and Edsta, B. (2015) *Störst, världsbäst, billigast: Det visste ni inte om fenomenet i Ullared.* Stockholm: Ultima Esperanza Books.

Newbold, K.B. (2010) *Population Geography: Tools and Issues.* Rowman and Littlefield.

Öner, Ö. (2014) *Retail Location.* Jönköping: Jönköping International Business School (Doctoral dissertation).

Porter, M.E. (1998) Clusters and the new economics of competition. *Harvard Business Review* 76 (6), 77–90.

Porter, M.E. (2000) Locations, clusters and company strategy. In G.L. Clark, M.P. Feldman and M.S. Gertler (eds) *The Oxford Handbook of Economic Geography* (pp. 253–274). Oxford University Press.

Postnord (2021) *E-barometern – Årsrapport* 2021. https://www.postnord.se/siteassets/pdf/rapporter/e-barometern-arsrapport-2021.pdf (accessed 1 December 2022).

Priporas C-V., Stylos, N. and Fotiadis, A.K. (2017) Generation Z consumers' expectations of interactions in smart retailing: A future agenda. *Computers in Human Behavior* 77 (Dec), 374–381.

Rabbiosi, C. (2011) The invention of shopping tourism: The discursive repositioning of landscape in an Italian retail-led case. *Journal of Tourism and Cultural Change* 9 (2), 70–86.

Retriever Business (2022) *Business Intelligence.* https://www.retrievergroup.com/sv/business-suite (accessed 6 December 2022).

Runyan, R.C. (2006) Tourist dependent small towns: Understanding competitive advantage. *Journal of Vacation Marketing* 12 (4), 329–343.

Runyan, R.C. and Huddleston, P. (2006) Getting customers downtown: The role of branding in achieving success for central business districts. *Journal of Product and Brand Management* 15 (1), 48–61.

Scott, A.J. (1998) *Regions and the World Economy: The Coming Shape of Global Production, Competition, and Political Order.* Oxford University Press.

Statistics Sweden (2022a) *Folkmängden efter region, civilstånd, ålder och kön. År 1968–2021.* https://www.statistikdatabasen.scb.se/pxweb/sv/ssd/START__BE__BE0101__BE0101A/BefolkningNy/ (accessed 6 December 2022).

Statistics Sweden (2022b) *Statistiska tätorter 2020, befolkning, landareal, befolkningstäthet.* https://www.scb.se/hitta-statistik/statistik-efter-amne/miljo/markanvandning/tatorter/pong/tabell-och-diagram/statistiska-tatorter-2020-befolkning-landareal-befolkningstathet-per-tatort/ (accessed 6 December 2022).

Statistics Sweden (2023) Befolkning 2013 – latest Population. https://zeus.slu.se/get/ (acessed 31 August 2023).

Stobart, J. and Howard, V. (eds) (2019) *The Routledge Companion to the History of Retailing.* Routledge.

Sullivan, P., Bonn, M.A., Bhardwaj, V. and DuPont, A. (2012) Mexican national cross-border shopping: Exploration of retail tourism. *Journal of Retailing and Consumer Services* 19, 596–604.

Sunley, P. (2000) Urban and regional growth. In E. Sheppard and T.J. Barnes (eds) *A Companion to Economic Geography* (pp. 187–201). Blackwell.

Svensk Handel (2018) *Det stora detaljhandelsskiftet* 2018. https://www.svenskhandel.se/globalassets/dokument/aktuellt-och-opinion/pressmeddelande/rapport_det-stora-detaljhandelsskiftet_2018-digital-version.pdf (accessed 8 December 2022).

Svensk Handel (2019) *Framtidens Handel – Nya företag som utvecklar Sveriges mest spännande bransch*. https://www.svenskhandel.se/globalassets/dokument/aktuellt-och-opinion/rapporter-och-foldrar/e-handelsrapporter/framtidens-handel-2019.pdf (accessed 15 December 2022).

Svensk Handel (2020) *Shoppingturism i Sverige* 2020. https://www.svenskhandel.se/globalassets/dokument/aktuellt-och-opinion/rapporter-och-foldrar/shoppingturism/shoppingturism-i-sverige-2020.pdf (accessed 8 December 2022).

Svenssons (2022) *Vår historia*. See: https://www.svenssons.se/om-oss/var-historia/ (accessed 15 December 2022).

SVT (2022) *Efter Svenssons varsel i Lammhult: Rädsla att butiken försvinner från orten*. https://www.svt.se/nyheter/lokalt/smaland/efter-svenssons-varsel-i-lammhult-radsla-att-butiken-forsvinner-fran-orten (accessed 15 December 2022).

Terry-Armstrong, N. (2012) Ikea: The homeware category killer. *Busidate* 20 (1), 7–10.

Timothy, D.J. (2005) *Shopping Tourism, Retailing and Leisure*. Channel View Publications.

Timothy, D.J. and Butler, R.W. (1995) Cross-border shopping: A North American perspective. *Annals of Tourism Research* 22 (1), 16–34.

Underhill, P. (2004) *Call of the Mall*. Simon and Schuster.

Verhoef, P.C., Lemon, K.N., Parasuraman, A., Roggeveen, A., Tsiros, M. and Schlesinger, L.A. (2009) Customer experience creation: Determinants, dynamics and management strategies. *Journal of Retailing* 85 (1), 31–41.

VK (2022) *Efter 100 år – nu avslöjar Häggströms hemligheten i Lögdeå*. https://www.vk.se/2022-04-21/efter-100-ar-nu-avslojar-haggstroms-hemligheten-i-logdea (accessed 15 December 2022).

Wang, S. and Du, P. (2021) *Retail Geography*. Routledge.

Warnaby, G., Bennison, D. and Davies, B. (2005) Retailing and the marketing of urban places: A UK perspective. *International Review of Retail, Distribution and Consumer Research* 15 (2), 191–215.

Williams, (2012) Re-theorising the role of retail shopping centres as tourist attractions in economic development. In C. McIntyre (ed.) *Tourism and Retail: The Psychogeography of Liminal Consumption* (pp. 27–44). Routledge.

Zaidan, E. (2016) Tourism shopping and new urban entertainment: A case study of Dubai. *Journal of Vacation Marketing* 22 (1), 29–41.

Zaidan, E. (2019) Shopping, tourism and hyper-development in the Middle East and North Africa. In D.J. Timothy (ed.) *Routledge Handbook on Tourism in the Middle East and North Africa* (pp. 365–377). Routledge.

10 Rethinking the Retail and Tourism Nexus as a Heritage-Making Performance: A European Perspective

Chiara Rabbiosi

Introduction

The economic crisis at the end of the first decade of the 2000s and the environmental crisis that was undoubtedly becoming apparent at more or less the same time, not to mention the ongoing warfare at the time of writing, have forced the relationship between retail and tourism to reposition itself to continue to be successful. The way consumerism was traditionally understood, in close relation to mass consumption (Baudrillard, 1970) and globalisation (Ritzer, 1993, 1999), is at odds with the current economic, political and social climate. The last two decades have been characterised by a return of the local dimension in consumer culture, which seemingly incorporates environmental awareness and economic and social sustainability principles in a configuration that has been called 'ethical' or 'conscious' consumption (Carfagna *et al.*, 2014; Thompson, 2011) and has even been labelled as consumer activism (Kuehn, 2017). From their side, in branding their products as local, producers can find an easy way to distinguish themselves from global competitors, as key drivers to community-building in a specific location, as well as a reduced environmental impact in terms of the commodity chain externalities at the basis of their business (Pike, 2011).

This chapter argues that, despite the crisis of the kind of consumerism that originally supported leisure shopping as a branding theme to attract a wide range of visitors and, consequently, foster urban and regional development, the nexus between retail and tourism remains successful. To continue this, however, retail and tourism need to be increasingly mediated by notions that rework the way consumer culture is commonly understood and practised. While research in this vein has focused on concepts of environmental sustainability or social awareness (Carfagna *et al.*, 2014; Thompson, 2011), this chapter highlights the notion of cultural heritage. There is general agreement that cultural heritage combines complex entanglements of practices, representations, expressions, knowledge and skills, with artefacts, infrastructures and sites, to identify a selective process of 'something from the past' that is valued anew in the present (Timothy, 2021; Vecco, 2010). The term 'value' here shall be understood as having both economic and cultural facets. With this acceptance, cultural heritage has also found a legitimised place among the so-called cultural and creative industries (CCIs) (Santagata, 2010).

The ways in which cultural heritage is nurtured by, and nurtures, space and place dynamics are multidimensional and performative (Rabbiosi, 2016b; Ren & Blichfeldt, 2011). This means that images, discourses, physical objects, spatialities and technologies may articulate a thematic selection of what from the past shall be saved for the future. As far as the retail and tourism nexus is concerned, this selection may stabilise in a tourismscape that includes the interior design of shops, the visual arrangement and display of commodities, the bodily performances of shopkeepers and tourists, and their narratives (Rabbiosi, 2016a, 2016b). In particular, the multifarious performances enacted in sites that combine retail, tourism and cultural heritage suggests that consumer and spatial imaginaries are combined in a bi-directional way. As such, branding commodities and branding places support a unique process that is consistent with the reproduction of the retail and tourism nexus in the current societal and economic context.

This chapter is descriptive in nature, with most of the empirical material coming from Europe, where information is most plentifully documented. First, the link between consumption and tourism by means of shopping will be described, with a specific focus on how the nexus between retail and tourism can be engineered by performing different geographical associations. Second, the notion of cultural heritage will be introduced as a novel element bridging retail and tourism. Four archetypical sites where this process is most visible will be described, including historical department stores, museum stores, productive sites for CCIs and urban food halls. In the conclusion, some avenues of future research are suggested by also providing examples that extend beyond the typology presented.

Retail, Tourism and the Role of Geographical Associations

Historically, there is a close link between consumption and tourism in the form of the practice of shopping, and this has been widely explored both in broad terms (Jin *et al.*, 2017; Timothy, 2005) and through diverse case studies (e.g. Gurova & Ratilainen, 2016; Jin *et al.*, 2020; Michalkó *et al.*, 2014; Rabbiosi, 2016a). Promoting leisure shopping sites, events and itineraries characterises late consumer capitalism, boosting the set of spatial imaginaries a place's brand image can be based on, emphasising a wide range of retail venues, from shopping malls located in metropolitan or regional areas (Rabbiosi, 2011) to historical city centres (Rabbiosi, 2015).

During the late 1990s and extending into the first decade of the present millennium, several cities began advertising shopping experiences to tourists, alongside more traditional themes for tourism promotion, such as heritage and cultural tours (Evans, 2003; Gotham, 2002). Elsewhere (Rabbiosi, 2015), I have proposed that cities have been branded as shopping venues in the context of a growing visitor economy through three main themes.

The first theme is 'fantasy city' and includes flagship developments, often driven by large real estate interests, and is based on globalised corporate brands as major retail assets. These may include large retail centres within historic city centres, and may be incorporated in the boosting of a renovated image of a city that may be attractive to both tourists and city users, as in the case of the redevelopment of Potsdamer Platz into a major shopping venue in post-1989 Berlin (Colomb, 2012). The same theme may be used to promote retail citadels in former industrial regions, as in the case of a luxury outlet village developed during the first decade of the 2000s in the former industrial triangle between the cities of Turin, Milan and Genoa in Italy (Rabbiosi, 2011).

The second theme – the 'cultural and creative city' – is characterised by cultural events associated with local productions in such a way that the creative process beyond the production of specific commodities could become profitable. Consider, for instance, the variety of events connected with CCIs that have become tourist attractors. A case in point is Milan's *Fuori Salone*, a collateral event to the Milan Furniture Fair, when designers open their shop doors to city visitors. The success of the initiative meant the event was able to be institutionalised, and today is considered one of the more profitable events for the city from a tourism point of view (Di Vita, 2022).

The third theme covers more vernacular commercial venues that have begun to be appealing specifically because they represent the opposite of downtown or especially regenerated areas. Local shopping streets have been defined as 'urban cultural ecosystems (…) formed by ordinary city dwellers interacting in vernacular spaces' (…) 'present[ing] a "face" of local social and cultural identity' (Zukin, 2012: 282). Ordinary shopping sites, as local shopping streets or street markets, are often at the centre of place branding projects more oriented toward the local community than to tourists (Ntounis & Kavaratzis, 2017).

Increasingly, playing with geographical associations to bridge retail and tourism emerges as a key driver for renewing the appeal of this nexus. The emphasis on the origin of consumer products and services has, paradoxically, increased with the radicalisation of globalisation, despite it being extremely difficult to link products and places to one location in this context. Commodities are mainly composed of a variety of components (either tangible or intangible) located in many different locations (including virtual locations). Despite this, or possibly a fortiori, geographical associations are increasingly used to differentiate commodities on the basis of the layers of cultural meanings of the places with which they are associated. The analysis of this phenomenon is certainly not new (Cook & Crang, 1996; Harvey, 1990); more recently, it has been argued that the 'origination' of a product is socially constructed by a set of actors, including producers, circulators, consumers and regulators, whose actions help 'connote, suggest and/or appeal to particular spatial references that embody and mean certain valuable things in specific market situations' (Pike, 2015: 73). Bringing new light to the 'place in product' (Molotch, 2002) aligns with new economic and symbolic values given to traditional production techniques and consumption rituals that are threatened by globalisation.

The local scale emerges as a particularly successful dimension in the association game, albeit a problematic dimension. On the one hand, previous coherent elements are disarticulated and old concepts of place are disrupted by new relationships with a world of global flows; on the other, new claims are made regarding the usually exclusive character of places and the people who belong to those places. Glocalisation (Swyngedouw, 1997) was the term used to stress the simultaneous emergence of the 'local' in the context of globalisation. In the same context, the emphasis on the places of origin of commodities and services has given rise to critical processes. This is the case, for example, for the so-called 'local trap' (Born & Purcell, 2006), that is essentialising and abstracting the local dimension from the flow of global social and spatial relations in which it is always immersed, in such a way as to give the local a positive value without scrutinising the relationship it is built from. These processes connote contemporary retail and tourism, which also indicate a sphere where geographical associations are performed.

Shopping for Cultural Heritage

As different as they are, the 'fantasy city', the 'cultural-creative city' and the 'local shopping street' themes all emphasise the role of retail services and the shopping experience in forging urban landscapes in a way that is particularly suitable for leisure and tourism. It is important to stress that these themes – or brands – are not self-excluding, or equally adopted in all parts of the world. They may also coexist in the same place. In addition, they are subject to change and can evolve. The 'fantasy city', for instance, has been particularly affected by changes in consumer culture and urban planning that have supported 'de-malling' processes, assigning new roles to shopping malls, sometimes totally disconnected from shopping (Coutinho Guimarães, 2019). The 'cultural-creative city' increasingly overlaps with the 'local shopping street', once the acceptance of the notion of CCIs is enlarged so as to include not only a wide range of productions but also the relationship they have with a place. This shift is increasingly enacted through the notion of cultural heritage, which always engages with space and place dynamics (Atkinson, 2005; Graham *et al.*, 2000; Lowenthal, 1998).

As already stated, the most basic definition of cultural heritage refers to what is inherited from the past that is considered of particular significance in the present. For a long time, the notion of cultural heritage related to artefacts, monuments, or sites, selected as the physical representations of an identity that embodied the moral values espoused by the culture surrounding them. Accordingly, it privileged monumentality and grand narratives. Cultural heritage is based on the understanding that the value of artefacts and sites is intrinsic, objective, tied to time and scientific or expert judgement often based on their aesthetic (Smith, 2006; Vecco, 2010). It is this understanding that informed the early definitions of heritage tourism, which may be reduced to 'people visiting heritage places or viewing historical resources' (Timothy, 2021: 4). However, the concept of cultural heritage currently includes intangible aspects – such as practices, representations, expressions, knowledge, skills – that can be understood from a subjective and emotional point of view (Waterton, 2014), and extending to 'mundane' heritage places (Atkinson, 2008). Therefore, a broader definition of heritage tourism includes 'travelers seeing or otherwise experiencing built heritage, living culture or manifestations of art' (Timothy, 2021: 4) and encompasses a multitude of motives, resources and experiences.

Commonly, the statistics provided by the tourist bureaus of cities and regions show distinct percentages of tourist activities or tourist spending. Based on these numbers, we know that shopping is quite a common tourist activity, as much so as visiting heritage sites. For instance, in a list of 18 activities, shopping ranked the fourth most practised activity by tourists in Paris in 2021, according to data provided by Visit Paris Region (https://pro.visitparisregion.com/chiffres-du-tourisme/chiffres-annuels/reperes-de-l-activite-touristique-a-paris-ile-de-france-2022/statistiques-detaillees/frequentation-touristique/La-frequentation-touristique-a-Paris-Ile-de-France/les-clienteles-touristiques-a-paris (last accessed 7 January 2023)). The most engaged in activity was visiting museums, permanent exhibitions, monuments and castles; the second was strolling around the city; and the third was a vaguely defined 'other' activity. The data portrayed by statistics are disentangled from materialities and bodies, and often present phenomena as separate while in practice they are embedded with one another. This chapter now turns to specific sites that epitomise the reworking of the retail and tourism nexus through diverse performances that play with geographical associations while simultaneously developing the notion of cultural heritage.

Historical department stores

Historical department stores are archetypical sites where retail and tourism are significantly entangled with notions of cultural heritage (Whitaker, 2011). The way cultural heritage is performed by historical department stores also plays with geographical associations, building a solid link with the cities where these are located. Consider the example of Paris's 19th-century department stores: the Galeries du Louvre opened in 1855 on the ground floor of the Grand Hôtel du Louvre, while the Hôtel Lutetia was built in 1910 opposite the Bon Marché on the initiative of the wife of the owner of the department store itself, showing the longstanding link between shopping and tourism in Paris (Miller, 1981). Today, there are five historic department stores still in operation that have become tourist attractions in the city – Galeries Lafayettes, Printemps Haussmann, Bon Marché, Samaritaine and the BHV Marais (formerly Bazar de l'Hotel de Ville). They have become part of the construction of the Paris imaginary as a destination for shopping, exploited by institutional tourist promotions as an expression of the city's culture.

The five historical department stores represent significant shopping sites because of the imaginative power associated with the production and consumption of certain commodities – namely luxury and fashion commodities – but they also intertwine with the world of cultural heritage in a broader sense. One on side, the buildings of these department stores are icons of a certain *epoque*, which was a golden age for Paris and France, namely the period between the 1870s and WWI. The link with this past time is shown by the exterior monumentality of the buildings, and the interior architectural details (staircases, domes, glass decorations, mosaics). A geographical association between the historical department store and the city, and the nation that the city is part of, is created by both retail managers and tourism stakeholders, including the tourists who visit the department stores as any other cultural heritage attraction. For instance, during the European Heritage Days – an initiative sponsored by the European Commission and the Council of Europe – guided tours are given at the Galeries Lafayette, at the Printemps and at the Samaritaine.

In visiting Paris, tourists visit the city's department stores not only to shop, but also to experience the atmosphere of the 19th-century consumer culture within that specific city. Within this frame, Paris replaces the postmodern fantasies (at best) of the commercial citadels created ad hoc in suburban areas. The city becomes the main sales point of department stores, while these, for their part, can offer an exclusive commercial arrangement made up of both French brands (enhanced by the interior layout of the department store) and an urban and cultural myth. In addition, the relationship that department stores have with the neighbourhoods in which they are located, which are rich in commercial activities as well as other heritage resources (e.g. cultural venues, architecture, and a specific urban atmosphere), contributes to the reproduction of a specific place brand for Paris, which renews a historical legacy through leisure shopping and tourism (Rabbiosi, 2015).

It is possible for historical department stores to be interpreted as museums, if we consider the commodities they sell as iconic items connected with fashion brands that people may be willing to admire and experience. Clothes, shoes and perfumes are the pieces both exposed and shopped for, whose economic and cultural value is mediated by the notion of heritage that intersects both consumer culture and the city.

Museum stores

While department stores are not museums but may be enjoyed as such by some visitors, it seems that museums are becoming closer to department stores. Museums and heritage sites have provided merchandise for visitors to purchase since their earliest incarnations as public attractions in the 19th century (Larkin, 2016). Retailing can be formalised in a variety of ways in front of museums – most simply with reference to the sale of tickets and guides – but it finds its most experiential form in museum stores, which increasingly offer a broad series of commodities 'to shop for'. Originally, museum stores were closer to the format of a bookshop, offering a range of books connected with what is exhibited in the museum. Another kind of museum shop simply included a set of souvenirs. Souvenirs are commodities that convey meanings while also being strictly connected with the commercial side of tourism (Swanson & Timothy, 2012). They are 'tangible reminders' of a place and its experience, but also have a symbolic association with the wider brand denoting a place. Therefore, while souvenirs have become an integral component of museum shops, they can be as much thought as part of museums' brands as cultural commodities conveying messages about the broader place in which a museum is located (Trabskaia et al., 2019).

There are examples of museums becoming part of place brands because of their own well-established brands, especially once they are part of the cultural heritage of a city. Museums are historical institutions in global cities, where they are depositaries of cultural capital connected with their economic and political history. Think of the British Museum in London, for instance, which exhibits collections strictly linked to the history of the British Empire and is inextricably associated with London as a tourist destination. In fact, museums may provide a form of 'imagineering' in which they act as a source of identity and a reference point for the place in which they are located, and are pivotal in attracting visitors from far afield. Museums and exhibition centres may also significantly contribute to a city's signposting on an international culture map to gain a global competitive advantage (Timothy, 2005; Paddison & Miles, 2007).

Museums have undergone a dramatic change in the way they are curated in the last two decades or so, and this is also true for their stores (Booth & Powell, 2016; Larkin, 2016; Shao *et al.*, 2019). Museum stores are now increasingly less confined to be bookshops or sellers of magnets, to become more sophisticated retail venues, elaborating the notion of cultural heritage with reference to the place in which they are located or the collections they include. Museum stores have become true shopping sites, sometimes serving customers well beyond the museum visitors and extending to a larger audience of urbanites in search of cultural commodities such as designer objects or artisan crafts. A case in point to explain this shift might be the museum stores of the Doge's Palace and Museo Correr in Venice (Figure 10.1). These museums, which line Piazza San Marco, are just two of a network of museums owned by the City of Venice. Until 2018, the bookshops of all these museums were managed by two of the main Italian publishing companies. Currently they are managed by a consortium composed of a leading company in Italy for the planning, organising and staging of art exhibitions, and a leading Spanish company specialising in the comprehensive management of shops in museums, art galleries and other cultural establishments, whose aim is 'to develop sales and to promote the cultural heritage in sites such as museums, palaces, cathedrals, monuments and all kinds of exhibitions, both permanent and temporary' (Palacios y Museos, 2024, n.p.).

Figure 10.1 Souvenirs created by a designer company from obsolete wooden pilings from the canals of Venice. Museums' store of the Doge's Palace and Museo Correr, Venice. Photo: Chiara Rabbiosi, 6 December 2022

This shift in museum stores' management lines up with a sophistication of museum stores as retail and tourist venues which results in a more complex network of actors, including retail managers who rub shoulders with exhibition curators, global commodities providers and local artisans and designers. For example, the team in charge of the museum store of the Doge's Palace and Museo Correr is in constant dialogue with their Spanish partner, which provides the museum with a set of template souvenirs, such as notebooks or fans. These are then printed in different patterns according to the specific collection of the Doge's Palace and Museo Correr or their temporary exhibitions. However, the team is also in contact with national designers and local artisans. For instance, souvenirs made from *briccole* wood can be found in the museum store. The design company that brands these objects is significantly named Pieces of Venice, as *briccole* are the dolphins serving as protective shields along the docks of Venetian canals. Tourists can also find miniature watercolour paintings produced by a Venetian artist that has a workshop in town, or the typical striped shirts of Venetian gondoliers branded with the logo of the local gondoliers' association.

These examples show the explicit intent of museum stores to provide souvenirs that are also cultural commodities strictly linked to the cultural heritage connected with the museum's location. In this way, souvenirs maintain their role of communicating a destination but also contribute to heritage-making performances, stabilising the tourism- and shopping-scape they entail.

Productive sites for creative industries

Other significant sites that are gaining a role in bringing together retail and tourism via cultural heritage are the productive sites for creative industries and artisans. Indeed, a wide range of locations, including small and medium-sized towns, are turning to tourism and enhancing their CCIs (Rabbiosi & Ioannides, 2022). It has been suggested that the 'industrial approach' focusing on the spatialised production of culture (Pratt, 2004) can be integrated with a 'territorial approach' (Lazzeroni et al., 2013; Santagata, 2010), highlighting the role that local assets may play in maintaining a set of intangible resources and sociocultural relationships, such as passions, affective components, moral sentiments, tacit awareness and trust. Increasing attention is given to the cultural landscape that CCIs represent. For instance, Stoke-on-Trent, a former industrial town in the UK, has self-designated itself as the Capital of Ceramics, and potteries are today considered a significant part of the city's cultural heritage, in both its tangible and intangible counterparts. Once there, visiting a pottery-making atelier that utilises materials from the surroundings and observing the actions of workers who create their art in a studio setting might be more attractive than simply visiting a pottery museum. At the end, a tourist may be willing to buy a piece of pottery that recalls the experience they have had, centring the fact that CCIs are part of the living cultural heritage of that specific place.

However, the nexus between retail and tourism does not always function to preserve the reproduction of CCIs and keep alive the cultural heritage they draw from. Consider Murano, a series of islands in the Venetian Lagoon well known as a glass-manufacturing site, which was established there at the end of the 13th century. Already by 1861 a Glass Museum was established with the aim of providing an archive of the long glass-making history. The production of glass in Murano was world-leading until recently. Then, the increasing competition at the end of the 20th century from foreign producers pushed an increasing number of Murano manufacturers to orient themselves to the local market, which is mainly the tourist market pivoting around Venice. Segre and Russo (2005) stated that the orientation to tourist demand was not without consequences, with those Murano producers that have chosen to come more directly in contact with tourism changing their nature from manufacturers to retailers, utilising abandoned factories as showrooms and substantially lowering the quality of the products. Murano's glass-making is part of the place brand of Venice and tourists do shop for Murano glass, often in the form of souvenirs. However, while the retail and tourism repositioning of this production contributes to heritagising glass-making, the makers have not been able to turn it into a thriving place-based CCI.

Urban food halls

Food and drinks have recently gained a proper place as tangible and intangible heritage, in particular as they are recognised as diets, agricultural and manufacturing techniques, or consumption rituals. For its part, gastronomy is increasingly understood as a part of CCIs. Food and drinks have always been something that travellers shop for, even if just for feeding themselves. However food and drinks now occupy a prominent position in regard to shopping, as part of culinary tourism (Everett, 2016a) or even heritage tourism (Timothy, 2021). For example, culinary trails and wine routes are often developed to thematically connect several productive locative sites. Culinary trails and wine routes have been examined to understand their contribution to both promoting the sales of specific products and to fostering heritage and place brands, by utilising the geographical associations they foster (Everett, 2016b; Hashimoto & Telfer, 2016).

Culinary trails and wine routes originated as tools to foster tourism for local development in rural areas. However, cities have recently turned to food and drinks to further develop the retail and tourism nexus, which also expands their branding campaigns. This is particularly evident with reference to the regeneration of urban food markets and market halls, which have become the object of urban policies in connection with the elicitation of discourses around environmental sustainability, food security, public space and, indeed, cultural heritage and tourism. Think, for instance, of Borough Market in London, located adjacent to London Bridge in an area made more attractive since the late 1990s thanks to a strong cultural heritage investment, with the opening of the Globe Theatre and the Tate Modern (Coles & Crang, 2011); or the Mercat de Santa Caterina in Barcelona, the subject of major architectural reinvestment as part of the redevelopment of the old district market in Barcelona's Ciutat Vella neighbourhood, also in the late 1990s (Cordero, 2014).

A further case in point is Mercato Centrale, a retail brand used to market food halls in a variety of Italian cities including Turin, Milan, Florence and Rome. As Bourlessas *et al.* (2022: 1340) suggest while studying the case of Turin, 'Mercato Centrale is not what is considered to be a traditional market, with plain stalls and goods simply on display; instead, its spatial organisation and aesthetics materialise the brand together with associated meanings and values'. The retail brand is constructed not only by showcasing food and drink commodities, but also through the exhibition of machines and other objects of production, the demonstration through screens, and the simulation of aesthetics typical of artisanal production, as in a museum. All these elements, and the performances entangled with them, contribute to heritagising food (see Chapter 14). At the same time, they are used to play with diverse geographical associations. While the rhetoric of the local is present in this kind of food market hall, especially drawing on the possibility of buying local food, thereby reducing food miles and supporting environmental sustainability, the national referent is used to construct the notion of Italianness, as a worldwide character that also guarantees quality. This may easily overlap with a specific 'otherness' in the same place, such as ethnic food, which can ensure a cosmopolitan touch that is consistent with global cities.

Lastly, it should be noted that food markets and food halls are increasingly understood as being part of foodification – a neologism combining 'food' and 'gentrification,' which may also intersect with 'touristification', that is used to criticise the food hype in urban cultures (Bourlessas *et al.*, 2022). The development of brand new food halls or the regeneration of existing ones emerge as a particularly ambiguous showcase for the staging of the widespread discourse on the cultural and creative city or on the heritage city from a neoliberal perspective, especially considering that most of the new urban food halls are controlled by major private actors (Gonzalez & Waley, 2013).

Conclusion

In showing a variety of examples through which retail and tourism are bridged by playing with geographical associations, this chapter focuses on an emergent theme used in shopping promotion in the last two decades or so, that of cultural heritage. Four typologies of sites where this theme emerges most significantly have been presented, and it has been suggested that in these sites a thematic reworking of the nexus between retail and tourism emerges through multifarious performances, which also contribute to smoothing the consumerist aspect, considering that this can be at odds with global economic and environmental crises or warfare. The notion of cultural heritage is employed in shopping venues, their commodities and their atmospheres in such a way as to hide the societal or cultural paradoxes that consumer capitalism may bring with it. In this way, the notion of cultural heritage plays a salvific role in regard to consumption. Although not examined in this chapter, traditional urban marketplaces, such as the souks of the Arab world or the vast produce markets of Southeast Asia, are another locality of heritage consumption that have become salient tourist attractions and are often looked upon by city dwellers, planners and tourist-consumers as key crucibles of heritage-making and heritage consumption (Zandieh & Seifpour, 2020).

This chapter is descriptive in nature and has suggested some analytical keys for further research development. On one hand, it is important to undertake deeper analysis through specific case studies to delve into the paradoxes and frictions resulting from heritage performances in specific settings, as well as beyond those settings. As mentioned throughout the chapter, the 'goodness' of heritage-making may be problematic in regard to the variety of processes it may generate. For instance, it may reduce CCIs to consumer entities disconnected from productive living cultures. On the other hand, it may boost gentrification processes, harbingers of social inequalities and exclusions. In this sense, it is important to move from case studies to extend the scope of research to wider economic and/or urban implications.

A second line for further research might be an enquiry into how retail and tourism play with multiscalar geographical associations, considering the transcalar dimension that the two phenomena – and even more so the nexus between the two – always include. On this point, my attention was recently struck by the label Fabriqué à Paris developed by the City of Paris, aimed at promoting the diversity and richness of Parisian manufacturing so as to 'meet a strong demand from Parisians and tourists for objects that embody Parisian identity and traditions' and 'highlight the richness of local production and the dynamism of Parisian crafts' (Parisjecoute, 2024, n.p.). In such branding, which is at the same time a form of retail, tourism and urban branding, how is the 'local' recrafted in such a way that inevitably dialogues with other scales that are not explicitly mentioned? And again, how is it connected with the notion of cultural heritage?

Lastly, a third research line may enlarge the discussion begun here to include themes that are increasingly marking the renewal of consumer practices, such as the notion of a circular economy. Consider the case of the designer's piece made from *briccole* wood sold at the Doge's Palace and Museo Correr in Venice, or the case of Chaussette Orphelines – a brand that produces socks by recycling worn out socks, and that has also been branded with the Fabriqué à Paris label. The notion of a circular economy may be productive of place performances (discursive, visual, material) that again imbue consumer culture, and the retail and tourism nexus as a consequence, with new, less consumerist, meanings.

To conclude, I wish to highlight that this chapter features examples, concepts and thoughts that reveal a European perspective. Future research on the retail and tourism nexus as a heritage-making performance in other parts of the world may bring to life diverse ways of understanding not only the agentic role of shopping, but also different notions of consumer culture and tourist experience at large. It may also acknowledge typological patterns that have not been considered here and that might offer a useful avenue of further research.

References

Atkinson, D. (2005) Heritage. In D. Atkinson, P. Jackson, D. Sibley and N. Washbourne (eds) *Cultural Geography: A Critical Dictionary of Key Concepts* (pp. 141–150). IB Tauris.

Atkinson, D. (2008) The heritage of mundane places. In B.J. Graham and P. Howard (eds) *The Ashgate Research Companion to Heritage and Identity* (pp. 382–395). Ashgate.

Baudrillard, J. (1970) *La Société de consommation: Ses mythes, ses structures*. Denoël.

Booth, E. and Powell, R. (2016) Museums: From cabinets of curiosity to cultural shopping experiences. In V. Katsoni and A. Stratigea (eds) *Tourism and Culture in the Age of Innovation* (pp. 131–143). Springer. https://doi.org/10.1007/978-3-319-27528-4_9

Born, B. and Purcell, M. (2006) Avoiding the local trap: Scale and food systems in planning research. *Journal of Planning Education and Research* 26 (2), 195–207. https://doi.org/10.1177/0739456X06291389

Bourlessas, P., Cenere, S. and Vanolo, A. (2022) The work of foodification: An analysis of food gentrification in Turin, Italy. *Urban Geography* 43 (9), 1328–1349. https://doi.org/10.1080/02723638.2021.1927547

Carfagna, L.B., Dubois, E.A., Fitzmaurice, C., Ouimette, M.Y., Schor, J.B., Willis, M. and Laidley, T. (2014) An emerging eco-habitus: The reconfiguration of high cultural capital practices among ethical consumers. *Journal of Consumer Culture* 14 (2), 158–178. https://doi.org/10.1177/1469540514526227

Coles, B. and Crang, P. (2011) Placing alternative consumption. Commodity fetishism in Borough Fine Foods Market, London. In T. Lewis and E. Potter (eds) *Ethical Consumption: A Critical Introduction* (pp. 87–102). Routledge.

Colomb, C. (2012) *Staging the New Berlin: Place Marketing and the Politics of Urban Reinvention Post-1989*. Routledge.

Cook, I. and Crang, P. (1996) The world on a plate: Culinary culture, displacement and geographical knowledges. *Journal of Material Culture* 1 (2), 131–153. https://doi.org/10.1177/135918359600100201

Cordero, A. (2014) *Gentrificaciòn comercial y mercados publicos: El Mercado de Santa Caterina, Barcelona*. Working Paper Series Contested_Cities.

Coutinho Guimarães, P.P. (2019) Shopping centres in decline: Analysis of demalling in Lisbon. *Cities* 87, 21–29. https://doi.org/10.1016/j.cities.2018.12.021

Di Vita, S. (2022) Cultural events and heritage policy for the Milan Expo 2015: Experimental intersections between mega-event and city. *European Planning Studies* 30 (3), 499–513. https://doi.org/10.1080/09654313.2021.1959723

Evans, G. (2003) Hard-branding the cultural city – from Prado to Prada. *International Journal of Urban and Regional Research* 27 (2), 417–440. https://doi.org/10.1111/1468-2427.00455

Everett, S. (2016a) *Food and Drink Tourism: Principles and Practice*. Sage.

Everett, S. (2016b) Iconic cuisines, marketing and place promotion. In D.J. Timothy (ed.) *Heritage Cuisines: Traditions, Identities and Tourism*. Routledge.

Gonzalez, S. and Waley, P. (2013) Traditional retail markets: The new gentrification frontier? *Antipode* 45 (4), 965–983. https://doi.org/10.1111/j.1467-8330.2012.01040.x

Gotham, K. (2002) Marketing Mardi Gras: Commodification, spectacle and the political economy of tourism in New Orleans. *Urban Studies* 39 (10), 1735–1756. https://doi.org/10.1080/0042098022000002939

Graham, B.J., Ashworth, G.J. and Tunbridge, J.E. (2000) *A Geography of Heritage: Power, Culture, and Economy*. Oxford University Press.

Gurova, O. and Ratilainen, S. (2016) From shuttle traders to middle-class consumers: Russian tourists in Finnish newspaper discourse between the years 1990 and 2014. *Scandinavian Journal of Hospitality and Tourism* 16 (1), 51–65. https://doi.org/10.1080/15022250.2016.1244507

Harvey, D. (1990) Between space and time: Reflections on the geographical imagination. *Annals of the Association of American Geographers* 80 (3), 418–434.

Hashimoto, A. and Telfer, D.J. (2016) Culinary trails. In D.J. Timothy (ed.) *Heritage Cuisines: Traditions, Identities and Tourism*. Routledge.

Jin, H., Moscardo, G. and Murphy, L. (2017) Making sense of tourist shopping research: A critical review. *Tourism Management* 62, 120–134. https://doi.org/10.1016/j.tourman.2017.03.027

Jin, H., Moscardo, G. and Murphy, L. (2020) Exploring Chinese outbound tourist shopping: A social practice framework. *Journal of Travel Research* 59 (1), 156–172. https://doi.org/10.1177/0047287519826303

Kuehn, K.M. (2017) Brand local: Consumer evaluations as commodity activism on Yelp. com. *Journal of Consumer Culture* 17 (2), 205–224. https://doi.org/10.1177/1469540515586866

Larkin, J. (2016) 'All museums will become department stores': The development and implications of retailing at museums and heritage sites. *Archaeology International* 19 (1), 109–121. https://doi.org/10.5334/ai.1917

Lazzeroni, M., Bellini, N., Cortesi, G. and Loffredo, A. (2013) The territorial approach to cultural economy: New opportunities for the development of small towns. *European Planning Studies* 21 (4), 452–472. https://doi.org/10.1080/09654313.2012.722920

Lowenthal, D. (1998) *The Heritage Crusade and the Spoils of History* (1st pbk. edn). Cambridge University Press.

Michalkó, G., Rátz, T., Hinek, M. and Tömöri, M. (2014) Shopping tourism in Hungary during the period of the economic crisis. *Tourism Economics* 20 (6), 1319–1336. https://doi.org/10.5367/te.2014.0387

Miller, M.B. (1981) *The Bon Marché: Bourgeois Culture and The Department Store 1869–1920*. Princeton University Press.

Molotch, H. (2002) Place in product. *International Journal of Urban and Regional Research* 26 (4), 665–688. https://doi.org/10.1111/1468-2427.00410

Ntounis, N. and Kavaratzis, M. (2017) Re-branding the High Street: The place branding process and reflections from three UK towns. *Journal of Place Management and Development* 10 (4), 392–403. https://doi.org/10.1108/JPMD-12-2015-0056

Paddison, R. and Miles, S. (eds) (2007) *Culture-led Urban Regeneration*. Routledge.

Palacios y Museos (2024) Comprehensive management for the cultural tourist. Online http://www.palaciosymuseos.com/v2/index.php/en/about-us/our-company (last access Jan. 15 2024).

Parisjecoute (2024) Le label 'Fabriqué à Paris', qu'est-ce que c'est? Online: https://www.paris.fr/pages/le-label-fabrique-a-paris-5152 (accessed January 10, 2024)

Pike, A. (2011) Placing brands and branding: A socio-spatial biography of Newcastle Brown Ale. *Transactions of the Institute of British Geographers* 36 (2), 206–222. https://doi.org/10.1111/j.1475-5661.2011.00425.x

Pike, A. (2015) *Origination: The Geographies of Brands and Branding*. Wiley.

Pratt, A.C. (2004) The cultural economy: A call for spatialized 'production of culture' perspectives. *International Journal of Cultural Studies* 7 (1), 117–128. https://doi.org/10.1177/1367877904040609

Rabbiosi, C. (2011) The invention of shopping tourism. The discursive repositioning of landscape in an Italian retail-led case. *Journal of Tourism and Cultural Change* 9 (2), 70–86. https://doi.org/10.1080/14766825.2010.549233

Rabbiosi, C. (2015) Renewing a historical legacy: Tourism, leisure shopping and urban branding in Paris. *Cities* 42 (b), 195–203. https://doi.org/10.1016/j.cities.2014.07.001

Rabbiosi, C. (2016a) Itineraries of consumption: Co-producing leisure shopping sites in Rimini. *Journal of Consumer Culture* 16 (2), 412–431. https://doi.org/10.1177/1469540516635429

Rabbiosi, C. (2016b) Place branding performances in tourist local food shops. *Annals of Tourism Research* 60, 154–168. https://doi.org/10.1016/j.annals.2016.07.002

Rabbiosi, C. and Ioannides, D. (2022) Cultural tourism as a tool for transformation in small and medium-sized towns. In H. Mayer and M. Lazzeroni (eds) *A Research Agenda for Small and Medium-sized Towns* (pp. 107–126). Edward Elgar.

Ren, C. and Blichfeldt, B.S. (2011) One clear image? Challenging simplicity in place branding. *Scandinavian Journal of Hospitality and Tourism* 11 (4), 416–434. https://doi.org/10.1080/15022250.2011.598753

Ritzer, G. (1993) *The McDonaldization of Society*. Pine Forge Press.

Ritzer, G. (1999) *Enchanting a Disenchanted World: Revolutionizing the Means of Consumption*. Pine Forge Press.

Santagata, W. (2010) *The Culture Factory: Creativity and the Production of Culture*. Springer.

Segre, G. and Russo, A.P. (2005) Collective property rights for glass manufacturing in Murano: Where culture makes or breaks local economic development. *Working Paper Series. International Centre for Research on the Economics of Culture, Institutions, and Creativity (EBLA)*, (5). Retrieved from https://www.fondazionesantagata.it/en/publications/collective-property-rights-for-glass-manufacturing-murano-where-culture-makes-or-breaks-local-economic-development/

Shao, J., Ying, Q., Shu, S., Morrison, A.M. and Booth, E. (2019) Museum tourism 2.0: Experiences and satisfaction with shopping at the National Gallery in London. *Sustainability* 11 (24), 7108. https://doi.org/10.3390/su11247108

Smith, L. (2006) *Uses of Heritage*. Routledge.

Swanson, K.K. and Timothy, D.J. (2012) Souvenirs: Icons of meaning, commercialization and commoditization. *Tourism Management* 33 (3), 489–499. https://doi.org/10.1016/j.tourman.2011.10.007

Swyngedouw, E. (1997) Neither global nor local: 'Glocalization' and the politics of scale. In K.R. Cox (ed.) *Spaces of Globalization: Reasserting the Power of the Local* (pp. 137–166). Guilford Press.

Thompson, C.J. (2011) Understanding consumption as political and moral practice: Introduction to the special issue. *Journal of Consumer Culture* 11 (2), 139–144. https://doi.org/10.1177/1469540511403892

Timothy, D.J. (2005) *Shopping Tourism, Retailing, and Leisure*. Channel View Publications.

Timothy, D.J. (2021) *Cultural Heritage and Tourism: An Introduction* (2nd edn). Channel View Publications.

Trabskaia, I., Shuliateva, I., Abushena, R., Gordin, V. and Dedova, M. (2019) City branding and museum souvenirs: Towards improving the St. Petersburg city brand: Do museums sell souvenirs or do souvenirs sell museums? *Journal of Place Management and Development* 12 (4), 529–544. https://doi.org/10.1108/JPMD-06-2017-0049

Vecco, M. (2010) A definition of cultural heritage: From the tangible to the intangible. *Journal of Cultural Heritage* 11 (3), 321–324. doi: 10.1016/j.culher.2010.01.006

Waterton, E. (2014) A more-than-representational understanding of heritage? The 'past' and the politics of affect: A more-than-representational understanding of heritage? *Geography Compass* 8 (11), 823–833. https://doi.org/10.1111/gec3.12182

Whitaker, J. (2011) *The World of Department Stores*. Vendome Press.

Zandieh, M. and Seifpour, Z. (2020) Preserving traditional marketplaces as places of intangible heritage for tourism. *Journal of Heritage Tourism* 15 (1), 111–121.

Zukin, S. (2012) The social production of urban cultural heritage: Identity and ecosystem on an Amsterdam shopping street. *City, Culture and Society* 3 (4), 281–291. https://doi.org/10.1016/j.ccs.2012.10.002

Part 3

Experiential and Niche Aspects of Shopping Tourism

11 Leisure Shopping, Retail Experiences and Destination Satisfaction

Tim Coles

Introduction

Shopping sells! Fifth Avenue in New York, Oxford Street in London, Kurfürstendamm in Berlin and the Champs-Elysees in Paris evoke alluring images of the outlets of sophisticated, exclusive high-end brands intended for metropolitan elites. Beyond the glitz and glamour of boutiques and department stores, the Grand Bazaar in Istanbul, the souk of Tangiers and the galleries of Milano engender both a sense of exoticism and experiences of shopping in yesteryear. As the latter-day cathedrals of consumptions, the gigantic Mall of America in Bloomington and the Mall of the Emirates in Dubai have become destinations in their own rights, each conjuring images of shopping paradises and, among those fortunate enough to have visited them, evoking distinctive memories of their trips.

These days there is hardly a town, city or rural locality (Hurst & Niehm, 2012), that does not extol to prospective visitors the virtues of its shopping offer or the local products and produce it has to satisfy visitors. Shopping is so widely embedded in the everyday lives of citizen-consumers in late modernity, not least the proliferation and popularity of the shopping mall (Warnaby & Medway, 2018). This is so much the case, especially in the developed world, that we sometimes lose sight of the fact that it is still in historical terms a relatively recent phenomenon (Stobart, 2010). Until the end of the 19th century, leisure shopping was the preserve of urban elites and the rise of the shopping embodied a shift in the nature of urbanism from the city of production to the city of consumption (Crossick & Jaumain, 1999; Shaw & Benson, 1992). The flânerie of well-to-do, petite bourgeois consumers and their increasing patronage of chain stores and department stores as innovations (Coles, 1999a) set them apart from the urban masses for whom the acquisition of goods was mostly restricted to buying household essentials from more modest establishments and artisans (Crossick & Haupt, 1984; Shaw & Benson, 1992).

Today shopping is also such a routine component in tourism episodes that it is often taken for granted by consumers but, as this volume attests, it has started to attract a significant but still evolving body of scholarly knowledge (Jin *et al.*, 2017). A distinctive aspect of this has been the positioning of tourist shopping within trips. Some commentators have identified shopping as a major attraction and motive for travel and tourism, arguing that it features strongly in destination selection and decision-making (Moscardo, 2004; Timothy, 2005). As alluded to above, this has been recognised by many destination managers and tourism 'thought leaders' who have argued that shopping facilities and retail assets form a major and essential part of the 'destination mix' that drives their attractiveness and ultimately their competitiveness (UNWTO, 2014). Others taking their cues from the Mobilities turn (Hall, 2005), have noted that trips away are increasingly complicated in nature with shopping just one among several diverse activities that comprise a leisure episode. Rather than being the sole or principal motivator for, or the exclusive defining feature of, trips, there are 'unplanned' as well as 'planned tourist shoppers' with differences in shopping behaviours and socio-demographics characteristics among the types (Baek & Park, 2024).

Of course, the emphasis of shopping in decision-making, as well as its conspicuous featuring in destination marketing, both result in the generation of expectations among visitors which, in principle at least, should be matched by their experiences, and it is the extent of that congruence that studies of destination satisfaction seek to assess. The chapter critically examines the relationship between leisure shopping by visitors with their retail experiences and destination satisfaction. Most, if not all of us, shop and come to a chapter (and even a volume) like this with our preconceptions of what shopping and retailing are. More likely than not, we also have our own experiences of shopping when away, perhaps on holiday or possibly when on a business trip, as well as our personal valorisations of whether we were satisfied or not. We (think we) know leisure shopping because we have done it. The purpose of this chapter is to question some of these assumptions and to argue that, despite the considerable scholarly progress made in recent years (cf. Choi *et al.*, 2016; Coles, 2004a; Jin *et al.*, 2017; Timothy, 2005, 2014), there is little room for complacency moving forward with notable challenges ahead for this particular focus of study. Just as chain stores and department stores were major disruptors that changed the nature of both retailing and shopping at the fin-de-siecle (Coles, 1999a, 1999b; Crossick & Jaumain, 1999; Shaw & Benson, 1992) or shopping malls (Warnaby & Medway, 2018) and designer outlets (Rabbiosi, 2011) at the fin-de-millennium, in their own right equally-radical, digital innovations look set to challenge what we know about tourism shopping and destination satisfaction and the enduring valence of this knowledge in the 21st century. The next section examines some of the established views on leisure shopping and destination satisfaction.

Shopping and Satisfaction: Established Positions

Destination satisfaction has been a cornerstone of tourism research for many years, as reviews and surveys of the state-of-the-art attest (Chen et al., 2013; Fuchs & Weiermaier, 2003). There is neither space nor scope in this chapter to revisit this substantial area of tourism studies. However, it is important to observe that satisfaction is a multi-layered and highly nuanced phenomenon. It is also dynamic and changing based on expectations of, and promises for, the delivery of service, product and experience at any given moment. Thus, as the nature of tourism and visitor behaviour have evolved in recent decades, there has been a longstanding, ever greater and arguably never-ending challenge in terms of how (and when) best to measure (and compare) satisfaction in destinations over time (cf. Pizam et al., 1978; Kozak, 2009).

Be this as it may, a great many studies have identified, to one degree or another, a positive relationship between tourism shopping experiences and destination satisfaction (see Chang et al., 2006; Choi et al., 2016; Heung & Cheng, 2000; Lee & Choi, 2020; Lin & Lin, 2006; McDowall, 2010; Parasakul, 2020; Song et al., 2011; Wong & Wan, 2013). Broadly writ, the orthodox thesis is, somewhat understandably, that the more positive the appraisal of shopping facilities the more satisfied a visitor has been or will be with a destination. An added dimension in more recent work has been that the more satisfied a visitor, the more loyal they are likely to be to the destination and return (McDowall, 2010; Song et al., 2011; Sirakaya-Turk et al., 2015; Suhartanto, 2018).

In general, there have been two main ways in which visitors' satisfaction with (retailing and) shopping have been investigated, in both cases through quantitative modes of enquiry. Several studies have approached the notion of satisfaction in a granulated manner exploring the performance of particular facets and attributes of service delivery (Ayeh & Chen, 2013; Lloyd et al., 2011; Yeung et al., 2004; Yüksel, 2004). For example, in a comparative study of Hong Kong and Singapore, 15 attributes of both shopping destinations were measured in surveys of visitors, with Singapore then out-performing Hong Kong 'notably in many areas, such as language ability, attitude and efficiency of service staff' (Yeung et al., 2004: 85). Employing importance-performance analysis (IPA), Ayeh and Chen (2013) also focus on Hong Kong as a shopping destination and explore attributes for shopping facilities there in a general, collective sense across the Special Administrative Region. Typical of the genre and also adopting survey instruments, they captured visitors' views on the importance and performance of, *inter alia*: the pleasantness of the shopping environment; the convenience of opening hours; quality of information on products and prices; range of choice; convenience of payment; and knowledge, honesty, communication skills and attitude of staff (Ayeh & Chen, 2013: 253). Standard two-by-two matrices revealed the correspondence for each attribute: those in the high-importance–high-performance quadrant suggesting strong satisfaction (and to be maintained in future deliberative management), those in high-importance–low-performance category suggesting lower satisfaction and, from a practical perspective, areas for uplift through future management interventions. In sum, Ayeh and Chen (2013: 259) noted the need for an urgent improvement in language and communication skills in the retail sector, with Asian visitors less satisfied overall than non-Asian visitors (Ayeh & Chen, 2013: 259).

Rather than inferring relative satisfaction from the closeness or distance between importance and performance of particular attributes, many studies view satisfaction as it relates to shopping as a more complex construct that defies straightforward measurement by a single index (i.e. a literal measure). Instead, through a composite view, satisfaction is revealed by measurement of a series of related indices (i.e. variables) and the strength and coherence of their relationship with one another. For instance, in the initial design of their study, Song *et al.* (2011: 85) argue that tourist satisfaction is a latent variable revealed through three broad indicators: overall satisfaction, confirmation of expectations and comparison with the ideal. In a structural equation modelling approach, measurements of satisfaction with the destination were related to perceived performance, expectations and assessed value for three sectors of the visitor economy, once more in Hong Kong: hotels, retailers and tour operators.

Not surprisingly then, if satisfaction is a construct and best revealed through multiple indices, questions arise of how best to measure satisfaction as it relates to the multifaceted nature of shopping and, more specifically, which particular variables are most appropriate for this purpose from a large array of potential options? Using data reduction techniques, like factor analysis, some studies have concentrated on developing scales to measure tourist shopping satisfaction (Wong *et al.*, 2013). In a widely cited contribution, Wong *et al.* (2013: 29) attempted to 'systematically develop a scale that conceptualizes tourists' shopping satisfaction as a four-dimensional construct that reflects tourists' satisfaction of service product and environment, merchandise value, staff service quality and service differentiation during their shopping excursion'. In the development of a series of hypotheses, they note that tourist shopping satisfaction is also related to a series of other services and attributes (i.e. safety, transportation, location, cleanliness, size). This reflects tourism shopping as operating in, and reliant on, destinations characterised by complex systems of provision and functional interdependencies, for instance with accommodation, hospitality, transport providers, attractions (Hall, 2005; Henseler *et al.*, 2022) and, in the technology-enabled smart city, telecommunications and virtual space (Buhalis *et al.*, 2023).

Although by no means exhaustive nor intended to be a systematic review, a closer inspection of this and related work reveals three important features regarding the nature of enquiry on leisure shopping and destination satisfaction. The first is the predominant geographical focus of the academic gaze. By far, the greatest volume of attention has been devoted to Asian destinations and Asian shoppers (Chang *et al.*, 2006; Lin & Lin, 2006; McDowell, 2010; Parasakul, 2020). This reflects the size of the markets, in particular China including Hong Kong and Macau (Ayeh & Chen, 2013; Choi *et al.*, 2016; Heung & Cheng, 2000; Lee & Choi, 2020; Lloyd *et al.*, 2011; Parasakul, 2020; Wong & Law, 2003; Yeung *et al.*, 2004), as well as the popularity of shopping as a leisure pastime. Second, much of the research has an inherently applied dimension, with the clearly intended and signposted outcomes of many a study being to inform (future) policy and practice beyond the academy in some way. For example, in the body of work on Hong Kong (Ayeh & Chen, 2013; Song *et al.*, 2011; Yeung *et al.*, 2004), a clear sub-text is the importance of satisfaction with shopping-related services, facilities and amenities as a means by which Hong Kong may retain and possibly grow its position in a congested global market. The implications of Ayen and Chen's (2013) work, as noted above, were consistent with earlier studies, not least those of Yeung *et al.* (2004) and Song *et al.* (2011). The former pointed to the need for improvement in five 'unsatisfactory attributes', including most crucially 'product reliability' if Hong Kong was not to lose ground against major competitors like Singapore (Yeung *et al.*, 2004: 95). The latter examined the performance of hotel, retail and tour operators as both discrete and integrated elements in destination experiences. Separately Chinese visitors to Hong Kong were most satisfied with the performance of the hotel sector followed by retail sector and least satisfied with tour operators (Song *et al.*, 2011: 82).

Findings that connect shopping and destination satisfaction in a positive relationship are clearly appealing and useful for end users in policy and practice, not least because they provide some basis of justifying past and future investment decisions. So, too, are the tools that generate the evidence, and a third feature of the literature is that data collection and data analysis largely concentrate on the experience and attributes of retailing and shopping in a 'physical' sense and in 'real world' retail environs. For example, Wong *et al.* (2013) consider the virtues of their scale and instrument as serving 'as a diagnostic tool to find out areas [sic] tourists are satisfied with and areas they are not happy with'. In their view, application of such an approach locally should furthermore offer potential users the attractive prospect of identifying 'places that are able to offer tourists excellent merchandise value, staff service quality, and differentiated service….' and of competing with major international shopping destinations like Honolulu, Hong Kong, Paris and Singapore (Wong *et al.*, 2013: 38).

In such an approach, the retail environment is aggregated and analysis speaks to satisfaction with the shopping experience at a particular destination *per se*, almost in a general, collective sense. Some limited attempts have been made to identify differences in satisfaction among different 'segments' or groups of shopping tourists. Usually though, segments are differentiated broadly on the basis of domestic and international (Egresi, 2017; Lloyd *et al.*, 2011; Yüksel, 2004) or by state citizenship (Song *et al.*, 2011) and in just a few cases those patronising particular retail spaces such as shopping malls (LeHew & Wesley, 2007). Conversely, segmentation of tourist shoppers based on their satisfaction with *types* (plural, emphasis intended) of store or space they frequent (i.e. a retail-based view) or the goods or items purchased and consumed appears to have been overlooked.

Notably, there appears to have been little research on the relative satisfaction with, or the contribution to overall destination satisfaction of, different types of retail space (e.g. small independent boutiques; chain stores; factory outlets; shopping malls; villages and centres; and street markets) or different locales within complex shopping destinations in a *comparative* sense. Cities such as London, Paris and New York have multiple centres of activity and the experience of tourism shoppers is not equal or even across all of them. For example, Berlin has two major focal points for shopping including Kufürstendamm in the west and Alexanderplatz in the east. Although these also functioned as the main shopping locales in the divided city, they were already distinctive growth poles before 1914 growing around their respective 'anchor stores', the major department stores: Kaufhaus des Westens and Hertie (Coles, 1999b). Zaidan's (2016, 2019) study of tourism shopping and new urban entertainment measures the perception of nine different shopping venues in Dubai, including: Dubai Mall, Mall of Emirates, Arab Historical Souks, Burjiman, Global Village, Jumeirah Plaza, hotel shops and Dubai Airport duty-free market. In one of the few studies of its type, McKercher (2020) examines the anatomy of successful shopping districts. Arguing for the importance of governance and creativity, his research emphasises the particular conditions in a locale that come together to result in a distinctive local offer, and because conditions vary across urban centres almost inevitably the nature of the experience (and hence satisfaction) will vary too. Drawing on the experience of Mouraria, a neighbourhood in Lisbon's city centre, Guimarães (2022: 221) argues that the way retail space incorporates community values, multiculturality and features of the townscape is diverse and intertwined with different perspectives on authenticity. This will, in turn, impact on satisfaction as will retail mix, specifically the countervailing forces of homogenisation (i.e. the gradual reproduction locally of store types and formats also elsewhere) and heterogeneity (i.e. to reflect local character and market dynamics) in engineering that (Guimarães, 2020). Focusing on Charleston (South Carolina, USA), Litvin and Rosene (2017: 821) point to the importance of achieving the right balance between chain merchants to what was 'once predominantly a local main street' in this heritage destination.

Unpacking Shopping and Satisfaction: Further Critical Reflections

What begins to emerge from the account so far is that there are distinctive ways in which the retail-shopping-leisure-tourism nexus have been studied and how destination satisfaction features within this. Existing reviews, such as that by Choi *et al.* (2016) point to the limited coverage of this theme both in absolute (viz. the number of studies and the number of researchers working in the space) and relative terms (viz. coverage of other subjects). Although there has been further, more recent progress since they compiled their appraisal, the application of a more critical lens suggests that there are still some significant gaps and major challenges ahead. For instance, from a basic scaled geographical perspective, aggregated views on the destination represent a macro-level view of satisfaction. Although the most frequent unit of analysis in existing studies, their pre-eminence masks that more differentiated meso-level (e.g. of particular shopping districts, centres or streets) and micro-level (e.g. of particular outlets, stores and markets) views are lacking by comparison. In the case of the latter, there appears to have little if any work to date on the relationship between the deliberative nature of service design (Hlee *et al.*, 2019) and the satisfaction of tourism shoppers with the spaces and experiences often carefully curated for them (Coles, 2004b, 2008; Pine & Gilmore, 1999). This is different from the assessment of the external environment surrounding, and providing context for, the shopping experience (cf. Yüksel, 2013), which may have developed 'organically', gradually over time and without the benefit of deliberate curation. Although this may appear to some as a fine point of distinction, it is important to praxis. More targeted practical research at the meso- and micro-levels is of greater relevance and potential assistance to managers in stores and retail organisations; that is, those with the greater prospect of introducing beneficial change to the retail environment and ultimately the destination including, but not restricted to, how their staff are trained for encounters with visitors (Lourenço *et al.*, 2022).

In more literal interpretations, the focus has been on the satisfaction of the consumer, the visitor, the shopper, the person experiencing the transaction first-hand and at the moment of exchange – in other words, a form of immediate satisfaction, there and then (cf. Wong *et al.*, 2013). As research on souvenirs and gift-giving demonstrates though, destination satisfaction may be deferred and even displaced. Satisfaction is deferred for the visitor who, comfortably back at home, looks fondly on their mementos of trips gone-by, some days, months or even years ago (Suhartanto, 2018; Vega-Vázquez *et al.*, 2017). Furthermore, if bought for others, destination satisfaction can be informed at some unspecified point in the future by the handing-over of gifts to their intended recipients.

The more substantive point is that, when distilled down further, further significant research questions remain for analysts around 'satisfaction for whom?' (i.e. retailer, shopper, family member, relative), 'satisfaction when?' (i.e. at the point of consumption, on return, later), and 'satisfaction where?' (i.e. where when away). Destination satisfaction is imbued with perceptions of items and, as Pine and Gilmore (1999) first argued persuasively, experiences and the ways in which they are performed – it is not just in the perceptions and recollections of the attributes and characteristics of retail spaces nor the economic valorisations of the transaction but also the memorable experiences (Sthapit *et al.*, 2018). Recollection may include sensory memories of the stall, the boutique or the store, its location, the ambience, the *patois*, sales pitch or scripting used, the temperature, the sounds and noises, the smells and aromas, the hustle-and-bustle and so on. Destination satisfaction is not only of the present but it may also be viewed retrospectively, overlain by nostalgia and complicated by sentiment and/or by selective revisiting of personal and familial events and episodes.

Viewing the destination experience and satisfaction in this manner makes it all the more surprising that there has not been more qualitative research in this subject area. Instead, in existing studies interpretations of constructs and findings are mostly confirmed by quantitative techniques or the intuitions of the particular researchers based on their interpretations of the statistical data on behalf of the recipients (e.g. Lee & Choi, 2020; Song *et al.*, 2011; Wong *et al.*, 2013). Seldom are the voices of consumers (i.e. tourist shoppers) heard. Arguably then, previous research has been somewhat 'safe' and 'sanitised'. Adding to the indictment, the de-emphasis of corporeal and sensory experiences questions the conceptual validity of existing survey work. Put another way, in asking whether visitors *feel* satisfied (emphasis intended) or not with their shopping experiences, such work is seeking to capture how an experience was sensed, and it raises the question of whether the not insignificant body of quantitative work adequately captures, even measures the phenomenon it set out to measure?

Admittedly, there are limits to surveying in terms of question numbers and what a survey can capture. No doubt for the quantifiers, there are counter-critiques of the usefulness in practice of the smaller sample sizes associated with qualitative modes of research and the perceived desirability among policymakers and practitioners of larger sample sizes as the bedrock of evidence-based decision-making. Be this as it may, the predominant orientation of research presents a limited view of the practice of shopping. In previous earlier work, I have argued that the relationship between tourism, shopping and retailing was considered almost axiomatic by many analysts (Coles, 2004a), and in the case of the latter two, this conflation of transaction and social practice represents a persistent issue. For a consumer, a successful or enjoyable trip does not necessarily have to result in the completion of a retail transaction (i.e. if the retailer offer does not convince the customer to make a purchase). Not just about the acquisition of goods, the practice of 'going shopping' may involve experiencing the range of attractions, services and facilities a destination has to offer visitors, including eating, drinking, taking in a show, visiting attractions and so on. Shopping may also involve 'haggling' over price where 'fun' and hence satisfaction derives not necessarily just from the buyer feeling as though they have got a good deal, but also from how the political act of negotiating a purchase is played out (Wu *et al.*, 2014).

The Future of Leisure Shopping and Destination Satisfaction

The previous section points to a series of lacunae within the existing research effort before the declaration of the worldwide Coronavirus pandemic in 2020. Without wishing to overly-summarise considerable public discourse on business, one of the main considerations at the time was the extent to which the pandemic would disrupt established economic systems and provide a basis for reset, a 'new normal' for the conduct of commerce (WEF, 2020). With this in mind, the obvious questions connected to the theme of this chapter were 'what is the future of shopping?', 'what does the future hold for shopping tourism?' and 'how will future changes in the nature of shopping impact on destination experience and satisfaction?'.

Although pertinent, there have been very few direct studies of the effect of the recent pandemic on shopping tourism. Notwithstanding, the pandemic has perhaps offered some tantalising clues to, and hints of, what a future may look like when physical access to shops and shopping locations is restricted, it is replaced by digital consumption, and shopping destinations are rendered temporarily redundant (García-Milon *et al.*, 2021). Locked-down at home for long periods of time, many consumers reverted to online channels for all but essential foodstuffs and medicines, but often for them too. In the UK as in many other economies, the future of shopping as well as that of the 'High Street' (or 'Main Street' to North American readers – Talen & Jeong, 2019) as the main locus for shopping (tourism), attracted much discussion. In particular the discussion focused on whether there would be a return to pre-pandemic levels of retail provision and, by extension, satisfaction with, shopping spaces and experiences.

Within the UK, some commentators viewed, and continue to take the position, that this is unlikely, arguing that new (hybrid) working patterns have changed the way city centres are perceived and used such that the future will be qualitatively different and there is little likelihood of being able to revert to the past (LGA, 2022). Resilience is to be found in greater flexibility to consumer trends, a 'greener', more environmentally-responsible approach to space, and anchoring in local offers and (community) contexts, not least relating to their predominant social and experiential uses as retail 'needs are increasingly met online' (LGA, 2022: n.p.). Others argue that, while 'large and successful centres have borne the brunt of the pandemic', the situation was not as bad as once feared and 'Covid-19 has far from killed high streets and city centres' (CfC. 2022: n.p.). The effects have been differential and, in relative terms, among the hardest hit 'low levels of footfall and consumer spending power were unable to sustain high street amenities' such that 'many of these places should shift away from being places of consumption towards production, and from over-reliance on retail space' (CfC, 2022: n.p.).

In parallel, other perspectives have taken a longer view of the future of shopping mainly involving digital innovation and its potentially transformative effects. Published more than a decade ago, several observations recorded by Rigby (2011) continue to resonate. Among the changes he foresaw (but which have yet to be played out fully), the future of shopping would involve overcoming 'an industry [i.e. retail] stuck in analog' and new approaches would be required to 'store economics, measurement systems, and incentives'. As conventional retailers had not 'had great experiences with breakthrough innovation', there was a case to 'redesign shopping from scratch' and to focus on the changing experience of shopping in an omni-channel environment where there is a degree of complementarity such that 'an omnichannel experience with stores… is superior to a purely digital retail strategy' (Rigby, 2011: 6, 12; see also García-Milon *et al.*, 2020). In other words, there is opportunity through the possibility of designing for, and more closely integrating, the digital and the physical worlds (Buhalis *et al.*, 2023), which would have obvious implications for retail satisfaction.

While some of these suggestions may appear somewhat tame and uncontroversial in retrospect, other more recent reviews suggest that their resonance endures. For instance, Adhi *et al.* (2021: n.p.) argue that 'the most successful retailers will be those that connect with consumers in new ways by leaning in on their digital, omnichannel, and in-store technology ambitions'. Broadly echoing this sentiment, Jain and Trivedi (2022: n.p.) argue for the importance of reimaging shopping, with a more complex ecosystem of 'players' involved in configuring experiences, including influencers, 'brand managers, gig workers, small business owners…and customer experience leads'. Beyond just a more data-driven approach for instance that involves the predictive advantages of artificial intelligence, the metaverse offers consumers increasing opportunities to order online and receive goods and services in real life. As Morgan (2022: n.p.) puts it, 'physical stores will still exist as showrooms to provide immersive experiences' and future shopping experiences will be personalised, benefitting from convenience, technology-driven, seamless and balanced between digital and physical world.

These views are not included because they are the outcome of extensive or structured searches. Yet, on the one hand, they cover many (but not all) the points made in several other 'think pieces' that readers can reveal by basic online queries using typical search engines. On the other hand and tactically more important, they were offered via outlets with highly credible reputations in the world of commerce for forecasting future trends. For instance, Rigby's views appeared in the *Harvard Business Review*, while Morgan's ideas were published in a *Forbes* blogspace. Adhi *et al*. (2022) argued for a 'New Normal' on the McKinsey website, while Jain and Trivedi's (2022) views were offered in the context of services that JP Morgan is able to offer prospective clients in this space. Put another way, these are ideas put into the public domain by organisations that consider their content to be reasonable possibilities for the future of shopping.

From a closer reading of these (and other similar) texts, two key absences are notable, in addition to the attractive ideas and opinions that are present. The first is that there is no obvious connection to, or teasing out of, the implications for shopping tourism or tourists and their destination satisfaction. This is perhaps not unexpected given the situation of authors and their backgrounds, as well as the positionings of the pieces for their intended audiences. Nevertheless, each has implications for tourism shopping. As far back as Rigby's (2011) study, he noted there was doubt among North American customers as to why they should buy in stores when buying online increasingly included delivery costs. Put another way, there was far less hassle associated with having goods delivered to home. In proposing the idea of integration of the digital and the physical world, there is potential to encourage far more enhanced tourism shopping experiences with commensurately higher levels of destination satisfaction. For instance, items shopped for in-destination when away, in 'real' or 'physical' space, may have commercial transactions and accompanying logistics and delivery arranged in 'virtual' or 'digital' space, and collection completed when at home, again in real space. Alternatively, browsing for, and selection of, items may take place in real space, when at home, followed ordering taking place online and through digital supply chains, and delivery taking place when away and in real space, as part of a special trip where experiencing the brand or commodity becomes part of the destination experience, too. Rather than challenge the viability and vitality of destinations, in this case future digital innovation can augment them (Morgan, 2022). As several of the 'think pieces' noted, the future of shopping relates to (bespoke) experience and, from a perspective of service design, there is the potential to carefully curate highly tailored products integrating offers from the destination nexus (e.g. accommodation, hospitality) with particular products or brands to enhance experiences of both in mutually-reinforcing ways. In one distinctive case, as early as the 2000s, German car brands like VW, Porsche and Mercedes were offering dedicated factory collection, often in the form of personalised packages and itineraries, to take advantage of factory facilities (e.g. test tracks) and/or Germany's Autobahn system where the vehicles could, in theory, be experienced to the full, especially on those stretches with no speed limits (Coles, 2004b, 2008).

Tourism scholars represent a second key absence from thought leadership in this space. This is somewhat ironic because, as noted above, many scholars of tourism shopping often seek to influence future policy and practice as outcomes of their work. Without greater engagement with the future of shopping, work on this topic may be depicted as a form of (very) contemporary history. In other words, it engages with extant forms of retailing and shopping considering changes that have (recently) taken place and waves of innovation that have broken. Here the use of the past tense is instructive; there is little if any evidence that tourism scholarship is engaging with possible emergent retail trends and shopping futures that have the potential to (or *will*) reconfigure the nature and value of shopping tourism in destinations.

In turn, this lack of engagement has the potential to result in maladapted policy and praxis. To date, the implication appears to have been that by understanding past trends and patterns in the development of retailing and shopping, there is an obvious opportunity to inform and even shape near-term future destination satisfaction by projecting forward. Implicit in such a view is that innovation is gradual and progressive, and that, in effect, the past is the key to the present and the future. Yet, innovation can often be highly disruptive leading to 'step-change' differences in production and consumption, and the future more radically departs from the past (Coles, 2022a). Muro-Rodriguez *et al.* (2020) explored shopping tourism and its use as a strategy option for the local development of cities in Spain. As they correctly note, 'with proper planning, tourism can be a part of the solution to cities' economic problems by helping to avoid closing businesses in cities that accelerate city depopulation problems and cause economic losses'. Yet, as examples of potential solutions, they invoke 'town centre management', 'business improvement districts' and 'tourist shopping villages' as possible inclusive, multi-stakeholder collaborative initiatives towards 'promoting quality and sustainable shopping tourism' (Muro-Rodriguez *et al.*, 2020: n.p.). Town centre management (Coca-Stefaniak *et al.*, 2009) and tourist shopping villages (Getz, 1993; Murphy *et al.*, 2011) are ideas that were first conspicuous in the 1990s while the notion of business improvement districts (BIDs) has a longer genealogy dating back to the 1970s (Kudla, 2022). In the UK, they were originally considered in the 1980s as one potential solution to combatting coastal resort malaise (Shaw & Williams, 1997). There is no doubt of the continued relevance of solutions rooted in the analogue era (Kudla, 2022; Thornley, 2020); current UK government policy continues to employ BIDs as ringfenced locations where additional local taxation is reinvested in projects to uplift resident businesses and commerce (HMG, 2014). However, the more salient questions are 'are solutions rooted in the analogue era still germane to the new digital age?' and 'how is such resource invested?'. In the case of the latter, in analogue technologies? As recent work on smart tourism makes clear (Buhalis *et al.*, 2023; Coles, 2022b), practical destination development and management solutions that eschew or de-emphasize, however inadvertently, the ever-greater pervasiveness of, and the distinctive practicalities associated with, digital are likely to be of very limited appeal to, or success among, contemporary visitors.

Conclusion

In a contribution two decades ago, I posed the question of whether tourism, shopping and retailing existed in an axiomatic relationship (Coles, 2004a). In other words, because of their apparent and perceived connections, it was a self-evident truth, an axiom, for many that they were connected, conceptually, methodologically and practically. As argued there much earlier and here, albeit far more briefly, there are important conceptual differences among them, and that those interested in the connections and possibilities should think through more carefully the analytical implications of conflating the concepts. That advice remains germane today.

Two decades on and the world has become even more complex. The ways in which many consumers shop have changed and continue to develop as both the retailing and distribution channels offered to them have been innovated and evolve. Digital technology pervades economy, society and culture in ever-increasing ways. Yet the relationship has been (and still continues to be) treated as largely axiomatic and, arguably far more dangerously, in a somewhat static sense. The view of retailing is still rooted for the most part in modes and spaces of distribution developed in the 20th century. Applying a critical lens suggests that, despite the appearance of progress, there remain significant gaps in our understanding. Shopping and shopping tourism are predominantly portrayed as pursuits conducted in-situ, in-destination, and through physical interactions between retailer and consumer. Taking a temporal perspective, many studies take a backwards-facing view to consider the implications for the future. Moreover, they do so at a moment in time when the nature and pace of retail innovation is faster than ever before and the expectations for experiential shopping and consumption are at their highest. Instead of dwelling on their analogue pasts, many established retailers are focusing on their digital futures. Inadvertently or otherwise, the predominance of retrospective research in academic analysis and discourse on tourism shopping and destination satisfaction preserves a particular view of the past and, extending the logic further, promotes futures as being necessarily rooted in, or disproportionately dependent on, the past.

Older, more established forms of tourist shopping will persist and continue to be important in many destinations for some time to come. Yet, the pandemic has pointed to the risks of being complacent and underestimating the transformative effects of digital technology. Thus, the perpetuation of traditional views of tourist shopping (and its links to destination satisfaction) does a disservice to, and rather defeats the otherwise professed object of, many a study on the topic towards informing future praxis involving policymakers, planners and destination managers as the intended audiences and users of the research. Empirically we know very little about how trends in shopping tourism are unfolding directly in response to recent radical innovations and major changes in market conditions and operation. This includes the extent to which shopping will continue to feature as a prime motive for, or feature in, contemporary tourism episodes in response to the likely changes to the future nature of shopping, broadly writ. In quite instrumental terms, as the nature of both distribution and shopping continues to evolve, we will know less about the value of shopping tourism (in older and/or newer, revised forms) to destinations unless the research effort is refocused.

Thankfully we are forewarned about this 'known–unknown' and the recalibration should extend to the future nature of the relationship between destination satisfaction and shopping. Within current academic discourse on this relationship, it seems that this too is treated as axiomatic and the research effort will continue much as it is now: shopping will be positively related to destination satisfaction, even to trust and loyalty. However, even from past research and in view of the methodological challenges in irrefutably demonstrating this relationship, it is difficult to be certain about whether it really is the self-evident truth as many contributions portray it. As retailing and shopping continue to evolve, the connections between tourist shopping and destination satisfaction are likely to become more complicated by digital technology, not simplified by it, and the sooner it is embraced, the sooner this research will re-establish its relevance.

References

Adhi, P., Hazan, E., Kohli, S. and Robinson, K. (2021) Omnichannel shopping in 2021. Four McKinsey leaders envision the future. Online: https://www.mckinsey.com/capabilities/growth-marketing-and-sales/our-insights/omnichannel-shopping-in-2030#/. Last accessed: 20/01/23.

Ayeh, J.K. and Chen, R.X. (2013) 'How's the service?' A study of service quality perceptions across sectors and source markets. *International Journal of Tourism Research* 15, 241–260.

Baek, E. and Park, S. (2024) Planned or unplanned? Understanding the shopping behaviours and experiences of tourists. *Journal of Vacation Marketing* 30 (3), 392–407.

Buhalis, D., O'Connor, P. and Leung, R. (2023) Smart hospitality: From smart cities and smart tourism towards agile business ecosystems in networked destinations. *International Journal of Contemporary Hospitality Management* 35 (1), 369–393.

Centre for Cities (CfC) (2022) How have two years of Covid-19 shaped the high street? Online: https://www.centreforcities.org/blog/how-have-two-years-of-covid-19-shaped-the-high-street/. Last accessed: 30/01/23.

Chang, J., Yang, B.-T. and Yu, C.-G. (2006) The moderating effect of salespersons' selling behaviour on shopping motivation and satisfaction: Taiwan tourists in China. *Tourism Management* 27, 934–42.

Chen, Y., Zhang, H. and Qiu, L. (2013) A review on tourist satisfaction of tourism destinations. In Z. Zhang, R. Zhang and J. Zhang (eds) *LISS 2012*. Springer.
Choi, M.J., Heo, C.Y. and Law, R. (2016) Progress in shopping tourism. *Journal of Travel and Tourism Marketing* 33 (1), 1–24. https://doi.org/10.1080/10548408.2014.969393
Coca-Stefaniak, J.A., Parker, C., Quin, S., Rinalduk, R. and Byrom, J. (2009) Town centre management models: A European perspective. *Cities* 26, 74–80.
Coles, T.E. (1999a) Department stores as innovations in retail marketing in Germany: Some observations on marketing practice and perception in the Wilhelmine Period. *Journal of Macromarketing* 19 (1), 34–47.
Coles, T.E. (1999b) Department stores as retail innovations in Germany: An historical-geographical perspective on the period 1870 to 1914. In G. Crossick and S. Jaumain (eds) *Cathedrals of Consumption: European Department Stores, 1850–1939* (pp. 72–96). Ashgate.
Coles, T.E. (2004a) Tourism, retailing and shopping: An axiomatic relationship? In A.A. Lew, A.M. Williams and C.M. Hall (eds) *A Companion to Tourism* (pp. 360–373). Blackwell.
Coles, T.E. (2004b) Tourism and retail transactions: Lessons from the Porsche experience. *Journal of Vacation Marketing* 10 (4), 378-389.
Coles, T.E. (2008) Tourism, commodity culture and international car manufacturers. In T.E. Coles and C.M. Hall (eds) *International Business and Tourism: Global Issues, Contemporary Interactions* (pp. 238-255). Routledge.
Coles, T.E. (2022a) Business models. In D. Buhalis (ed.) *Encyclopaedia of Tourism Management and Marketing*. Edward Elgar.
Coles, T.E. (2022b) Hidden in plain sight? AR apps and the sustainable management of urban heritage tourism. In F. Nevola and D. Rosenthal (eds) *Hidden Cities: Urban Space, Geolocated Apps and Public History in Early Modern Europe* (pp. 203–224). Routledge.
Crossick, G. and Haupt, M.-G. (eds) (1984) *Shopkeepers and Master Artisans in Nineteenth-century Europe*. Methuen.
Crossick, G. and Jaumain, S. (eds) (1999) *Cathedrals of Consumption. The European Department Stores, 1850–1939*. Ashgate.
Egresi, I. (2017) Tourists' satisfaction with shopping experience based on reviews on TripAdvisor. *Tourism* 65 (3), 330–345.
Fuchs, M. and Weiermaier, K. (2003) New perspectives of satisfaction research in tourism destinations. *Tourism Review* 58 (3), 6–14.
García-Milon, A., Juaneda-Ayensa, E., Olarte-Pascual, C. and Pelegrín-Borondo, J. (2020) Tourist shopping and omnichanneling. In S.J. Teixeira and J.M. Ferreira (eds) *Multilevel Approach to Competitiveness in the Global Tourism Industry* (pp. 87–97). IGI Global.
García-Milon, A., Olarte-Pascual, C. and Juaneda-Ayensa, E. (2021) Assessing the moderating effect of COVID-19 on intention to use smartphones on the tourist shopping journey. *Tourism Management* 87, 104361.
Getz, D. (1993) Tourist shopping villages: Development and planning strategies. *Tourism Management* 14, 15–26. https://doi.org/10.1016/0261-5177 (93)90078-Y
Guimarães, P.P.C. (2020) Retail change in a context of an overtourism city: The case of Lisbon. *International Journal of Tourism Cities* 7 (2), 546–564.
Guimarães, P.P.C. (2022) Unfolding authenticity within retail gentrification in Mouraria, Lisbon. *Journal of Tourism and Cultural Change* 20 (1-2), 221–240. https://doi.org/10.1080/14766825.2021.1876079
Hall, C.M. (2005) *Tourism: Rethinking the Social Science of Mobility*. Harlow: Pearson.
Henseler, M., Maisonnave, H. and Maskaeva, A. (2022) Economic impacts of COVID-19 on the tourism sector in Tanzania. *Annals of Tourism Research Empirical Insights* 3, 100042. https://doi.org/10.1016/j.annale.2022.100042.

Her Majesty's Government (HMG) (2014) Business Improvement Districts. Information and Guidance on Business Improvement Districts. Online: https://www.gov.uk/guidance/business-improvement-districts#what-is-a-business-improvement-district. Last accessed: 27/01/23.

Heung, V.C.S. and Cheng, E. (2000) Assessing tourists' satisfaction with shopping in the Hong Kong Special Administrative Region of China. *Journal of Travel Research* 38 (4), 396–404.

Hlee, S., Yhee, Y., Chung, N. and Koo, C. (2019) Service innovation by design thinking methods: A case of Seoul children's grand park (SCGP). *e-Review of Tourism Research* 17 (2), 271–291.

Hurst, J.L. and Niehm, L.S. (2012) Tourism shopping in rural markets: A case study in rural Iowa. *International Journal of Culture, Tourism and Hospitality Research* 6, 194–208.

Jain, M. and Trivedi, H. (2022) Reimaging shopping is more important than ever. Online: https://www.jpmorgan.com/solutions/treasury-payments/insights/future-of-shopping-reimagined. Last accessed: 20/01/23.

Jin, H., Moscardo, G. and Murphy, L. (2017) Making sense of tourist shopping research: A critical review. *Tourism Management* 62, 120–134.

Kozak, M. (2009) Measuring tourist satisfaction with multiple destination attributes. *Tourism Review* 14 (3), 229–240.

Kudla, D. (2022) Fifty years of Business Improvement Districts: A reappraisal of the dominant perspectives and debates. *Urban Studies* 59 (14), 2837–2856.

Lee, J.-S. and Choi, M. (2020) Examining the asymmetric effect of multi-shopping attributes on overall shopping destination satisfaction. *Journal of Travel Research* 59 (2), 295–314.

LeHew, M.L.A. and Wesley, S.C. (2007) Tourist shoppers' satisfaction with regional shopping mall experiences. *International Journal of Culture, Tourism and Hospitality Research* 1 (1), 82–96.

Lin, Y.H. and Lin, K.Q. (2006) Assessing Mainland Chinese visitors' satisfaction with shopping in Taiwan. *Asia Pacific Journal of Tourism Research* 11 (3), 247–68.

Litvin, S.W. and Rosene, J.T. (2017) Revisiting main street: Balancing chain and local retail in a historic city's downtown. *Journal of Travel Research* 56 (6), 821–831.

Lloyd, A.E., Yip, L.S.C. and Luk, S.T.K. (2011) An examination of the differences in retail service evaluation between domestic and tourist shoppers in Hong Kong. *Tourism Management* 32, 520–533.

Local Government Association (LGA) (2022) Creating resilient and revitalised high streets in the 'new normal'. Online: https://www.local.gov.uk/publications/creating-resilient-and-revitalised-high-streets-new-normal#pestle. Last accessed: 30/01/23.

Lourenço, F., Li, Z., Ren, L. and Cheng, R. (2022) What retail experts say about tourism retail education? A case of Macao using an integrated Bloom-Kolb Learning Design Canvas. *Journal of Quality Assurance in Hospitality and Tourism* 23 (1), 275–297.

McDowall, S. (2010) International tourist satisfaction and destination loyalty: Bangkok, Thailand. *Asia Pacific Journal of Tourism Research* 15 (1), 21–42.

McKercher, B. (2020) Anatomy of successful shopping districts. *International Journal of Tourism Cities* 6 (4), 831–846.

Morgan, B. (2022) Inside the shopping experience of the future. Online: https://www.forbes.com/sites/blakemorgan/2022/12/15/inside-the-shopping-experience-of-the-future/?sh=56a44e006221. Last accessed: 20/01/23.

Moscardo, G. (2004) Shopping as a destination attraction: An empirical examination of the role of shopping in tourists' destination choice and experience. *Journal of Vacation Marketing* 10 (4), 294–307.

Muro-Rodríguez, A.I., Pérez-Jiménez, I.R. and Sánchez-Araque, J.A. (2020) Impact of shopping tourism for the retail trade as a strategy for the local development of cities. *Frontiers in Psychology* 11, 67.

Murphy L., Moscardo G., Benckendorff, P. and Pearce, P. (2011) Evaluating tourist satisfaction with the retail experience in a typical tourist shopping village. *Journal of Retailing and Consumer Services* 18 (4), 302–310.

Parasakul, L. (2020) Assessing Chinese tourists' satisfaction with their shopping experiences in Bangkok metropolis. *Journal of China Tourism Research* 16 (2), 279–296.

Pine, B. and Gilmore, J.H. (1999) *The Experience Economy*. Harvard Business School Press.

Pizam, A., Neumann, Y. and Reichel, A. (1978) Dimensions of tourist satisfaction with a destination area. *Annals of Tourism Research* 5 (3), 314–322.

Rabbiosi, C. (2011) The invention of shopping tourism: The discursive repositioning of landscape in an Italian retail-led case. *Journal of Tourism and Cultural Change* 9 (2), 70–86.

Rigby, D. (2011) The future of shopping. *Harvard Business Review*, December 2011. Online: https://hbr.org/2011/12/the-future-of-shopping. Last accessed: 20/01/2023.

Shaw, G. and Benson, J. (eds) (1992) *The Evolution of Retail Systems, c1800–1914*. Leicester University Press.

Shaw, G. and Williams, A.M. (eds) (1997) *The Rise and Fall of British Coastal Resorts: Cultural and Economic Perspectives*. Pinter.

Sirakaya-Turk E., Ekinci, Y. and Martin, D. (2015) The efficacy of shopping value in predicting destination loyalty. *Journal of Business Research* 68 (9), 1878–1885.

Song, H., Li, G., van der Veen, R. and Chen, J.L. (2011) Assessing Mainland Chinese tourists' satisfaction with Hong Kong using tourist satisfaction index. *International Journal of Tourism Research* 13, 82–96.

Sthapit, E., Coudounaris, D.N. and Björk, P. (2018) The memorable souvenir-shopping experience: Antecedents and outcomes. *Leisure Studies* 37 (5), 628–643.

Stobart, J. (2010) Explorations and insights. A history of shopping: The missing link between retail and consumer revolutions. *Journal of Historical Research in Marketing* 2 (3), 342–349.

Suhartanto, D. (2018) Tourist satisfaction with souvenir shopping: Evidence from Indonesian domestic tourists. *Current Issues in Tourism* 21 (6), 663–679.

Talen, E. and Jeong, H. (2019) What is the value of 'main street'? Framing and testing the arguments. *Cities* 92, 208–218.

Thornley, A. (2020) *Urban Planning under Thatcherism: The Challenge of the Market*. Routledge. (Republished, originally 1991).

Timothy, D.J. (2005) *Shopping Tourism, Retailing and Leisure*. Channel View Publications.

Timothy, D.J. (2014) Trends in tourism, shopping, and retailing. In A.A. Lew, C.M. Hall and A.M. Williams (eds) *The Wiley Blackwell Companion to Tourism* (pp. 378–388). Wiley.

Vega-Vázquez, M., Castellanos-Verdugo, M. and Oviedo García, M.Á. (2017) Shopping value, tourist satisfaction and positive word of mouth: The mediating role of souvenir shopping satisfaction. *Current Issues in Tourism* 20 (13), 1413–1430.

Warnaby, G. and Medway, D. (2018) Marketplace icons: Shopping malls. *Consumption Markets and Culture* 21 (3), 275–282.

Wong, I.A. and Wan, Y.P.K. (2013) A systematic approach to scale development in tourist shopping satisfaction: Linking destination attributes and shopping experience. *Journal of Travel Research* 52 (1), 29–41.

Wong, J. and Law, R. (2003) Difference in shopping satisfaction levels: A study of tourists in Hong Kong. *Tourism Management* 24, 401–410.

World Economic Forum (WEF) (2020) The Great Reset. Online document. Available from: https://www.weforum.org/great-reset/. Last accessed: 25/01/23.

World Tourism Organization (UNWTO) (2014) *Global Report on Shopping Tourism*. Madrid: UNWTO. Online: https://www.e-unwto.org/doi/book/10.18111/97892 84416172. Last accessed: 25/01/23

Wu, M.-Y., Wall, G. and Pearce, P.L. (2014) Shopping experiences: International tourists in Beijing's Silk Market. *Tourism Management* 41, 96–106. http://dx.doi.org/10.1016/j.tourman.2013.09.010

Yeung, S., Wong, J. and Ko, E. (2004) Preferred shopping destination: Hong Kong versus Singapore. *International Journal of Tourism Research* 6, 85–96. https://doi.org/10.1002/jtr.474

Yüksel, A. (2004) Shopping experience evaluation: A case of domestic and international visitors. *Tourism Management* 25, 751–759. https://doi.org/10.1016/j.tourman.2003.09.012

Yüksel, F. (2013) The streetscape: Effects on shopping tourists' product/service quality inferences and their approach behaviours. *Journal of Quality Assurance in Hospitality and Tourism* 14, 101–122. https://doi.org/10.1080/1528008X.2013.749387

Zaidan, E. (2016) Tourism shopping and new urban entertainment: A case study of Dubai. *Journal of Vacation Marketing* 22 (1), 29–41. https://doi.org/10.1177/1356766715589426

Zaidan, E. (2019) Shopping, tourism and hyper-development in the Middle East and North Africa. In D.J. Timothy (ed.) *Routledge Handbook on Tourism in the Middle East and North Africa* (pp. 365–377). Routledge.

12 The Ancillary Role of Shopping in Other Types of Tourism

Jennifer Frost and Warwick Frost

Introduction

On a river cruise down the Nile in 1990, an unexpected shopping opportunity presented itself to one of the authors when local merchants in small boats surrounded the vessel with goods to be purchased, such as a *galabeya*, a loose-fitting garment traditionally worn by men and women, and small statues of ancient Egyptian gods. When guests on the boat asked about an item, it was pulled up to the deck by ropes, and the haggling over price began. If the price was agreed, the money was sent back via rope to the merchant in question; otherwise, the item was returned in the same manner and the process began again. The items themselves were a fun souvenir to take home but it was the interaction with locals, with lots of laughter, and the unique way of buying gifts that formed the real tourist experience. It was an example of shopping that was *ancillary* or secondary (Timothy, 2014) to a specific form of tourism, in this case cruise tourism.

There are other examples of this phenomenon that we discuss in this chapter, such as shopping that is ancillary to wine tourism, food tourism, cultural heritage tourism, spa tourism and sports tourism. Overall, this is an under-researched area in the tourism literature, which focuses primarily on shopping as a main attraction. We have reviewed current literature in the field and suggest areas that may represent gaps in knowledge that would benefit from future academic study.

Winery Cellar Doors

Operating in a crowded international marketplace with complex and expensive distribution chains, winemakers generally aim to maximise their revenue by selling directly to visitors through their cellar doors (Mitchell & Hall, 2001). Accordingly, these cellar doors are often highly visible, with a great deal of strategic planning to establish them as distinctive identifiers of the winemaker's brand and a unique marketing proposition. This may be manifested in fantasy architecture – ranging from the historical to the modern – set in enticing gardens and a rural setting that is carefully designed to draw in potential visitors (Danielmeier, 2014; Pan *et al.*, 2008). Such tourism fantasies are epitomised in the faux French chateaux found in New World wine regions such as California and Australia, and even in Asia (Frost & Laing, 2014).

Wineries offer a wide range of tourism activities, including tastings, tours, festivals, concerts, restaurants, cafes, exhibitions and the staging of private functions such as weddings. (Note: We need to acknowledge that not all wineries offer all of these tourist activities and that there is a great deal of variation around the globe. For example, in parts of the USA, local laws prohibit restaurants and weddings being provided at wineries, whereas in Australia these are quite common.) Ancillary shopping is primarily the sale of wine after tastings, but also includes souvenirs, homewares, clothing, art and crafts and non-wine produce, such as preserves. Tourist shopping is focused on the cellar-door or tasting room, which is the main space for engagement with tourists. Tastings are curated, with cellar-door staff guiding visitors through a range of wine varieties and vintages, typically in the most efficacious order, and these provide a *front-stage* experience for tourists (Frost *et al.*, 2020). (Particularly as winery tours are becoming less common, often for health and safety reasons.) Such tastings may be free, or have a fixed charge and there is ongoing debate within the industry as to which approach is the best for wineries (McNamara & Cassidy, 2015). Following the tasting, the visitor has the option of whether or not to make a purchase (O'Mahony *et al.*, 2006). (Boutique breweries and distilleries follow the same model and their numbers have expanded in recent decades. Research, however, is limited on their operations.)

Several studies have focused on the role and effectiveness of winery cellar doors. These are predominantly set in Australia, suggesting that the cellar door is of paramount importance to the industry in that country. In a study using 'mystery' shoppers, it was found that:

> The main difference reported between large and small wineries was the clearly commercial focus of the larger ones. This was in part a result of the aesthetics and appearance of the bigger establishments ... There were often large number of visitors in the tasting rooms of large wineries also which, coupled with limited numbers of tasting room staff, meant there was very little personal attention or sense of welcome. (Charters *et al.*, 2009: 127)

This study also found conflicts between the different types of ancillary shopping, particularly at larger wineries, where 'the presence of a large quantity of merchandise, often in the form of foods, clothing, and other wine-related accessories, at times detracted from the winemaking message' (Charters *et al.*, 2009: 127). This study concluded that in some respects smaller wineries had a competitive advantage, as:

> Comments about the personal and 'real' nature of the interaction between the visitor and staff of smaller tasting rooms were frequently made, and there was a sense that at these smaller wineries, participants were made to feel special ... The 'connection' between the staff and the wine they were selling, and the passion many small winery operators demonstrated for their product also added to the sense of a genuine interaction. This experience was often described as 'authentic'. (Charters *et al.*, 2009: 128)

A study of tourists at a range of South Australian wineries emphasised the importance of skilled cellar door staff, finding that:

> Regardless of the quality and value of their wine products, the behavior and knowledge exhibited by cellar door staff during a service interaction may be an independent source of visitor satisfaction. This finding suggested that improving cellar door staff interpersonal skills can improve customers' overall experience at that cellar door ... Given this evidence, cellar door managers should recognize that, to establish a competitive advantage, resources should be dedicated not only to wine products, but also to those elements that comprise a service encounter. (Chen *et al.*, 2016: 88)

In another study of winemakers, Frost *et al.* (2020) emphasise how storytelling and personal encounters were important at the cellar door, arguing that:

> Tourist interactions were often with winemakers, family members and experienced staff, who provided an encounter that was unscripted and personalised. This led to a strong sense that tourists were gaining real insights into the personal stories and lived experiences of those who worked at the wineries. (2020: 7)

Given the variations that occur across the globe, future research should focus on comparisons between practices in different countries in terms of retail sales at wineries. There is also a need to examine new developments. A major one is the potential 'social contract' that exists between visitors and wineries, being the idea that a free tasting requires a reciprocal purchase. With the growth of payments for tasting, is such a social contract now redundant, resulting in less sales? Furthermore, is the trend towards charging for tasting leading towards wineries become de facto wine bars and changing the socioeconomic mix of visitors? Another new development is the push by some wineries to restrict tastings to appointments, which may affect the spontaneity of touring around a wine region. Future research could also consider the effect of shopping opportunities that do not involve wine on the cellar door experience. Does this affect the perception of wine as a premium product and consequently the branding of the winery?

Food Tourism

Food tourism, also referred to as gastronomic or culinary tourism, is a broad term that encompasses the wide spectrum of tourist activities or experiences connected to food at a destination, including visits to farms, food markets, farmers markets, food festivals and food-related businesses; taking part in food-related tours, food trails and cooking demonstrations or classes; or local dining. Although all tourists must eat while away, whether self-catered or bought at a destination, and not all tourists regard food as the 'major aspect of the holistic holiday experience' (Björk & Kauppinen-Räisänen, 2016: 1263), we are focusing in this chapter on individuals who are actively seeking a food-related experience while on holiday, and the role that shopping plays in these experiences.

Food purchased during travel can of course be consumed in situ, but much of it will be taken home as a souvenir (Stone, in press) (see Chapter 14). Björk and Kauppinen-Räisänen (2016) found that 40.5% of their respondents (Finns attending a travel fair) reported that they had brought home local food items as souvenirs, yet surprisingly, 'the issue [of souvenirs] has received scarce attention in terms of food' (2016: 1268). Stone *et al.* (2019) found an even higher percentage of the leisure travellers they studied intended to take food home for themselves to enjoy (67.4%) or as gifts (65.1%). Examples of food souvenirs might include foodstuffs or ingredients that can be used in cooking back home or given as presents to family and friends, such as spices, canned goods, oil or dry food, or food-related gifts such as cookbooks, aprons, utensils, dishes or drinking vessels. Sims (2009) interviewed tourists and producers of food in the English Lakes District and noted that some visitors planned their food souvenirs before they left home, with one interviewee commenting: 'I've heard fudge is quite a good thing to take back, and I'm hoping to get something in the Scrumpy cider line for my neighbour who is doing little jobs – you know, gardening' (2009: 328).

The research by Björk and Kauppinen-Räisänen (2016), while not squarely about food tourists, suggests two broad motivations for buying food as souvenirs. First, there was a desire to either extend or *relive* the holiday experience, with the souvenir acting as an expression of memories of the places that have been visited. Second, some respondents wanted to 'share the food and food culture they had experienced on a destination' (2016: 1274) with others back home, thus building social connections and relationships, a phenomenon which has also been observed by Sims (2009) and Horng and Tsai (2010). Both types of behaviour are often encouraged by destination marketing organisations (DMOs), with Horng and Tsai (2010: 78) referring to the following example on the Korea Tourism Organization website: 'Cooking Korean dishes can bring back many pleasant culinary memories. This is a new and delightful way to surprise your family and friends, whether you are cooking for a formal or informal gathering'. This website goes on to suggest the purchase of Korean cookbooks in English, to facilitate this trip down memory lane. Thus, a further consequence of taking food home from a holiday is that it may lead to greater consumption of these kinds of foods thereafter, with an 'economic impact beyond on-site consumption' (Stone et al., 2019: 149). Another motivation for bringing food-related souvenirs home as a gift for friends and family might be that they are seen as a cultural obligation (Bernardo & Rodrigues, 2020).

Food souvenirs could also constitute a form of *external validation* of a tourist's sophistication and taste (Bernardo & Rodrigues, 2020; Björk & Kauppinen-Räisänen, 2016), an example of what Bourdieu (1984) refers to as *cultural capital*. This can be defined as 'embodied or objectified forms of cultural knowledge and competence [that] are socially esteemed in themselves and facilitate participation in other socially-esteemed activities' (Smith Maguire & Lim, 2015: 232). Goolaup et al. (2018) have studied extraordinary food travel experiences and created a typology of tourists based on their food cultural capital, building on the work of Bourdieu (1984); however, their participants do not mention any shopping connected to these experiences *in situ*. This is the case even though those participants identified as possessing high levels of food cultural capital are characterised by 'a large collection of cookery, [and] wine books, subscriptions to several food magazines, and some fancy cooking equipment' (Goolaup et al., 2018: 223). Their study did not consider whether these *cultivated food tourists* prefer to purchase their food-related consumables at home, rather than while on holiday. The link between food shopping and cultural capital is an area that is ripe for further research and could be extended to consider shopping based on ethical choices. For example, Sims (2009: 328) argues that buying local or regional food products to take home is associated with being a 'good' traveller and aligns with values 'such as being better for the environment, conserving "traditional" rural landscapes and supporting the local economy'.

There are issues of authenticity with respect to food souvenirs, which may heighten their prestige or status when brought back home. Some food may, for example, carry 'national or international certification or distinction' (Bernardo & Rodrigues, 2020: 132) that links it with a destination and makes it by definition unique. Bernardo and Rodrigues (2020) studied the purchase of sugared almonds by tourists in Moncorvo, a municipality in the Douro region of north Portugal. This product has been given a 'protected geographical indication' (PGI) by the European Union and its key ingredient, local almonds, has received 'protected designation of origin' (PDO) certification in connection with the Douro. Their study finds that these types of certification programmes, while aimed at safeguarding economic benefits to a region, may exclude individuals who have been traditionally involved in this food production as part of their heritage yet do not wish to stick to rigid rules as to how it must be created and sold. This may have the consequence of 'stripping [these individuals] of the legitimacy to continue to innovate and reinvent their own assets' (Bernardo & Rodrigues, 2020: 138). The authors of this study have suggested that further research be conducted to explore how tourists feel about the certification process and its importance (or not) to their decision to buy these sugared almonds.

A purchase of local food while on holiday has also been linked in the literature to existential authenticity, or the *search for meaning* (Wang, 1999). This is about finding one's authentic self through experiences that engage us. Sims (2009: 333) interviews a bakery owner in the Lake District, UK, who refers to their wares in these terms: 'you literally get it in a bag, walk out and eat it, and it can still actually sometimes be warm. And people aren't just buying that – they're buying the whole package'. The tourist feels a connection with the local producer through the consumption of their product and 'thus, tourists choosing to consume [buy] local products may not just be enjoying the physical taste of the food. Instead, they are also consuming the meaning behind it' (Sims, 2009: 333).

Another area that lacks research relates to the carriage of food or food products across borders by tourists. Countries such as Australia and New Zealand have strict quarantine laws administered by customs officials, both at the airport for individual travellers or for freight shipped home by tourists. A popular television show, *Border Security: Australia's Front Line* (2004–present) has run for 15 seasons and regularly depicts tourists being searched and charged with failing to declare food products, notably those that are illegal to bring into the country. It is surprising that this issue has not received more attention in the tourism literature given its ubiquity and the money spent annually on its detection and control. Extant research on illegal importation of food products that considers the efficacy of biosecurity measures at borders (e.g. Noordhuizen *et al.*, 2013; Soon, 2020) is currently in fields other than tourism such as veterinary studies, agriculture and biology.

There is also the problem of maintaining the integrity of food products during transport home by a tourist. The sugared almonds mentioned above are required to be sold in 'a small transparent plastic bag that allows the consumer to check its integrity, for no broken or incorrectly produced almonds can have the PGI label' (Bernardo & Rodrigues, 2020: 136). The tourist must pack the souvenir carefully to avoid it breaking during their journey, but if they take it out of its original packaging, it may lose its appeal when shown to or provided as a gift to others back home. This dilemma represents an opportunity for research.

Cruise Tourism

Cruise tourism is starting to grow in popularity again after a drop in demand (and supply) during the COVID-19 pandemic. Tourist shopping is so inextricably linked with cruising that it could be argued that it is hardly an ancillary experience at all. In this chapter, we have assumed that shopping is not the main reason for taking a cruise, based on research that suggests that escape/relaxation are the major motivations for cruise tourism (Hung & Petrick, 2011), although shopping is clearly of high interest to many passengers and some market segments in particular. A review of the Chinese cruise industry by Sun *et al.* (2014: 76) noted that 'To meet the desire of Chinese cruisers to shop, Costa [foreign-owned cruise company based in China] dedicated more than 50% of the shopping space by removing the sunbathing area on their Costa Allegra ... and offered more duty-free items onboard and shopping opportunities during excursions ashore'. Other studies (e.g. Brida *et al.*, 2013) have also highlighted the importance of shopping while cruising, although they found that there are different levels of interest and expenditures.

Research on shopping during cruise tourism tends to fall into five categories. The first (and most prevalent) focuses on the economic benefits of shopping for local economies when a cruise ship is in port. These studies quantify the economic impact and/or provide recommendations as to how this might be increased for destinations (e.g. Gouveia & Eusébio, 2019; Lee & Lee, 2017; Sciortino *et al.*, 2022). The economic rationale put forward in many of these studies is that 'The purchase of local products [by cruise passengers] is of utmost relevance to the economic development of a destination, given that these products use local resources in the production process which will lead to an increase in the multiplier effects of cruise tourism' (Gouveia & Eusébio, 2019: 431). The nature of the destination may affect the importance that shopping plays for cruise visitors and thus the impact for local economies (Andriotis & Agiomirgianakis, 2010; Teye & Paris, 2011; Timothy, 2005, 2014).

However, there is a growing body of work that has criticised the welcoming of cruise liners as contributing to overtourism or as promising much but delivering little to local businesses (Hritz & Cecil, 2008; Kumar *et al.*, 2022; Larsen & Wolff, 2016). This research does not support the premise that cruise tourism is a positive economic benefit for small businesses. For example, Kumar *et al.* (2022: 11) found in a Fijian context that 'air passengers display a more discerning and sustainable souvenir purchasing behaviour than cruise ship visitors', while Larsen and Wolff's (2016) study of cruise visitors to Bergen, Norway, refers to the fact that many cruise passengers are only at a destination for a very short period of time and return to their ships for meals, thus limiting shopping opportunities and the economic impact of this form of tourism. Concerns about the way these visits are being managed may also contribute to negative attitudes towards cruises by local residents and communities and ultimately to community discord (McCaughey *et al.*, 2018), which affects the sociocultural sustainability of cruise tourism. Further studies are needed to tease out these issues.

A second category of research considers the nature of the cruise shopping experience. For example, Buzova and colleagues (2019: 367) examine satisfaction with shore excursions and found that the mass cruise passengers they studied evinced concerns that adequate time be allocated for shopping, alongside the availability of a selection of different retail opportunities. In contrast, passengers undertaking a premium or luxury cruise focused more on the 'expressive components of the tour' such as the quality of the guiding. Another study by Li (2019) of on-shore excursions undertaken by Chinese cruise passengers found that the experience was spoilt by too much time being allocated to shopping compared to attractions; visits to shops that sold similar products and having to wait for 'tardy' passengers to return to the bus. These criticisms relate to the organisation of these excursions by tour companies (see discussion below on the supply side of shopping) and may be attributed in part to the receipt of commissions for taking visitors to certain retail outlets (Li, 2019). Cave *et al.* (2012) consider souvenirs bought during cruises as 'mementos of place' and found that perceived authenticity was an important part of the experience, as well as 'retail atmospherics'. Perceptions of an aggressive retail environment at a cruise destination may conversely have a negative impact on the shopping experience and traveller intentions to return for a future visit (Henthorne *et al.*, 2013).

The third area of academic focus relates to understanding the cruise tourist. Some studies have segmented these tourists in terms of their shopping habits (Brida *et al.*, 2013) or satisfaction with the onshore shopping experience (Sorrentino *et al.*, 2019), explored their purchase intentions (Militz *et al.*, 2021) and analysed their shopping behaviour. Ahn and Kwon (2022) consider shopping by cruise passengers based on impulsive buying tendency and found that this individual trait had a stronger influence on impulsive purchase behaviour than emotions. The researchers recommend ways for cruise companies to 'enhance' this behaviour in their customers or even to target them in their pre-cruise marketing, which seems to fly in the face of the sustainability concerns referred to previously, although they do acknowledge the potential for such behaviour to lead to addiction and threaten well-being. The influence of other passengers on shopping behaviour was noted by Li (2019: 224): 'Although some passengers did not originally intend to buy anything, observing other passengers purchasing items encouraged them to make purchases'. Li's (2019: 227) study also confirms the cultural pressure to buy gifts for those back home, in this case for Chinese people visiting Japan and Korea: 'Reciprocity is valued in Chinese society and one way to maintain it is by giving gifts'. Research has also considered the mobility of cruise tourists. Henthorne *et al.* (2013) found that cruise tourists to Jamaica who were travelling with others or who were repeat visitors spent more time shopping and bought more from local vendors than the solo or first-time traveller, which may be associated with perceptions of risk. Sciortino *et al.*

(2022) also found that greater cruise passenger mobility on shore was associated with higher passenger expenditure.

Fourthly, there are studies that examine the supply side of onshore shopping. We have already mentioned Li (2019) who notes concerns with the way the shopping element of onshore excursions is structured. Sun et al. (2021) analyse characteristics of these excursions and how local resources can be bundled up to create onshore products. Shopping is mentioned as an opportunity for cruise tourists but is not specifically considered in depth. More in-depth research would be useful as to how shopping can be accommodated within on-shore excursions to enhance the tourist experience, bearing in mind the sustainability of local businesses and the destination.

A fifth body of work aims to link the demand and supply sides of shopping during cruise tourism. Gutberlet (2019, 2024) has studied German-speaking cruise tourists and their perceptions of 'Otherness' while visiting a *souq* (market) in Oman and buying souvenirs. She also examines the guide as a cultural broker for these cruise passengers and the way that frankincense in particular is promoted heavily to passengers as an exotic and mystical purchase, linked to tourist imaginaries such as the Arabian Nights and Christian identities. The shop becomes a stage for the guide's performance that in turn mediates the tourist experience. Likewise, Kumar et al. (2022) explore perceptions of souvenirs from the angle of both cruise tourists and micro-entrepreneurs in Fiji, and note that it is not just the tourist who derives meaning from the purchase of souvenirs. While the micro-entrepreneurs acknowledge the importance of these products economically, 'producing and selling souvenirs was [also] perceived as a way to connect with and sustain vendors' cultural identity' (Kumar et al., 2022: 9).

Cultural Heritage Attractions: Museums, Art Galleries and Historic Buildings

The cultural heritage tourism sector features a range of visitor attractions including museums, art galleries and historic buildings. At the top end of the scale, these typically anchor large destinations such as major cities and employ professional staff covering management, curating, preservation, interpretation and visitor services. Examples include the Louvre and Versailles in Paris, the National Gallery, British Museum and Tower of London in London and the Metropolitan Museum of Art and the American Natural History Museum in New York. Products available in their shops include books, clothing and homewares and often feature reproductions of art and artefacts housed within the collections.

Research into shopping at cultural heritage attractions has been primarily confined to the marketing and museum studies fields, with surprisingly little consideration in tourism studies (see Chapter 10). Generally, selling souvenirs at historic sites helps meet financial objectives. Cultural heritage institutions only gain part of their revenue from government funding and private donations and so need to charge admission fees and earn profits from ancillary operations such as shopping (Timothy, 2021). Originally, gift shops usually started as small-scale fundraisers run by volunteer groups, but in recent decades they have become professionalised and run on a larger scale (Cameron, 2007; Komarac *et al.*, 2019). As one of a range of visitor services, it is argued by some researchers that a professional museum shop is now expected and valued by visitors as an integral part of the museum experience (Komarac *et al.*, 2019; McIntyre, 2010).

Products for sale are often based on both permanent and travelling exhibitions, with the latter at times leading to their own distinct 'pop-up' shop. Whether or not other products were appropriate was sometimes a matter of contention. A study of visitors to English museums and galleries, found that:

> Respondents questioned the sales of unrelated merchandise ... but most expressed a desire for merchandise that was not only restricted to the exhibitions. It was stated, for instance, that a museum and/or art gallery could function as a general arts and crafts shop, if such a facility was felt to be lacking in the locality. (McIntyre, 2010: 188)

A common issue in these studies was that the opposing realms of culture and commerce frequently collided. One study of the Vancouver Art Gallery (VAG) found that:

> Shop staff members identify much more strongly with their artistic identities than with their employment in the service industry. They are overwhelmingly from fine art and art history backgrounds, grew up in middle-class families, and many belong to local 'art' social networks. Staff members possess a great deal of cultural capital, if not economic capital, and identify more with the VAG as an institution than with their positions in the Gallery Store. (Cameron, 2007: 558)

This led to a workplace culture of disdain for certain shoppers. As one employee stated:

> Over and over again I see people connecting with the consumption of the image rather than the image itself. Someone, for example, will spend two minutes looking at an Impressionist painting on the Gallery wall and twenty minutes in the giftshop looking at a reproduction of the same painting wondering whether to buy it or not. (Quoted in Cameron, 2007: 560–561)

A contrasting view was advanced by Kent (2010) in a study of visitors to the museum shop at the Imperial War Museum in London, UK. Kent argues that visitors utilised the shop to extend and deepen their experience, often purchasing items that they view as furthering the knowledge they had gained during their visit. Books were a good example of products that added to the experience. It may be that this more engaged utilisation of this museum's shop is attributable to its nature as a war memorial and to potential family connections among visitors and suggests that this is a topic that requires further research.

Dark tourism museums and heritage attractions are particularly problematic when it comes to achieving the appropriate balance between respect, remembrance and commerce. In a study of Auschwitz, Brown (2013) found that this was a major consideration in the management of commercial services and had led to the shop being renamed as a bookshop, as:

> The universal horror surrounding the Holocaust dictates the way in which it can be depicted and sold at Auschwitz. The book shop presents itself as a worthy place of learning, which reconfirms the memorialising message and sober atmosphere of the museum, but also meets the needs of the visitors by providing the educational material they expect. (Brown, 2013: 275)

The tension between commerce and culture is often manifested in other ways. A study of Australian heritage attractions found that government funding agencies promoted grants for building ancillary business operations to such an extent that managers felt pressured to conform to expectations of greater commercialism. Such pressure led to problems when ancillary financial targets were not met and local custodians reported that they were dispirited and even 'burnt out' by having to focus on commercial operations that they were not interested in rather than the heritage that had attracted them (Carlsen *et al.*, 2008; Timothy, 2005).

The increased *adaptive re-use* of heritage buildings for tourist shopping also leads to problems. A good illustration of this is the case of a *caravanserai* (guesthouse for merchants on the Silk Road) in Qazvin, Iran. Abandoned and rapidly deteriorating, the building was restored and re-opened with a mix of handicraft and artisanal shops aimed at tourists. While this project gained a great deal of community support, local critics viewed the shops as inauthentic and unrelated to the original function of the caravanserai (Rezaei *et al.*, 2022).

Shopping in a heritage setting raises issues of what may broadly be termed *cultural sensitivity*. While this is often recognised in the tourism literature, coverage tends towards the descriptive, with little in the way of empirical studies. This perhaps reflects a widespread held attitude that cultural sensitivity (like, say sustainability) is important, but this may not necessarily parlay into action and strategies and, indeed, might be ignored by both tourists and managers when convenient.

Cultural sensitivity plays out in four main ways in the context of shopping at heritage operations, yet little empirical work has been carried out, particularly in the tourism sphere. The first regards what is appropriate for sale in a giftshop. As noted above, this has attracted interest in terms of museums dealing with atrocities or tragedies (Brown, 2013; Kent, 2010). This may also be an issue where themes are less intense, but authenticity is still important for the tourist experience. An example of this are giftshops at medieval castles, which commonly have clothing, souvenirs and toys related to knights and princesses, which may be viewed by some as tacky and even no longer culturally relevant. Going even further, such shops stock items and imagery that are either mythical (for example, King Arthur, dragons, fairies) or fictional (*The Lord of the Rings, Game of Thrones, World of Warcraft*)

A second issue is *cultural appropriation*, where the artworks and artefacts of a particular culture are purchased by people from another culture, who do not fully respect or understand that their consumption/use/display may cause offence. This appropriation typically occurs where tourists from modern, affluent cultures purchase items from more traditional societies, which may have religious significance and limits on whom may utilise these items. An example is the sale of First Nation headdresses in museums and art galleries across the American West.

A third issue is where the sale of cultural items in a museum giftshop or similar tends to reinforce potential negative imagery about a culture. This might commonly entail the presentation of that culture as primitive or frozen in a past era and result in more modern arts and handicrafts of that culture being ignored or devalued.

The fourth issue concerns the provenance of products for sale. This is manifested by artefacts being represented as produced by traditional artists or makers, whereas they are actually cheaply mass-produced in another country. This is a common issue across the globe, with Indigenous art and handicrafts. While a well-publicised issue, research into this is more likely to appear in the museum studies and intellectual property law literatures, rather than in tourism studies.

Spa/Wellness Tourism

Spa tourism used to be regarded as a niche activity but is now a global phenomenon (Kucukusta *et al.*, 2013). Many accommodation providers offer in-house spa services such as massages and facials, while various destinations are known for their spa traditions such as the Japanese *onsen* or hot springs and Turkish *hammams* or therapies such as Thai massage, Japanese reiki (energy healing) and Indian Ayurvedic massage. While there has been a growing body of research on spa tourism, including motivations and the spa tourism experience, there is a dearth of literature on shopping connected to spa tourism. This is curious, given that many spas promote their own range of products such as oils, scrubs, massage brushes and loofas or carry branded products that could be categorised as high-end or luxury in terms of cost, uniqueness, packaging and ingredients

(Kucukusta *et al.*, 2013), and these kinds of sales can be lucrative for a spa business (Buxton, 2017; Hjalager & Konu, 2011; Monteson & Singer, 1992). It is also the case that many spa employees receive a commission on the sale of these products (Buxton, 2017; Joppe, 2010), which supplements what can often be a meagre remuneration.

Hjalager and Konu (2011: 879) have considered cosmeceuticals, a cross between cosmetics and pharmaceuticals, 'which embrace anti-aging creams, vitamin enriched moisturizers, bath salts, lip balms, and many other products that have or are claimed to have a beneficial health effect compared to other cosmetics' in tourism. They refer to concerns about regulation or certification of these products, which may have a quasi-medical association in the minds of consumers that may not be able to be substantiated. The role of cosmeceuticals in spa branding is also discussed by Hjalager and Konu (2011), with co-branding between a supplier of these products and the spa facility. This sometimes takes the form of 'signature products', which have a direct connection to place and/or the local community and their traditions and are thus used to distinguish a provider from their competitors (Smith & Puczkó, 2015) and bestow authenticity. Tourists are informed about products used during their treatments and may be encouraged to purchase these products, as 'the first step towards loyalty to the particular product' (Hjalager & Konu, 2011: 882), which is an example of the *mutual* or reciprocal effects of co-branding. Some of the branding is subtle. For example, in one of the spas studied by Hjalager and Konu (2011: 887) 'the manager's expectation is that placing the name of the spa on gifts and products placed in private bathrooms will create some business in the spa'. Spa businesses need to take care that the therapist is not conducting a 'hard sell' to tourists, which runs counter to 'the client's need to relax and escape' (Tabacchi, 2010: 114). Purchasing a spa product, like other ancillary retail activities discussed in this chapter, may assist in extending the tourist experience, by aiding memory post-travel (Buxton, 2017; Monteson & Singer, 1992; Wuttke & Cohen, 2010).

The importance of employee knowledge and training about the retail products they sell to spa tourists has been noted by Smith and Wallace (2019: 10) in their study of key issues in spa management. They observe, '[Employees] should also know their products and brand well enough to create customer trust'. A spa studied by Hjalager and Konu (2011: 887) highlighted a problem with a lack of staff training about products, which affected retail sales: 'The manager envisages a need for a more general focus on new employee skills before full exploitation of co-branding'. Some employees may feel there is a disconnection between their role as a therapist and feelings of pressure to upsell products to customers: 'This has met resistance from some therapists who consider themselves to be caring practitioners rather than salespeople' (Wuttke & Cohen, 2010: 209). In a study one of the authors conducted of spa therapists in Australia, one of the interviewees (Joanna) discussed how this was a source of emotional labour at work:

> We have challenges every month to get a target and if you don't reach the targets, you're sort of in the bad books. This [spa] has to make a certain amount of money each month and if you don't, it's losing money for the company. I find it hard because not everybody in here wants to buy, and that puts me down as well. Because everyone's like, 'keep selling stuff, keep trying to sell to people'. They try to put a positive spin on it and then you're like 'well, I try, but everyone doesn't say yes'. That puts me in a negative frame of mind.

Another interviewee (Marie) made it clear that she would not compromise her professional integrity to increase retail sales to spa tourists: 'A big part of our job is retail. I don't sell people stuff that they don't need and I don't sell them stuff I don't like'. The association between this stance and authenticity of emotions at work warrants further research.

Sports Tourism, Events and Stadia

Sports tourism is a rapidly expanding market, with fans attracted by the prospect of viewing their team (or champion players) in person and experiencing the atmosphere and excitement of a live event. The range of sports tourism experiences includes attending mega event competitions between various countries (such as the Olympic Games and various World Cups and championships), regular international contests between two country teams (for example, the Ashes in cricket between England and Australia), tournaments featuring individual players (the various Opens in tennis and golf), club competitions (the National Basketball Association in the USA and the English Premier League), visits to famous stadia, often with associated museums (Lord's in London and the Melbourne Cricket Ground) and purpose-built sporting halls of fame and museums (such as the National Football Museum in Manchester, UK and the National Baseball Hall of Fame and Museum in Cooperstown, New York, USA). One recent trend is 'Fan Zones', which are precincts outside stadia with entertainment and shopping (Richards *et al.*, 2022).

Surprisingly, there has been almost no research into shopping for merchandise and sports tourism. Equally surprisingly, there has been little academic examination of the linkages between sporting events and tourist shopping across destinations, even though this is often used as a key argument in favour of public funding of event bids. In regards to stadia, halls of fame, sports museums and fan zones, tensions with shopping and other commercial activities are often identified as issues, but not considered in detail (see for example Frost, 2005; Ramshaw *et al.*, 2019; Richards *et al.*, 2022).

In light of this paucity of research, we have identified three issues that require further examination. The first is the link between merchandise and fan identity, which may be constructed in terms of players, clubs or national teams. What are the psychological benefits that fans gain from wearing gear featuring their team's colours, mascots or logos? Are their issues with certain imagery? For example, teams in the Australian Football League regularly feature Indigenous uniforms, which are designed by Indigenous artists and are popular with fans. How does purchasing and wearing such designs make fans feel? Another example is the appropriateness of sponsors' logos on uniforms. (Recently Cricket Australia has been criticised for having a coal-based electricity supplier as its main sponsor on its uniforms.) A second issue is the sale of 'throwaway' items and souvenirs, including cheap headwear, signs and plastics. Is such merchandising appropriate in times of increasing concern about sustainability? Third, there may be a tension between older and more serious fans and the commodification of a sport and its players aimed at the younger or more casual attendee.

Conclusion

Shopping is not always the primary purpose of travel. It can also be a significant ancillary function of the travel experience in a wide variety of tourism contexts, as described in this chapter. Regardless of the type of journey undertaken or the motivations for traveling, shopping is one of the most pervasive activities tourists undertake while away from home. Our review of the literature on shopping that is ancillary to different forms of tourism has elucidated gaps in knowledge that are ripe for further examination. While in some cases, ancillary shopping augments the main tourist experience, in others it may be a negative element, perhaps lessening the feeling of authenticity or perceived as an example of commodification. There are also nuances linked to cultural capital, memory making and sustainability that make this a fascinating and rich area of future study. These and other concepts have considerable potential for additional research, especially in tourism contexts that have not heretofore been well examined, such as pilgrimage/religious tourism, visiting friends and family (VFR), agritourism and solidarity travel.

References

Ahn, J. and Kwon, J. (2022) The role of trait and emotion in cruise customers' impulsive buying behavior: An empirical study. *Journal of Strategic Marketing* 30 (3), 320–333.
Andriotis, K. and Agiomirgianakis, G. (2010) Cruise visitors' experience in a Mediterranean port of call. *International Journal of Tourism Research* 12 (4), 390–404.
Bernardo, E. and Rodrigues, V. (2020) Buying sweet memories: The heritagization of food souvenirs in northern Portugal. *Journal of Gastronomy and Tourism* 4 (3), 129–140.
Björk, P. and Kauppinen-Räisänen, H. (2016) Exploring the multi-dimensionality of travellers' culinary-gastronomic experiences. *Current Issues in Tourism* 19 (12), 1260–1280.

Bourdieu, P. (1984) *Distinction: A Social Critique of the Judgment of Taste*, translated by R. Nice. Cambridge University Press.

Brida, J.G., Pulina, M., Riaño, E. and Aguirre, S.Z. (2013) Cruise passengers in a homeport: A market analysis. *Tourism Geographies* 15 (1), 68 bakery owner in the Lake District 87.

Brown, J. (2013) Dark tourism shops: Selling 'dark' and 'difficult' products. *International Journal of Culture, Tourism and Hospitality Research* 7 (3), 272–280.

Buxton, L. (2017) Selling the total spa product. In S. Rawlinson and T. Heap (eds) *International Spa Management: Principles and Practice* (pp. 99–113). Goodfellow Publishers.

Buzova, D., Sanz-Blas, S. and Cervera-Taulet, A. (2019) 'Tour me onshore': Understanding cruise tourists' evaluation of shore excursions through text mining. *Journal of Tourism and Cultural Change* 17 (3), 356–373.

Cameron, E. (2007) Exhibit and point of sale: Negotiating commerce and culture at the Vancouver Art Gallery. *Social and Cultural Geography* 8 (4), 551–573.

Carlsen, J., Hughes, M., Frost, W., Pocock, C. and Peel, V. (2008) *Success Factors in Cultural Heritage Tourism Enterprise Management*. Sustainable Tourism Centre for Co-operative Research.

Cave, J., Jolliffe, L. and De Coteau, D. (2012) Mementos of place: Souvenir purchases at the Bridgetown cruise terminal in Barbados. *Tourism Culture and Communication* 12 (1), 39–50.

Charters, S., Fountain, J. and Fish, N. (2009) 'You felt like lingering ...': Experiencing 'real' service at the winery tasting room. *Journal of Travel Research* 48 (1), 122–134.

Chen, X., Bruwer, J., Cohen, J. and Goodman, S. (2016) A wine tourist behaviour model for Australian wine cellar doors. *Tourism Analysis* 21 (1), 77–91.

Danielmeier, T. (2014) Winery architecture: Creating a sense of place. In M. Harvey, L. White and W. Frost (eds) *Wine and Identity: Branding, Heritage, Terroir* (pp. 213–229). Routledge.

Frost, W. (2005) The sustainability of sports heritage attractions: Lessons from the Australian Football League Hall of Fame. *Journal of Sports Tourism* 10 (4), 295–305.

Frost, W. and Laing, J. (2014) Old World winemakers in the New Worlds of California and Victoria. In M. Harvey, L. White and W. Frost (eds) *Wine and Identity: Branding, Heritage, Terroir* (pp. 17–28). Routledge.

Frost, W., Frost, J., Strickland, P. and Smith Maguire, J. (2020) Seeking a competitive advantage in wine tourism: Heritage and storytelling at the cellar-door. *International Journal of Hospitality Management* 87, 102460.

Goolaup, S., Solér, C. and Nunkoo, R. (2018) Developing a theory of surprise from travelers' extraordinary food experiences. *Journal of Travel Research* 57 (2), 218–231.

Gouveia, A.X. and Eusébio, C. (2019) Assessing the direct economic value of cruise tourism in a port of call: The case of Funchal on the island of Madeira. *Tourism and Hospitality Research* 19 (4), 422–438.

Gutberlet, M. (2019) Staging the Oriental other: Imaginaries and performances of German-speaking cruise tourists. *Tourist Studies* 19 (1), 110–137.

Gutberlet, M. (2024) *Overtourism and Cruise Tourism in Emerging Destinations on the Arabian Peninsula*. Routledge.

Henthorne, T.L., George, B.P. and Smith, W.C. (2013) Risk perception and buying behavior: An examination of some relationships in the context of cruise tourism in Jamaica. *International Journal of Hospitality and Tourism Administration* 14 (1), 66–86.

Hjalager, A.M. and Konu, H. (2011) Co-branding and co-creation in wellness tourism: The role of cosmeceuticals. *Journal of Hospitality Marketing and Management* 20 (8), 879–901.

Horng, J.S. and Tsai, C.T.S. (2010) Government websites for promoting East Asian culinary tourism: A cross-national analysis. *Tourism Management* 31 (1), 74–85.

Hritz, N. and Cecil, A.K. (2008) Investigating the sustainability of cruise tourism: A case study of Key West. *Journal of Sustainable Tourism* 16 (2), 168–181.

Hung, K. and Petrick, J.F. (2011) Why do you cruise? Exploring the motivations for taking cruise holidays, and the construction of a cruising motivation scale. *Tourism Management* 32 (2), 386–393.

Joppe, M. (2010) One country's transformation to spa destination: The case of Canada. *Journal of Hospitality and Tourism Management* 17 (1), 117–126.

Kent, T. (2010) The role of the museum shop in extending the visitor experience. *International Journal of Nonprofit and Voluntary Sector Marketing* 15 (1), 67–77.

Komarac, T., Ozretic-Dosen, D. and Skare, V. (2019) The role of the museum shop: Eliciting the opinions of museum professionals. *International Journal of Arts Management* 21 (3), 28–41.

Kucukusta, D., Pang, L. and Chui, S. (2013) Inbound travelers' selection criteria for hotel spas in Hong Kong. *Journal of Travel and Tourism Marketing* 30 (6), 557–576.

Kumar, N., Trupp, A. and Pratt, S. (2022) Linking tourists' and micro-entrepreneurs' perceptions of souvenirs: The case of Fiji. *Asia Pacific Journal of Tourism Research* 27 (1), 1–14.

Larsen, S. and Wolff, K. (2016) Exploring assumptions about cruise tourists' visits to ports. *Tourism Management Perspectives* 17, 44–49.

Lee, G. and Lee, M.K. (2017) Estimation of the shore excursion expenditure function during cruise tourism in Korea. *Maritime Policy and Management* 44 (4), 524–535.

Li, N. (2019) Mainland Chinese cruise passengers' onshore experience. *Tourism Recreation Research* 44 (2), 217–231.

McCaughey, R., Mao, I. and Dowling, R. (2018) Residents' perceptions towards cruise tourism development: The case of Esperance, Western Australia. *Tourism Recreation Research* 43 (3), 403–408.

McIntyre, C. (2010) Designing museum and gallery shops as integral, co-creative retail spaces within the overall visitor experience. *Museum Management and Curatorship* 25 (2), 181–198.

McNamara, N. and Cassidy, F. (2015) Wine tasting: To charge or not to charge? *International Journal of Hospitality Management* 49, 8–16.

Militz, T.A., Kershler, D.A. and Southgate, P.C. (2021) Informing artisanal pearl and pearl-shell handicraft production for the cruise tourism market through analysis of intended purchase behavior. *Tourism in Marine Environments* 16 (1), 45–58.

Mitchell, R. and Hall, C.M. (2001) Lifestyle behaviours of New Zealand winery visitors: Wine cub activities, wine cellars and place of purchase. *International Journal of Wine Marketing* 13 (3), 82–93.

Monteson, P.A. and Singer, J. (1992) Turn your spa into a winner: Spas are a popular amenity in upscale and destination hotels. They can also be a revenue center. Here are some of the considerations for marketing and operating your spa. *Cornell Hotel and Restaurant Administration Quarterly* 33 (3), 37–44.

Noordhuizen, J., Surborg, H. and Smulders, F.J. (2013) On the efficacy of current biosecurity measures at EU borders to prevent the transfer of zoonotic and livestock diseases by travellers. *Veterinary Quarterly* 33 (3), 161–171.

O'Mahony, B., Hall, J., Lockshin, L., Jago, L. and Brown, G. (2006) Understanding the impact of wine tourism on post-tour purchasing behaviour. In J. Carlsen and S. Charters (eds) *Global Wine Tourism: Research, Management and Marketing* (pp. 123–137). CABI.

Pan, F., Su, S. and Chiang, C. (2008) Dual attractiveness of winery: Atmospheric cues on purchasing. *International Journal of Wine Business Research* 20 (2), 95–110.

Ramshaw, G., Gammon, S. and Tobar, F. (2019) Negotiating the cultural and economic outcomes of sport heritage attractions: The case of the National Baseball Hall of Fame. *Journal of Sport and Tourism* 23 (2–3), 79–95.

Rezaei, N., Ghaderi, Z. and Ghanipour, M. (2022) Heritage tourism and place making: Investigating the users' perspectives towards Sa'd al-Saltaneh Caravanserai in Qazvin, Iran. *Journal of Heritage Tourism* 17 (2), 204–221.

Richards, J., Spanjaard, D., O'Shea, M. and Garlin, F. (2022) The changing carnival: Reimagining and recreating the match-day experience in multi-purpose stadiums. *Journal of Sport and Tourism* 26 (3), 269–284.

Sciortino, C., Ferrante, M., De Cantis, S. and Gyimóthy, S. (2022) Tracking cruise passengers' consumption: An analysis of the relationships between onshore mobility and expenditure. *Annals of Tourism Research Empirical Insights* 3 (2), 100059.

Sims, R. (2009) Food, place and authenticity: Local food and the sustainable tourism experience. *Journal of Sustainable Tourism* 17 (3), 321–336.

Smith, M. and Puczkó, L. (2015) More than a special interest: Defining and determining the demand for health tourism. *Tourism Recreation Research* 40 (2), 205–219.

Smith, M. and Wallace, M. (2019) An analysis of key issues in spa management: Viewpoints from international industry professionals. *International Journal of Spa and Wellness* 2 (3), 119–134.

Smith Maguire, J. and Lim, M. (2015) Lafite in China: Media representations of 'wine culture' in new markets. *Journal of Macromarketing* 35 (2), 229–242.

Soon, J.M. (2020) Application of Bayesian network modelling to predict food fraud products from China. *Food Control* 114, 107232.

Sorrentino, A., Risitano, M., Del Chiappa, G. and Abbate, T. (2019) Profiling cruise passengers in a Mediterranean port-of-call. *Anatolia* 30 (2), 279–290.

Stone, M.J. (in press) Not just another trinket: Defining unique attributes of food souvenirs. *Tourism Recreation Research* 1–6.

Stone, M.J., Migacz, S. and Wolf, E. (2019) Beyond the journey: The lasting impact of culinary tourism activities. *Current Issues in Tourism* 22 (2), 147–152.

Sun, X., Feng, X. and Gauri, D.K. (2014) The cruise industry in China: Efforts, progress and challenges. *International Journal of Hospitality Management* 42, 71–84.

Sun, X., Kwortnik, R., Xu, M., Lau, Y.Y. and Ni, R. (2021) Shore excursions of cruise destinations: Product categories, resource allocation, and regional differentiation. *Journal of Destination Marketing and Management* 22, 100660.

Tabacchi, M.H. (2010) Current research and events in the spa industry. *Cornell Hospitality Quarterly* 51 (1), 102–117.

Timothy, D.J. (2005) *Shopping Tourism, Retailing and Leisure*. Channel View Publications.

Timothy, D.J. (2014) Trends in tourism, shopping, and retailing. In A.A. Lew, C.M. Hall and A.M. Williams (eds) *The Wiley Blackwell Companion to Tourism* (pp. 378–388). Blackwell.

Timothy, D.J. (2021) *Cultural Heritage and Tourism: An Introduction* (2nd edn). Channel View Publications.

Teye, V.B. and Paris, C.M. (2011) Cruise line industry and Caribbean tourism: Guests' motivations, activities and destination preference. *Tourism Review International* 14 (1), 17–28.

Wang, N. (1999) Rethinking authenticity in the tourist experience. *Annals of Tourism Research* 26, 349–370.

Wuttke, M. and Cohen, M. (2010) Spa retail. In M. Cohen and G. Bodeker (eds) *Understanding the Global Spa Industry* (pp. 234–246). Routledge.

13 Shopping Festivals: High Impact Events for Shopping Tourism

Sangeeta Peter and Victor Anandkumar

Introduction

Shopping plays a part in almost every trip taken, and its share in total global tourism participation and expenditures is growing (Timothy, 2005; UN Tourism, 2014). Retail opportunities attract tourists to certain destinations and frequently serve as an ancillary attraction in a destination that is mostly popular for other reasons (see Chapter 12). Tourist shopping has a significant impact on the destination economy, the tourism industry and the retail sector (UN Tourism, 2014). This synergistic relationship between shopping and tourism has caused many destinations to focus much of their promotional effort on retail tourism, including shopping events (Peter, 2013; Timothy, 2005). Shopping festivals are promoted by destination management organizations (DMOs) for their ability to attract tourists, employ local residents, boost visitor expenditures in a concentrated time and space, enhance place image and contribute generally to the local economy.

All festivals essentially have a salient shopping/retail element. For instance, at harvest/agricultural festivals, one can buy fresh produce, honey, farm-fresh eggs, or homemade bread and cookies. Attendees at cultural heritage festivals regularly purchase traditional costumes, food items, musical instruments, arts and handicrafts, and religious icons (Chhabra *et al.*, 2002; Swanson & Horridge, 2004). Thus, the connection between retail and organized festivals is not new (Way & Robertson, 2013; Zaidan, 2016). Despite this connection between shopping and general festivals, this chapter is concerned particularly with festivals and events that focus on retail and shopping as their raison d'être. There are many small-town or regional festivals throughout the world associated with periodic markets, local heritage celebrations or Indigenous festivities, and these are key area activities that help support local artisans, farmers and merchants. However, larger-scale retail events occur in several international cities – places that are already known for their shopping appeal in the global tourism marketplace. This chapter describes the phenomenon of shopping events and festivals. It first examines the role of shopping festivals in the retail tourism milieu of places, the important role of tourist motivations and behaviors, and the role of governments in creating successful festivals. In the end, the chapter provides a conceptual framework for understanding the points of success and planning elements of large-scale shopping festivals.

Shopping Festivals in the Tourism Landscape

Destinations such as Singapore, Dubai and Hong Kong have positioned themselves as retail tourism hubs with a well-established retail presence. Real estate analyst CBRE reported that 62% of the 334 international luxury brands were present in Dubai with an estimated 4.6 million square meters of retail space in 2020, making it the world's premier retailing hub both in terms of size and brand concentration (FirstGroup, 2023). These leading positions in the world of shopping tourism has stimulated the growth of shopping festivals in various locations.

Festivals are periodic celebrations that focus on some element of place which is commemorated through public events and planned activities. Festivals typically celebrate something unique to a place, or for which the place is particularly known, such as a famous person or event, a religious occurrence, a specific food item or a unique cultural tradition (Ezeuduji, 2023). They may be small in scale at a village or local level, or regional in scale, attracting celebrants from other nearby localities. National festivals are often located in large cities that are centrally located or easily accessible and cater to domestic visitors. International festivals are large-scale events that are heavily promoted in a country's market source regions to encourage first-time and repeat visits.

Most festivals are concentrated in a bounded geographic space with marked entrance and exit points for counting attendees and issuing entrance passes (Allen *et al.*, 2021). They involve displays, activities for youth and others, merchandise for sale, concerts, art displays, auctions and various other shows and showcases. Special food vendors and transport options are set up temporarily for the duration of the event. Most shopping festivals, however, use existing retail spaces, food services and transportation infrastructure, although some additional dining services and shuttles are commonly established to cater to the needs of the influx of shoppers. Larger-scale shopping festivals tend to be scattered throughout the host city (Jakob, 2013), with the participation of existing retailers and catering services, and feature street events and other festivities in public open spaces. Thus, shopping festivals differ somewhat from other sorts of festivals in their physical layout and urban design. They typically involve mass participation by retailers throughout the city, rather than being contained in a consolidated area as many other festivals are.

At the international level, there are relatively few mega-events that could be classified as shopping festivals, but the number is growing. Several destinations that are already famous for their retail offerings have developed shopping festivals during the past 30 years to draw increased attention to their retailscapes. Shopping events emphasize a destination's shopping potential, building global awareness of opportunities for high-end retail and employing thousands of people during the festival and afterwards. They aim to attract tourists to the destination and are used by several known shopping destinations as additional branding tools to ensure that their retail image is enhanced in the global marketplace.

Capitalizing on their retail presence, destinations offer different categories of retail promotions during shopping festivals, including raffles and drawings, bundled offers, sales discounts and reduced-price coupons. Table 13.1 displays information about several of the world's biggest shopping events and festivals. Most of these activities are organized by shopping centers/malls, independent retailers, master franchisers, and brand and trade bodies – efforts that usually lead to an increase in footfall and merchandise sales (Peter & Anandkumar, 2013). Sales discounts ranging from 25% to 75% are typically offered on products such as gold, electronics, fashion, apparel, watches, jewelry, cosmetics, automobiles, home furnishings, real estate, sportswear and sports equipment (Henderson et al., 2011).

Established in 1994 by the Singapore Retail Association, in collaboration with the Singapore Tourism Board and UnionPay, the Great Singapore Sale was the first shopping mega-event in modern times (Table 13.1). It took place annually during the summer months, and according to the Singapore Tourism Board (2019), tourist arrivals grew tremendously year on year, largely owing to the festival, until 2019, after which the COVID-19 pandemic decimated the island's tourism economy, including the Great Singapore Sale. For two years, the sale was closed, and the Singapore Retail Association held its final event in 2022.

The Dubai Shopping festival was the second global mega-event to be organized. This annual celebration of shopping was founded in 1996 by the Dubai Festivals and Retail Establishment (Alhosani & Zaidan, 2014; Anwar & Sohail, 2004; Mehta et al., 2014). Although it suffered considerable planning and attendance setbacks from COVID-19, prior to the pandemic, the event accounted for nearly one third (30%) of annual retail sales in the emirate and helped solidify Dubai's reputation as a global shopping destination (Dubai Tourism, 2019). The revenue generated by this mega-event increased from its initial impact of $1.6 million USD in 1996 to over $50 million in 2018 (ValuStrat, 2019). The event in Dubai regularly draws between 3–4 million attendees each year (Timothy, 2018). The 2023 event was considered a post-pandemic success, and plans are underway to continue holding the event in future years.

Seeing the success of these early events, several other shopping festivals were initiated during the 2000s, including the Amazing Thailand

Table 13.1 A selection of international shopping festivals, general information

Shopping festival	Year started	Country	Duration of the festival	Organizer	Meta events
Great Singapore Sale	1994	Singapore	June–August (ceased in 2022)	Singapore Retailers Association, Supported by Singapore tourism Board and UnionPay	discounts, music concerts, food outlets, food festival, drawings and instant reward games
Dubai Shopping Festival	1996	UAE	January–February	Dubai Festivals and Retail Establishment	fireworks display, Global Village, daily DSF raffles, discounts and freebies, live concerts, DSF Carnival, 12-hour super sales, gold raffle, night market
Amazing Thailand Grand Sale	2002	Thailand	June–August	Tourism Authority of Thailand	special deals, discounts, Amazing Flash Sale, 'Expat Amazing Week', lucky draw
Hong Kong Shopping Festival	2004	Hong Kong	July–August	Hongkong Tourism Board and the Hong Kong Chinese Manufacturers' Association of Hong Kong	entertainment, gourmet events, music concerts, Grand Lucky Draw
Korea Grand Sale	2011	South Korea	January–February	Ministry of Culture, Sports and Tourism and Visit Korea Committee	cultural heritage experience programs, World Heritage of Korea 'experience zone', hands-on arts and crafts class, raffles, discounts on Korea-bound flights, along with discounts on railway services
Istanbul Shopping Festival	2011	Turkey	July–August	Ministry of Culture and Tourism and Ministry of Development in conjunction with the Istanbul governorship, the Istanbul metropolitan municipality, the Turkish Exporters Assembly, Turkish Airlines and Istanbul Chamber of Commerce	special discounts, raffles, concerts, competitions, fashion shows, cultural activities, theatrical performances, parades, children's games, competitions
Andorra Shopping Festival	2013	Andorra	November	VisitAndorra/Andorra Tourism	children's games and activities, street music, concerts, fine dining, parades, parallel food festival, sales and giveaways

Source: Compiled by the authors from multiple sources

Grand Sale, the Hong Kong Shopping Festival, the Korea Grand Sale, the Istanbul Shopping Festival and the Andorra Shopping Festival (Agarwal *et al.*, 2022; Alhamwi, 2020; Arslan & Kendir, 2019; Azmi *et al.*, 2020; Kütük-Kuriş, 2020; Timothy, 2018, 2021; Wu & Lo, 2018). All of these countries and cities regard shopping festivals as a significant boost to their tourism economies and have continued to hold the events each year because of its immediate profitability, as well as its lingering retail image derived from mass promotion and branding efforts. The Amazing Thailand Grand Sale 2023 was estimated to have attracted over 200,000 foreign shopper-tourists to the country, generating at least 2.8 million USD in direct sales in participating stores and at least 2 million USD in other tourism spending (Tourism Authority of Thailand, 2023).

Formal and informal shopping opportunities generally have a role to play in the 'festivalization' of retail tourism, including airport shops, flea markets, souvenir shops, bazaars, fairs, shopping malls and farmers markets (Hazlan *et al.*, 2019). Shopping events tend to encompass a mix of formal-structure, controlled retailing such as shops and malls, and informal-structure opportunities, such as less controllable and impermanent features (e.g. street vendors, flea markets, farmers markets and fairs) that tend to add more ludic and hedonic value to the event (Sherry, 1990a, 1990b).

Shopping festivals are systematically planned, citywide events organized by destinations aimed at attracting tourists and increasing local spending. They are usually carried out through the combined efforts of government agencies (e.g. tourism ministries), DMOs, retail associations and individual retailers (Henderson *et al.*, 2011; Vel *et al.*, 2014). Most of them are held annually for a period of one to three months (Table 13.1), often organized to coincide with holiday periods with the aim of attracting international tourists. Dubai promotes other shopping events during the low season, such as Dubai Summer Surprises (July/August), to promote retail and tourism during the hottest part of the year (Zaidan, 2019).

As noted earlier, shopping festivals involve many activities that pair well with shopping, such as dining, outdoor activities, concerts, shows and other entertainment events (Table 13.1). These activities are typically non-retailing in the traditional sense but complement shopping and are used to create a mega event with an appealing atmosphere that attracts tourists, increases total expenditures, and extends tourists' stay in the destination.

Components and Considerations of Shopping Festivals

As previously noted, shopping fetes are uniquely positioned to build public awareness of a destination's shopping tourism potential and to help create a shopping brand. The success of such events requires considerable knowledge about tourists' motives and behaviors and government participation so that effective planning can take place to create positive retail environments for out-of-town consumers. This section focuses on the shopping experience, shopper-tourists' behaviors and the role of government in carrying out successful retail extravaganzas.

The shopping experience

The shopping experience involves much more than simply acquiring goods; it involves interactions between tourists, the shopping environment, services and service providers, product attributes and the sociocultural and natural environments of the destination (Tosun *et al.*, 2007). The shopping experience, however, is also impacted by the retail milieu, with shopping in a diversified environment leading to more leisure encounters and greater satisfaction. Shopping environments range from new luxury malls to old street markets and ancient traditional markets (see Chapter 8), with local markets and street vendors, which used to be the province of the poor, having seen a resurgence in recent years as part of the holistic destination retail experience (Correia & Kozak, 2016; Furlan & Faggion, 2015; Lee & Pearce, 2020; Major & Tannous, 2020). These are all key features enveloped within shopping festivals.

Likewise, food and drink, as well as elements of local culture, play an important role in many shopping festivals. In the Middle East and North Africa, traditional markets (*souq*s and bazaars) and markets in other parts of the world play an important role in the heritagization of shopping festivals and retail heritage in general (Furlan & Faggion, 2015; Major & Tannous, 2020). In Andorra, shopping festival attendees are encouraged to partake of the country's cultural heritage and enjoy its natural landscapes (Visit Andorra, 2024).

Shopping mega-malls have undergone significant transformations since the mid-1900s to become more tourism-oriented, offering accommodations, attractions and non-shopping entertainment (Timothy, 2005). For example, the Dubai Mall includes hotels, added attractions (e.g. ice rink, aquarium and underwater zoo), retail stores and multiple dining options (Peter *et al.*, 2013). Traditional markets, such as bazaars, souqs, farmers markets and floating markets, help tourists connect with the destination's heritage (Hall, 2016; Timothy, 2005; Wattanacharoensil & Sakdiyakorn, 2016), whereas high street shopping adds an element of convenience and luxury as those avenues function as one-stop retail clusters that cater to the needs of the tourist. All of these venues generally have a role to play in large-scale shopping festivals, as they add a local flavor and element of culture and sense of place to what might otherwise be a mundane event that could be held anywhere.

In festivals, good physical connectivity between shopping venues within the city by local transportation, as well as other retail variables such as store attributes, mall atmospherics, safety conditions, mall amenities, product attributes and staff service positively influence tourists' shopping experience in general but are particularly important during festival times (Peter & Anandkumar, 2016a, 2016b). Tourists, however, differ in their preferences of shopping venue. For example, most tourists from Western countries prefer traditional markets and gravitate towards souvenir buying, whereas those from developing countries tend to choose modern shopping malls and are more inclined to buy brand-name merchandise (Egresi & Arslan, 2016). For Westerners, the pull of traditional marketplaces exudes a sense of heritage, authenticity and cultural gravitas that cannot be experienced in other retail venues (González, 2020). For consumers from the Global South, modern shopping centers and other luxury retail spaces are associated with the 'idealized' West, affluence and socioeconomic development, and therefore project the characteristics of retail places they want to be (Egresi & Arslan, 2016: 215). Many tourist-shoppers from the Global North eschew malls and shopping centers for these very same reasons – that they can be visited any time at home and do not truly represent 'authentic' destination conditions; they are simply symbols of globalization that could exist almost anywhere with the standardization, predictability and depersonalization that follow the McDonaldization of the world (Ritzer, 1996).

Stimulation of the senses, including colors, smells, sounds and interactive displays help heighten the retail experience (Timothy, 2005), which is something festival planners have to consider in how streetscapes are designed, signage is erected, and temporary displays are displayed. Such details can help create an environment conducive to increased expenditures, or conversely, they can hamper people's desires to shop (Bustamante & Rubio, 2017; Chatterjee & Shukla, 2020; Yakhlef, 2015). While most shopping festivals are physically distributed throughout the community, signage and other forms of physical connectedness are typically present to show which retailers are involved in the event and to guide consumers from one point to the next. In areas where retail is concentrated, it is much easier to plan for signage, displays and retail landscaping (Yoshimura *et al.*, 2021), but marked routes, printed maps and mobile phone apps are now common tools for connecting retailers and vendors participating in retail events.

Shopping value is the benefit derived from the shopping experience. Utilitarian shopping value is generally work-oriented and is the cognitive outcome of shopping, whereas hedonic shopping value reflects the entertaining, ludic and emotional aspects of shopping. Shopping festivals are planned to emphasize the hedonic side of shopping. Consumers, who view shopping as an escape enjoy bargaining and exploring new products; in essence, they seek hedonic value from their shopping experiences (Holbrook & Hirschman, 1982). Shopping value influences satisfaction, word-of-mouth, repatronage intentions and loyalty (Jones *et al.*, 2006) and can be enhanced through the conversion mechanism of shopping festivals that turns visitors into buyers. Shopping festivals also provide shoppers with both utilitarian and hedonic value. The experiential value of shopping may be derived from the act of shopping itself (including browsing), the overall environment in which the tourist shops or engaging with new retailers and products. Shopping value positively impacts both revisit intentions and satisfaction with shopping activities (Peter & Anandkumar, 2013).

Shopping festivals and tourists' behavior

Successful shopping festival management requires an understanding of guests' interests and behaviors. Various enabling decisions need to be taken by policymakers, the success of which depends on an understanding of what matters to the tourist. For instance, decisions about what will draw tourists to a shopping festival require an understanding of what motivates people to undertake a trip. Similarly, decisions regarding tourists' experiences and product development require an understanding of dimensions such as activities, cultures and points of satisfaction. Naturally, shopping is expected to be the primary motive for visiting a shopping festival. However, relatively few tourists travel with the primary motive of shopping but, as noted earlier, the majority of tourists do shop once at the destination (Egresi & Arslan, 2016). Multiple travel motives such as exploring new places and cultures, and visiting friends and family are common purposes, with motives often varying by nationality. The cultures of East Asia and the Middle East, for example, are known for their high propensity to shop while traveling and their propensity to travel for the purpose of shopping (Bui & Trupp, 2020; Kim *et al.*, 2011; Park & Reisinger, 2009; Xu & McGehee, 2012). The type of products purchased during the shopping festival also tend to differ across nationalities (Peter & Anandkumar, 2014).

The role of government

Owing to their economic and promotional potential, shopping festivals tend to have strong government support, with local governments and national tourism agencies actively promoting shopping as a tourist attraction. Most shopping festivals are an outcome of public-private partnerships (Table 13.1), with some government divisions being dedicated to promoting shopping events (Peter & Anandkumar, 2013). Dubai Festivals and Retail Establishment (DFRE) and a division of Singapore's Tourism Board are dedicated to planning and implementing shopping and tourism in partnership with retail associations. These initiatives help integrate shopping festivals within official policies and promotional efforts.

Most large-scale shopping festivals require government support for a variety of reasons (Litvin, 2013). First, some festivals offer special prices, which may include tax incentives. This obviously needs government action to waive certain taxes or reduce local consumer taxes during the period of the event. Second, local law enforcement is also involved. Heightened security and overcrowding are significant concerns, so an additional police presence is often needed to help alleviate congestion, enforce driving and parking laws, and ensure the safety of spenders and merchants. Third, extra effort is sometimes needed for city clean-up to prepare for the event and afterwards. This requires coordination between various public works offices (e.g. traffic control, garbage collection and utilities services). Fourth, special training is sometimes necessary to enhance local hospitality and to help vendors understand the unique retail needs of certain cultural groups. Finally, increased destination marketing is usually the responsibility of DMOs, which may be a specific government agency or a coalition of government agencies and private sector interests. These educational and marketing efforts often fall within the purview of ministries of tourism or local tourism agencies.

Shopping Festival Successes

Although it ended its 30-year run in 2022, the Great Singapore Sale was considered a beacon of success among the world's large shopping festivals. It successfully promoted Singapore as a shopping destination and drew in tens of thousands of retail consumers from throughout Asia and elsewhere every year. Likewise, it established a strong network of retailers and retail organizations, and had a significant spillover effect into other sectors of the tourism industry, including hotels, food services, cultural attractions and transportation. Ultimately, these are the goals of the promoters of shopping events.

According to one consultant familiar with these sorts of events (Yu, 2023), to achieve continued success, shopping festivals should ensure several actions. First, they should foster stronger relationships with government tourism authorities to improve marketing and raise their global visibility. Second, organizers should use current technology to improve digital marketing and e-commerce, where possible. Although ideally, DMOs would want to see increased numbers of tourists on site, even online shopping associated with a festival can increase expenditures in the destination's shops, malls and retail districts. Success also comes by creating long-lasting memories and satisfying retail experiences through entertainment, events and various interactive activities. Third, extending the period of the sales event can obviously increase footfall in the destination and increase sales. This will enable people to come whose schedules might not necessarily allow them to visit during shorter periods of time. Finally, creating partnerships with hotels and airlines to organize promotional packages and special deals can incentivize people to come to the event and spend more time and money in the destination (Yu, 2023: n.p.).

These and other factors that facilitate the success of shopping festivals are summarized in Figure 13.1. These include an assessment of tourists' behaviors and characteristics (e.g. travel motives, shopping behavior and demographic characteristics), the role of government, product promotions, events and various destination attributes. The government can assist in creating a supportive environment, through policies that support retail and tourism. Sales promotions can be used as a communicative tool, and help in meeting the short-term and long-term objectives of the brand (Lee, 2002). By inference, the same is applicable to shopping festivals. While price-based promotions help in meeting short-term objectives such as increase in sales and market share for the retailer, non-price promotions would help the DMO and shopping center managers in achieving long-term objectives such as enhancing the brand image and increasing loyalty and long-term profits. Tourists vary in their shopping motives which encompass both hedonic and utilitarian motives. These motives play a key role in bringing shoppers into stores, where their shopping behavior is affected by emotion. Shoppers motivated by hedonic elements, will pay greater attention to elements of the shopping environment such as atmospherics and window displays, while those motivated by utilitarian motives will be motivated by price-based promotions. The potential for shopping festivals to be developed, promoted and sustained as a tourism product would depend upon what the tourists derive from their shopping trip, rather than merely the sale of merchandise. Understanding the sources of satisfaction or dissatisfaction with shopping festivals is important to the stakeholders so as to be able to develop and improve the shopping and destination experience. The increase in the number of destinations organizing shopping festivals is leading to unprecedented competition, with destinations offering similar shopping facilities and

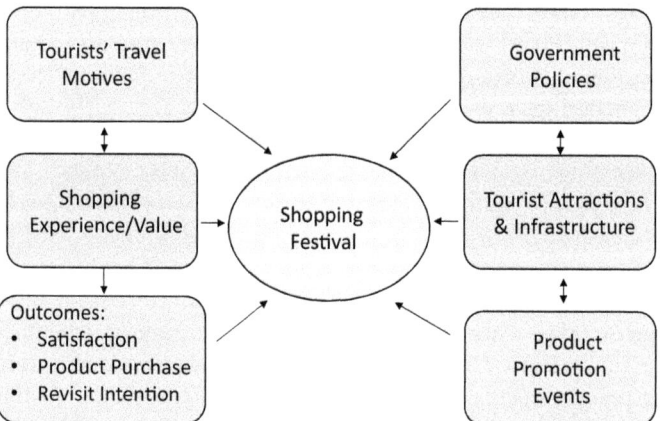

Figure 13.1 Framework for understanding and planning shopping festivals

various added attractions. In such a case, shopping festivals can be substituted and destinations may struggle to maintain their competitive advantage. Hence, destinations will have to constantly reinvent their offerings and upgrade the retail environment to attract tourists to the shopping festival.

Conclusion

Shopping festivals are becoming increasingly common for their ability to enhance a destination's retail image, increase tourists' expenditures and provide an overall economic stimulus for the destination. Successful shopping events have several things in common, including understanding the market (experiences and behaviors), collaboration between tourism stakeholders and service providers, government assistance and effective promotion to enhance a retail image through events that draw tens of thousands of consumers every time an event is held. Several global cities have seen considerable growth in shopping tourism as a result of their efforts to brand themselves as ultimate retail destinations through shopping events and festivals.

A very recent trend in shopping festivals is to include an online presence, where consumers no longer have to visit the destination to immerse themselves in the retail festivities. Online components are becoming an increasingly common part of shopping festivals, as they open additional revenue streams for retailers. Although most DMOs desire to increase the physical presence of shoppers because this has significant spinoff effects for other sectors of tourism, including lodging, food services and transportation, online components are nonetheless a critical part of the overall success of today's shopping festivals. Likewise, fully online shopping festivals, or 'retail extravaganzas', are now in some cases taking the place of traditional onsite retail events (Chen & Li, 2020; Wu & Ai, 2016; Xie et al., 2023), thereby spreading the positive economic impacts to destinations without them having to increase their security presence and physically set up for the event. While deterritorializing the retail sense of place that shopping events were traditionally meant to enhance, online shopping festivals can have a positive economic impact without the physical overcrowding and upfront costs associated with planning and executing onsite festivals. Despite this increase in online events, the corporeal experience of being in a famous shopping destination will likely continue far into the future as it presents enjoyable leisure experiences that will build customer loyalty, visitor satisfaction and draw consumers back for future shopping experiences and to appreciate other elements of the destination beyond only retail.

More research is needed to highlight different outcomes and impacts associated with different retail events, both large and small. Most work has focused on urban retail-based events, with good reason given cities' retail clusters and critical mass of shopping opportunities. However, the number of rural shopping festivals is growing, and we need to understand them better and their relationships with what makes rural environments appealing. There is also a dearth of knowledge about the physical displacement ('deplacing') associated with online retail festivals, and the potential placebo effect that could occur when or if online shopping were to replace retail opportunities in situ. Much more work is needed about these retail events a multitude of other shopping contexts and situations that effect retail mobility or prevent it.

References

Agarwal, S., Koo, K.M. and Qian, W. (2022) Consumption response to temporary price shock: Evidence from Singapore's annual sale event. *Journal of Financial Intermediation* 51, 100966.

Alhamwi, A. (2020) Istanbul Shopping Festival 2020! A magical experience for shopping enthusiasts. Online: https://move2turkey.com/istanbul-shopping-festival-2020-a-magical-experience-for-shopping-enthusiasts/

Alhosani, N. and Zaidan, E. (2014) Shopping tourism and destination development: Dubai as a case study. *The Arab World Geographer* 17 (1), 66–81.

Allen, J., Harris, R. and Jago, L. (2021) *Festival and Special Event Management Essentials*. Wiley.

Anwar, S.A. and Sohail, M.S. (2004) Festival tourism in the United Arab Emirates: First-time versus repeat visitor perceptions. *Journal of Vacation Marketing* 10 (2), 161–170.

Arslan, E. and Kendir, H. (2019) Evaluation of the effect of festival tourism on urban economy: The case of Van Shopping Fest. *Turizm Akademik Dergisi* 6 (2), 299–306.

Azmi, A., Abdullah, A., Nurhidayati, S.E. and Shaw, G. (2020) Shopping and tourism: A state-of-the-art review. *PalArch's Journal of Archaeology of Egypt/Egyptology* 17 (5), 1220–1239.

Bui, H.T. and Trupp, A. (2020) Asian tourism in Europe: Consumption, distinction, mobility, and diversity. *Tourism Recreation Research* 45 (1), 4–18.

Bustamante, J.C. and Rubio, N. (2017) Measuring customer experience in physical retail environments. *Journal of Service Management* 28 (5), 884–913.

Chatterjee, R. and Shukla, Y.S. (2020) Examining the role of culture, retail environment, and emotions in determining festival shopping engagement: An emerging market perspective. *International Journal of Indian Culture and Business Management* 20 (3), 385–408.

Chen, C. and Li, X. (2020) The effect of online shopping festival promotion strategies on consumer participation intention. *Industrial Management and Data Systems* 120 (12), 2375–2395.

Chhabra, D., Sills, E. and Rea, P. (2002) Tourist expenditures at heritage festivals. *Event Management* 7 (4), 221–230.

Correia, A. and Kozak, M. (2016) Tourists' shopping experiences at street markets: Cross-country research. *Tourism Management* 56, 85–95.

Dubai Tourism (2019) Annual Visitor Report, 2019. Online: https://dubaitourism.getbynder.com/m/3e56c8625ed93ce0/original/DTCM-ANNUAL-REPORT-2019-EN.pdf

Ezeuduji, I.O. (2023) Cultural events and tourism in Africa. In D.J. Timothy (ed.) *Cultural Heritage and Tourism in Africa* (pp. 67–84). Routledge.

Egresi, I. and Arslan, S. (2016) Shopping and tourism in Turkey: The perfect combination. In I. Egresi (ed.) *Alternative Tourism in Turkey: Role, Potential Development and Sustainability* (pp. 211–228). Springer.

FirstGroup (2023) Dubai: The world's most exciting retail destination. Online: https://www.thefirstgroup.com/en/news/dubai-the-world-s-most-exciting-retail-destination/

Furlan, R. and Faggion, L. (2015) The Souq Waqif heritage site in Doha: Spatial form and livability. *American Journal of Environmental Engineering* 5 (5), 146–161.

González, S. (2020) Contested marketplaces: Retail spaces at the global urban margins. *Progress in Human Geography* 44 (5), 877–897.

Hall, C.M. (2016) Heirloom products in heritage places: Farmers' markets, local food and food diversity. In D.J. Timothy (ed.) *Heritage Cuisines: Traditions, Identities and Tourism* (pp. 88–103). Routledge.

Hazlan, H.A.M., Ismail, H.N. and Jaafar, S.M.R.S. (2019) Flea market tourism: A review of motivation and characteristics of specialised tourist segmentation. *International Journal of Built Environment and Sustainability* 6 (1–2), 55–61.

Henderson, J.C., Chee, L., Mun, C.N. and Lee, C. (2011) Shopping, tourism and retailing in Singapore. *Managing Leisure* 16 (1), 36–48.

Holbrook, M.B. and Hirschman, E.C. (1982) The experiential aspects of consumption: Consumer fantasies, feelings, and fun. *Journal of Consumer Research* 9 (2), 132–140.

Jakob, D. (2013) The eventification of place: Urban development and experience consumption in Berlin and New York City. *European Urban and Regional Studies* 20 (4), 447–459.

Jones, M.A., Reynolds, K.E. and Arnold, M.J. (2006) Hedonic and utilitarian shopping value: Investigating differential effects on retail outcomes. *Journal of Business Research* 59 (9), 974–981.

Kim, S.S., Timothy, D.J. and Hwang, J. (2011) Understanding Japanese tourists' shopping preferences using the Decision Tree Analysis method. *Tourism Management* 32 (3), 544–554.

Kütük-Kuriş, M. (2020) Piety, fashion and festivity in a modest fashion shopping mall in Istanbul. *International Journal of Fashion Studies* 7 (2), 167–191.

Lee, C.W. (2002) Sales promotions as strategic communication: The case of Singapore. *Journal of Product and Brand Management* 11 (2), 103–114.

Lee, D. and Pearce, P. (2020) Shining a light on Asian night markets: vendors' and visitors' views. *International Journal of Tourism Cities* 6 (2), 467–484.

Litvin, S.W. (2013) Festivals and special events: Making the investment. *International Journal of Culture, Tourism and Hospitality Research* 7 (2), 184–187.

Major, M.D. and Tannous, H.O. (2020) Form and function in two traditional markets of the Middle East: Souq Mutrah and Souq Waqif. *Sustainability* 12 (17), 7154.

Mehta, S., Jain, A. and Jawale, R. (2014) Impact of tourism on retail shopping in Dubai. *International Journal of Trade, Economics and Finance* 5 (6), 530.

Park, K.S. and Reisinger, Y. (2009) Cultural differences in shopping for luxury goods: Western, Asian, and Hispanic tourists. *Journal of Travel and Tourism Marketing* 26 (8), 762–777.

Peter, S. (2013) 'What is in it for a tourist shopper?': An assessment of the shopping value of tourists in an international shopping festival. *Journal of Tourism* 14 (2), 47–54.

Peter, S. and Anandkumar, V. (2013) A study of sales promotions during Dubai Shopping Festival. *International Journal of Tourism, Hospitality and Catering* 5, 1–12.

Peter, S., Anandkumar, V. and Peter, S. (2013) Role of shopping festivals in destination branding: A tale of two shopping festivals in the United Arab Emirates. *Anatolia* 24 (2), 264–267.

Peter, S. and Anandkumar, V. (2014) Dubai Shopping Festival: Tourists' nationality and travel motives. *International Journal of Event and Festival Management* 5 (2), 116–131.

Peter, S. and Anandkumar, V. (2016a) Deconstructing the shopping experience of tourists to the Dubai Shopping Festival. *Cogent Business and Management* 3 (1), 1199109.

Peter, S. and Anandkumar, V. (2016b) Travel motivation-based typology of tourists who visit a shopping festival: An empirical study on the Dubai Shopping Festival. *Journal of Vacation Marketing* 22 (2), 142–153.

Ritzer, G. (1996) The McDonaldization thesis: Is expansion inevitable? *International Sociology* 11 (3), 291–308.
Sherry, J.F. (1990a) A sociocultural analysis of a Midwestern American flea market. *Journal of Consumer Research* 17 (1), 13–30.
Sherry, J.F. (1990b) Dealers and dealing in a periodic market: Informal retailing in ethnographic perspective. *Journal of Retailing* 66 (2), 174–200.
Singapore Tourism Board (2019) Third consecutive year of growth for Singapore tourism sector 2018. Online: https://www.stb.gov.sg/content/stb/en/media-centre/media-releases/third-consecutive-year-of-growth-for-singapore-tourism-sector-in-2018.html
Swanson, K.K. and Horridge, P.E. (2004) A structural model for souvenir consumption, travel activities, and tourist demographics. *Journal of Travel Research* 42 (4), 372–380.
Timothy, D.J. (2005) *Shopping Tourism, Retailing and Leisure*. Channel View Publications.
Timothy, D.J. (2018) Shopping tourism. In S. Agarwal, G. Busby and R. Huang (eds) *Special Interest Tourism: Concepts, Contexts and Cases* (pp. 134–144). CABI.
Timothy, D.J. (2021) *Tourism in European Microstates and Dependencies: Geopolitics, Scale and Resource Limitations*. CABI.
Tosun, C., Temizkan, S.P., Timothy, D.J. and Fyall, A. (2007) Tourist shopping experiences and satisfaction. *International Journal of Tourism Research* 9 (2), 87–102.
Tourism Authority of Thailand (2023) 'Shopping challenge' kicks off Amazing Thailand Grand Sale 2023. Online: https://www.tatnews.org/2023/06/shopping-challenge-kicks-off-amazing-thailand-grand-sale-2023/#:~:text=The%20Amazing%20Thailand%20Grand%20Sale%202023%20with%20the%20concept%20of,at%20over%2010%2C000%20participating%20stores.
UN Tourism (2014) *Global Report on Shopping Tourism*. Madrid: UN Tourism.
ValuStrat (2019) How does Dubai Shopping Festival affect the economy of UAE? Online: https://valustrat.com/how-does-dubai-shopping-festival-affect-the-economy-of-uae.
Vel, K.P., Suhail, L. and Dokhan, A. (2014) Events marketing model of Dubai shopping festival. *Brazilian Journal of Marketing* 13 (6), 138-147.
Visit Andorra (2024) Andorra Shopping Festival. Online: https://visitandorra.com/en/agenda/andorra-shopping-festival/
Wattanacharoensil, W. and Sakdiyakorn, M. (2016) The potential of floating markets for creative tourism: A study in Nakhon Pathom Province, Thailand. *Asia Pacific Journal of Tourism Research* 21 (1), 3–29.
Way, K.A. and Robertson, L.J. (2013) Shopping and tourism patterns of attendees of the bikes, blues and BBQ festival. *Journal of Hospitality Marketing and Management* 22 (1), 116–133.
Wu, H.C. and Ai, C.H. (2016) A study of festival switching intentions, festival satisfaction, festival image, festival affective impacts, and festival quality. *Tourism and Hospitality Research* 16 (4), 359–384.
Wu, S.S. and Lo, S.M. (2018) Events as community function of shopping centers: A case study of Hong Kong. *Cities* 72, 130–140.
Xie, J., Yoon, N. and Choo, H.J. (2023) How online shopping festival atmosphere promotes consumer participation in China. *Fashion and Textiles* 10 (1), 1–19.
Xu, Y. and McGehee, N.G. (2012) Shopping behavior of Chinese tourists visiting the United States: Letting the shoppers do the talking. *Tourism Management* 33 (2), 427–430.
Yakhlef, A. (2015) Customer experience within retail environments: An embodied, spatial approach. *Marketing Theory* 15 (4), 545–564.
Yoshimura, Y., Santi, P., Arias, J.M., Zheng, S. and Ratti, C. (2021) Spatial clustering: Influence of urban street networks on retail sales volumes. *Environment and Planning B: Urban Analytics and City Science* 48 (7), 1926–1942.
Yu, A. (2023) What role does the Great Singapore Sale play in attracting tourists to Singapore, and how can it be made more impactful? Consultancy summary. Online:

https://www.quora.com/What-role-does-the-Great-Singapore-Sale-play-in-attracting-tourists-to-Singapore-and-how-can-it-be-made-more-impactful

Zaidan, E.A. (2016) Tourism shopping and new urban entertainment: A case study of Dubai. *Journal of Vacation Marketing* 22 (1), 29–41.

Zaidan, E. (2019) Shopping tourism and hyper-development in the Middle East and North Africa. In D.J. Timothy (ed.) *Routledge Handbook on Tourism in the Middle East and North Africa* (pp. 365–377). Routledge.

14 Food Retail and Food Souvenirs in Tourism: Bringing Home a Taste of Place

Matthew J. Stone

Introduction

Travelers make retail food and drink purchases for many reasons. They may be for immediate consumption, but frequently they purchase food and drink to help them remember a trip, to relive an experience and to give as gifts. In most of these instances, these food purchases can be defined as *souvenirs*. Food and drink have become more important to travelers both as a motivator to travel and as experiences to engage in (Stone *et al.*, 2020). At the same time, destinations have increasingly used food and drink to differentiate themselves and attract travelers. Despite this increased importance of food in the tourism industry, the topic of food souvenirs has generally been overlooked by researchers (Lin & Mao, 2015).

This chapter explores the phenomenon of food and drink retail purchases (including souvenirs) while traveling. Following a brief overview of food tourism, this chapter considers food and retail purchases within the literature on souvenir purchases. The research on souvenirs and their authenticity is presented, with a focus on authenticity of food souvenirs. The following sections explore what makes a food souvenir (unique characteristics, typology and attributes), why people buy food souvenirs (i.e. purchase motivations), what impacts the selection of a food souvenir, and where food souvenirs are purchased. Finally, ideas for future research are presented. Because this chapter focuses on food retail, it does not investigate food travel activities (e.g. visits to restaurants, wineries, cooking classes) in detail. However, many activities, like visits to wineries, are accompanied by retail purchases. Therefore, food travel experiences are introduced throughout the chapter because of their relationship to food retail purchases while traveling. Everett (2016a) and Dixit (2019) have written extensively about food and drink tourism activities. For ease of reading, the word 'food' is often used to refer to both 'food and drink' items. This follows the World Food Travel Association, which uses the term 'food tourism' for 'food and drink tourism' (Wolf, 2014; WFTA, 2022).

What is Food Tourism/Culinary Tourism?

Individuals have always consumed food during their travels, but recently it has become much more common as a motivator for travel and as a desired touristic activity. A World Food Travel Association study of leisure travelers in six countries revealed that over half (53%) had recently been motivated to take a trip to eat the local or regional food or drink (Stone et al., 2020).

Culinary tourism (or food tourism) includes both tourism motivated by food and drink and the activity of participating in an intentional food or drink activity while traveling (Stone, 2022). Thus, retail purchases of food and drink while traveling often fit within the category of 'culinary tourism'. The term culinary tourism was first defined by Lucy Long (1998: 21) as 'the intentional, exploratory participation in the foodways of an other – participation including the consumption, preparation and presentation of a food item, cuisine, meal system, or eating style considered to belong to a culinary system not one's own'. A more recent definition (Smith & Xiao, 2008) noted that culinary tourism can also occur within one's own culture, provided the experience is 'intentional and reflective'.

The World Food Travel Association defines food tourism as 'the act of traveling for a taste of place to get a sense of place' (WFTA, 2022, n.p.). Tourists make food and drink retail purchases to get a taste of place and to bring these tastes home with them. This is where retail food and drink purchases fit into food tourism. The terms 'food tourism', 'gastronomic tourism' and 'culinary tourism' are closely related, and they are often used interchangeably (UN Tourism, 2012; Wolf, 2014). However, Everett (2016a) notes some perceived differences in the terms. For simplicity, the term food tourism and food travel will be used in this chapter to include food and drink tourism.

Food and drink retail within food tourism

Many, but not all, food retail purchases while traveling can be considered food tourism. It is not easy to specify exactly which activities are considered food tourism, but some guidelines can be presented. Food retail purchases may be part of food tourism, provided they are intentional and exploratory. For example, everyday activities like picking up a pre-packaged pastry at a convenience store for breakfast would typically not be considered food tourism. However, browsing the unusual or local pre-packaged foods and choosing some unusual ones to sample or to take home as gifts could be considered food tourism. It is important to note that not all travelers who participate in food travel activities can be considered 'culinary tourists'. Culinary tourists must not only participate in food activities while traveling, but food or drink must be a primary motivator when they choose a destination (Stone *et al.*, 2016). Over half (53%) of leisure travelers can be considered culinary travelers (Stone *et al.*, 2020).

Culinary tourism is a subset of cultural tourism, as it allows visitors to explore and experience a destination like a local. Food and drink can be some of the most distinctive elements of a destination's offering, and food and drink are often associated with destination image (Kivela & Crotts, 2006), as they become symbolic of the place (Lin & Mao, 2015). A robust food and beverage scene, with authentic and local products can encourage visitation, enhance the visitor experience and lead to economic benefits with food and drink purchases. Local food enhances the visitor experience by connecting travelers to a region's heritage and culture (Sims, 2009; Timothy, 2016). For example, whisky production and consumption is a part of the identity of Scotland, as well as a major tourist draw and source of retail revenue. Purchases of food souvenirs also increase destination awareness and promote destination image (Wilkins, 2011).

Travelers' local food purchases may also contribute to sustainability goals (Sims, 2009; UN Tourism, 2012). Food is 'part of the local culture that is consumed and can be promoted, contributing to the economic development of a tourism destination' (Bernardo & Rodrigues, 2020: 131). Travelers' food purchases contribute to economic sustainability, which extends beyond the individual journey. After returning home, travelers frequently repurchase food and drink they first experienced on a trip (Stone *et al.*, 2019). Food and drink support environmental sustainability when purchases benefit local producers, sustain local agriculture and reduce food miles. Sustainable sociocultural benefits of local craft food and drink experiences include 'a renewed sense of pride, identity and unity and allow local entrepreneurs to be creative, forge their own path, and earn a living through their passion' (Simone-Chateris, 2019: 438). It is easy to see how this extends to food and beverage retail. Exploratory food and drink consumption while traveling can be an antidote to globalization, as being on holiday exposes travelers to new foods (Altintzoglou *et al.*, 2016).

Beyond simple sustenance, food and drink also have value to the traveler. Researchers on consumption values have found that, aside from taste and quality, food consumption while traveling can provide emotional value (such as pleasure), prestige value (such as higher social status), epistemic value (such as learning more about a culture) and interaction value (social) (Choe & Kim, 2018; Stone & Zou, 2023). These seem to be more pronounced in a travel setting than at home. Food and drink given as gifts also provide a social value in maintaining relationships (Lin & Mao, 2015).

Souvenirs: An Overview

Quite simply, souvenirs are mementos collected on a journey. Swanson and Timothy (2012: 493) wrote that 'souvenirs are material commodities that are produced, sold, and consumed'. However, a souvenir can also be a free or priceless memento from a trip, such as a ticket stub or matchbook from a restaurant visited (MacCannell, 1976/1999). Tourism researchers (e.g. Wilkins, 2011) have also considered retail products purchased for others to be souvenirs, even if they are intended as a gift rather than being kept as a memento. This chapter acknowledges that both personal items and gifts can be classified as souvenirs.

The most commonly cited typology of souvenirs comes from Gordon (1986), who classified five types of souvenirs: pictorial images; piece-of-the-rock; symbolic shorthand; markers (e.g. t-shirts); and local product souvenirs (e.g. indigenous foods, food paraphernalia, local clothing and local crafts). Choi et al. (2008) categorized souvenirs into three types: local products (e.g. arts and crafts, foods); products marked with graphics; and clothing and jewelry.

Most tourism research focuses on physical objects as souvenirs, and food occupies a small corner of the souvenir literature. More often, researchers have considered souvenirs such as crafts (e.g. Littrell et al., 1993) and tchotchkes (e.g. Lasusa, 2007). For example, MacCannell's (1976/1999) seminal work, *The Tourist*, did not address food souvenirs in detail, instead mentioning items like necklaces, blown glass, boomerangs and Hawaiian shirts. The references to food-adjacent souvenirs are matchbook souvenirs from restaurants, rather than food or drink.

Food and drink have long been considered souvenir purchases. Sthapit (2017, 10–11) wrote that souvenir purchasing is 'almost essential' for travelers. Generally, 'souvenirs' are associated with being a 'commercially produced and purchased object to remind the purchaser of [a travel] experience' (Swanson & Horridge, 2006: 671), and souvenirs are often viewed as tangible means to encapsulate intangibly emotional experiences (Gordon, 1986). Researchers also include gifts for others under the heading of souvenirs (Wilkins, 2011). Food and drink are popular souvenirs for travelers. About two-thirds (67.4%) of leisure travelers bring back food and beverage from a trip to enjoy at home, and a similar percentage (65%) bring back food or beverage to give as gifts (Stone et al., 2019). Food and drink souvenirs may include everything from packaged chocolate to honey, jam, cheese, and local beer and wine. However, retail food and drink purchases while traveling extend beyond souvenir buying. Aside from souvenirs, tourists buy retail food for everyday consumption, to explore and experiment with local foods, and for later consumption.

Although food souvenirs are common in tourism, food has been very late to the souvenir literature. 'Despite its considerable importance, the role and meaning of food as a souvenir in tourism have rarely been discussed in the literature' (Lin & Mao, 2015: 20), and it has frequently been lumped together with unrelated items. While Gordon (1986) mentioned San Francisco sourdough bread, Greek olive oil, jam and liquor as food souvenirs, the food category was just a subcategory of local product souvenirs, which included local clothing and local crafts. Likewise, Choi *et al.* (2008) lumped food souvenirs with 'arts and crafts' in their local product category. Stone (2024), however, argues that food souvenirs should be classified separately from other souvenirs owing to their unique attributes.

The meanings and importance of souvenirs

Souvenirs have meaning beyond their material value. It is probably easy to think of a travel souvenir that holds special meaning to its owner. Swanson and Timothy (2012: 491) propose that a souvenir is not an 'ordinary item', but instead a holder of 'heightened meaning and symbolic transcendence'. Tourists actively create meaning in travel experiences (Littrell *et al.*, 1993), and souvenirs help to embody this meaning.

Souvenirs are often used to capture a moment in time for the traveler or share a moment in time with a gift. Through a souvenir, a person can share, enhance or re-live an event or experience. Swanson and Timothy (2012: 492) proposed that the object can 'transport the [unique destination] qualities home as reminders of what made the place special'. Food items that are taken home from a trip can embody the same meaning as 'traditional' souvenir purchases, like destination-branded paperweights, magnets and mugs.

'Touristic souvenirs are found in every corner of daily life' (MacCannell, 1976/1999: 147), not just from retail purchases. Any item may become a souvenir; Swanson and Timothy (2012) recognize that an ordinary commodity may become a souvenir, even 'by accident'. It may be that the more personal the experience is, the more meaningful a souvenir might be. For example, an Eiffel Tower statue may have less personal meaning than an empty bottle of wine from where a couple dined on their honeymoon.

Souvenirs given as gifts also have meaning beyond their material value, in that they possess a social value (Lin & Mao, 2015). Although these gifts will not carry the same value in recalling the journey or reliving the experience (as the recipient was not on the journey), souvenirs given as gifts help individuals 'express prestige, self-esteem and pride through the processes of souvenir giving and sharing' (Lin & Mao, 2015: 19). This topic is discussed more in a later section.

Souvenir Authenticity and Perceived Authenticity

Authenticity (or perceived authenticity) plays a key role in understanding tourists' retail purchases, including souvenirs. While a few paragraphs are insufficient to address authenticity, the concept must be mentioned. There are many reasons that tourists want to bring home 'authentic' souvenirs (Littrell *et al.*, 1993), and the meaning of a souvenir to the traveler may depend on its authenticity or perceived authenticity.

The concept of authenticity is ambiguous (Pizzichini *et al.*, 2020), and it is difficult to know what is actually authentic (MacCannell, 1976/1999), or even if true, objective authenticity exists. Determining authenticity in tourism is complicated by the proposition that much of tourism is 'staged' (MacCannell, 1976/1979). Authenticity is socially constructed (Cohen, 1988), personally constructed, contextual and changing (Littrell *et al.*, 1993). There are subjective ways to compare levels of authenticity (e.g. was an item crafted in a historical manner?; was it created locally?). However, what seems authentic to one person may not be authentic to another. This is important to remember when considering a traveler's personal decision to purchase a souvenir. In a food purchase scenario, one consumer may place more value on the heritage of the recipe, rather than the company producing the item. One may find only historic foods to be most authentic, while another may value modern creations that are associated with the area. Thus, 'perceived authenticity' is an important concept – acknowledging that authenticity may differ between individuals.

Food and drink may often be more authentic than other retail purchases, because food and drink are consumed by locals in their daily lives. Many tourists have the desire to participate in the 'real life' of the place and to see how locals live (MacCannell, 1976/1999), which helps to create an experience richer in authenticity. Therefore, purchasing and experiencing local food can be perceived as one of the best ways to get off of MacCannell's (1976/1999) proposed 'stage set' for tourists.

Although there is no singular classification of authenticity, Soukhathammavong and Park (2019) provide one example. In a survey of Laotian souvenir sellers, they identified many attributes of souvenir authenticity. These included: '1) integrating culture and history to represent place identity; 2) created in the city or nearby; 3) handmade or handicrafts; 4) a unique, attractive presentation; and 5) requiring specific skills of local artisans' (2019: 110). Bernardo and Kastenholz (2023) provide a comprehensive analysis of research on authenticity and meanings of souvenirs.

Establishing authenticity of food and drink can be complicated, and this also applies to food and drink retail purchases during travel. Björk and Kauppinen-Räisänen (2016) state that travelers value authenticity in local foods for food specialties in each region represent the region's unique identity (Everett, 2016b; Lin & Mao, 2015). In considering craft distilling, Simone-Chateris (2019) proposed many reasons that beverages may be thought of as authentic. The product may be linked to regional or local history, or it may use local ingredients. However, to be authentic, a food does not need to be indigenous (Lin & Mao, 2015). Owing to global movements of people and foods, it is often difficult to find foods that are truly indigenous (or native) to an area. For example, tomato sauces in Italy are authentic, but not indigenous, as they were introduced from the Americas. Lin and Mao (2015: 22) note that indigenous is a subset of authentic: 'All that is indigenous is authentic, but not all that is authentic is indigenous'.

Local and authentic are frequently related, but are also different concepts. Like the term 'authenticity', defining the term 'local' is also contested, but seeking local or traditional foods is part of a tourist's quest for authenticity (Sims, 2009). Eastham's (2019) research helps creates a structure for evaluating local foods. Local food supply chains can be categorized as foods grown in the region, foods processed in the region, foods grown and processed in the region, or foods produced elsewhere but representing traditional regional products (Eastham, 2019). Despite their differences, all of these may possess elements of authenticity. Many local foods are authentic. However, there are also many locally produced food retail products that may not be considered authentic. For example, a local artisan chocolate made and purchased in Chicago may represent a local craft food product. However, we cannot propose that chocolate is an authentic specialty of Chicago.

It appears that discussing authenticity in food souvenirs may result in more questions than answers. How authentic are foods brought by immigrants? How authentic are foods produced by recent entrepreneurs, even if they use heritage recipes? How authentic are locally-produced foods (e.g. jam or salsa) made by craft producers if the raw ingredients are from elsewhere or the recipes are not traditional? Likewise, how authentic are mass-produced foods, even if they are traditionally associated with an area? There may be a continuum of authenticity upon which to categorize food souvenirs, and here are several questions that can be asked to categorize authenticity, keeping in mind that different people may prioritize different issues in this categorization:

- Is the food or drink certified from a certain area of origin?
- Are the ingredients sourced locally?
- Is it produced locally?
- Is it a traditional food or traditional recipe?
- Is it a branded product of the area?
- Is it available at the traveler's home or only at the destination?
- Is it commonly consumed locally (regardless of the origin)?
- Is it perceived as authentic by locals?
- Is it a familiar product but prepared in a unique or special way?

What frequently makes a food souvenir purchase unique is that it is part of the authentic lived experience of the tourist. If a tourist eats a delicious barbeque meal in New York City and purchases a bottle of that barbeque sauce to relive the experience at home, then the sauce may have authentic meaning in the tourist's lived experience – even if the product was not original or authentic to that area. Thus, we may even categorize food purchases on a continuum from 'food from the tourist's lived experience' to 'food not experienced by the tourist'. On this scale, a snack purchased as a gift from an airport shop may not accompany the same depth of tourist experience as a bottle of wine picked up from a winery visited for a tasting.

What are Food Souvenirs? Typologies and Attributes of Food Retail Products

Food souvenirs are a subset of souvenirs (Gordon, 1986), often with authentic or local provenance. Food and drink brought home may have special meanings for tourists, helping them remember and even re-create experiences from their travels (Sthapit, 2017). There are many distinctions between food items purchased while traveling and other souvenirs, such as a handicraft, logoed item or tchotchke. Because of this, the study of food and drink retail tourism is likely different from retail tourism in general, and food souvenirs differ from other local souvenirs. For example, food and drink souvenirs are most often intended for consumption, rather than sitting on a shelf. Food retail products may be consumed while at the destination or while at home. Further, travelers may be more likely to repurchase food souvenirs they have enjoyed.

There are many different types of food, drink and culinary products that travelers purchase. This section explores ways of categorizing food souvenirs and their attributes. Buczkowska (2014) generally classified culinary souvenirs into four types: perishable food and beverage products; kitchen utensils and gadgets; recipes and photographs. This shows interest in culinary items, far beyond food retail. While this chapter concerns only food and drink products, culinary-related items are often sold alongside food.

Lin and Mao (2015) studied food specialty souvenirs. Through content analysis, they identified three important dimensions and 15 attributes which accompany the dimensions, of which the thirteen most common are described here. The *sensory dimension* includes flavor, texture, procedural innovation and visual appearance. The *utility dimension* includes characteristics of natural/organic, environmental sustainability, health, convenience and quality. The *symbolic dimension* includes elements of authentic/indigenous, traditional and craftsmanship. In a survey of Indonesian tourists, Sosianika *et al.* (2018) identify four general dimensions of food souvenirs: tangible, brand and packaging, value and food quality. Horodyski *et al.* (2014) (cited in Medeiros *et al.*, 2017: 350) proposed that a food souvenir must have a distinguishable identity, identification of origin, adequate packaging and easy transport.'

These categories do not seem to adequately encapsulate the many ways of categorizing food and drink souvenirs. Here I propose additional ways (though likely not exhaustive) of understanding and grouping food and drink souvenirs. Many of these attributes can be placed on a continuum.

- *Intended for tourists vs. intended for locals.* The ubiquitous Mozart chocolate balls sold all over Vienna, or Devonshire fudge 'wrapped in sunshine' (Cleave, 2013) seem to have tourists as their primary audience. However, a farmer's market or craft brewery's retail store may produce foods targeting local consumers. Either category can be souvenirs, as 'items that were not intentionally manufactured to be tourist mementos can be considered souvenirs if they fulfill the functions of memory holder and tradable commodity among tourists' (Swanson & Timothy, 2012: 493).
- *Mass production vs. craft production.* This divides the food or drink based on the size of the operation. Hershey, of the world's largest chocolate producers, can be contrasted with a local chocolate shop in France. Marshmallow Peeps (a mass-produced American candy most often associated with holidays like Easter) can be contrasted with craft production of local candies.
- *Local specialty vs. national specialty.* Some food and drink is particular to a certain area, while other foods are thought of in a larger scope. Here, one could think about what foods are 'American' compared to what foods are 'Southern' or available primarily in a particular city.
- *Produced onsite vs. produced elsewhere.* A tourist could purchase the same local beer from the brewery (point of production) or from a retail store. It may mean more for a tourist to make a food and drink purchase where it was produced. Lin and Mao (2015: 27) wrote that 'tourists appreciate craftsmanship when they are able to personally observe how talented local people make food specialties'.
- *Protected designation of origin (PDO) and protected geographical indication (PGI).* Many local food products contain a certification from a protected geographic origin or similar labeling, and certified products may be especially sought important in travel food retail. Some governments and agencies certify that agricultural products or processed foods are grown, processed and/or produced in a specific region. These certifications – protected designation of origin (PDO) and protected geographical indication (PGI) – are key indicators of local foods. 'The PGI is a differentiator that identifies and distinguishes characteristics of the product's geographical origin, whether related to its method of production, the community's know-how, or simply the quality of the raw materials. Its implementation is often justified as a guarantee of its origin or quality' (Bernardo & Rodrigues, 2020: 133). For example, certifications are important among food and wine products from designated Italian regions, and these certifications can lead to a higher cost to the consumer.

A few studies have shown that consumers perceive quality differences between certified and non-certified items and a higher likelihood to purchase the certified product (e.g. Supeková *et al.* (2008) in Slovakia; and Likoudis *et al.* (2016) in Greece). Therefore, it might be that PGI products may be especially important in travelers' food purchases because this designation adds an element of authenticity.

However, many areas (including the USA) do not have PGIs. In addition, Bernardo and Rodrigues (2020) argue that the 'heritagization' of food has a downside. The strict rules around certified products may favor certain producers and allow only some of a region's producers to benefit. 'Some local elements stand out and monopolize the official narrative around a vast heritage made singular and particular' (Bernardo & Rodrigues, 2020: 138). Thus, it is best to accept that PGI is advantageous, but not required, for food retail.

Why Do People Purchase Food Souvenirs and Retail Food Products?

Quite simply, food and drink can be bought while traveling for one's self or as a gift (Buczkowska, 2014). Among Chinese visitors to Taiwan, Lin (2017) found that food souvenirs were purchased for the tourist ('myself') (23.5%), family (32.6%), friends (24.4%) and work colleagues/supervisors (19.4%). Although this finding is unlikely to be universal, it shows that the consumer may be different from the purchaser. Regarding souvenir purchases in general, Wilkins (2011) listed three basic motivations for souvenir purchases: souvenirs as gifts; souvenirs as memory-keeper; and souvenirs as evidence of having been somewhere. In considering food souvenirs among Chinese tourists in Taiwan, Lin (2017) found that giving food as a gift was a stronger motive than food as memory-holder, but this was the only statistical difference in tourists' motivation to purchase.

Because food is also physically consumed, not just gifted or displayed, there seem to be additional reasons for food and drink retail purchases. These include food souvenirs as gifts, as memory-makers and evidence of having 'been there' (Wilkins, 2011). These may relate to the use of the food, rather than its social or emotional meanings. For food and beverage retail purchases, I add two additional categories of motivations to purchase food souvenirs: souvenirs to be consumed at home and food retail for consumption while traveling. The last category fits within the discussion of food retail, but not necessarily within discussions of souvenirs.

Food souvenirs are given as gifts in many cultures, particularly in societies with a sociocultural obligation of gift-giving (Swanson & Timothy, 2012). This is especially important in Asian cultures (Xu & McGehee, 2012), including the concept of *omiyage* in Japan and *sunmul* in Korea (investigated in detail by Park, 2000), where food and drink are among the most popular gifts. A recent Cebu, Philippines, tourism report states that a majority of tourists picked food as a homecoming gift (*pasalubong*) for friends and family (Cacho, 2022). In a Brazilian study on Serro cheese, 55% of travelers bought the cheese as a gift (Medeiros *et al.*, 2017). Food souvenirs allow travelers to share a 'taste of your trip' with family and friends (Lin & Mao, 2015).

Food souvenirs as memory refers to the way a souvenir can remind someone of, or help them recollect, an experience. Buczkowska (2014) found that three-quarters of respondents in her study bought souvenirs for themselves to remind them of the places visited. In a Finnish study, Sthapit (2017) found that buying food souvenirs prolonged the memorability of tourists' food experiences at the destination.

Food souvenirs as evidence focuses more on displaying or showing off than remembering. Souvenirs as evidence is to have proof of having been to a destination or having an experience. Buczkowska (2014) acknowledged that some food and drink products are never opened or consumed, while some consumers may retain the packaging of already-consumed products.

In the case of *food souvenirs for consumption at home*, souvenirs as memory and souvenirs as evidence seem to suggest a level of permanence. While food souvenirs may be perceived by their owners to have permanence, the primary role of food is consumption, and foods have a limited shelf life. Buczkowska (2014) found that 63% of her participants consumed food souvenirs 'some time' after returning from a trip, with about 36% consuming the items right after arriving at home. While some products may have an element of memory (e.g. displaying 'canned bear meat' at home), most food products are intended to be eaten.

Food retail for consumption while traveling suggests that retail food is not exclusively for souvenirs, so purchasing for immediate consumption is an additional category to consider. Much of the retail food expenditure, especially from grocery stores, is consumed immediately. As it has become more likely for travelers to stay in accommodations with kitchens (e.g. peer-to-peer rental accommodations), preparing food has become a more common travel activity. Part of the novelty of being in a new destination is intentionally experimenting with new foods or brands. Trying and cooking new foods while traveling may impact a tourist's decision to bring home these products for themselves or others. Thus, the everyday grocery shopping experience may also impact souvenir purchases.

Product Attributes Affecting Selection of Retail Food Souvenirs

Food attributes are a primary reason to select a food souvenir. Altintzoglou *et al.* (2016) found that taste, quality and local production to be the three most important factors in choosing food souvenirs in

Norway. Lin and Mao (2015) identify attributes of foods that may impact food souvenir buying. These include natural and organic, flavor versatility, convenience, authenticity and indigenousness, and craftsmanship. However, their analysis includes only attributes of award-winning food products rather than consumer behavior. In a very localized study, focusing on brown sugar steamed cake in the Penghu Islands, Taiwan, Chen *et al.* (2022) found that souvenir food image, the destination brand of brown sugar cake, and perceived tourist value accounted for a majority of the variance in purchase intentions.

Practical (transport) considerations also influence food retail purchases, as not all foods are suitable to be taken home as souvenirs. Travelers consider the suitability and ease of transport (Buczkowska, 2014), accounting for the popularity of dry goods, snack food and candy. Behind taste, quality and local production, the next most important factors in food souvenir buying in Norway were practical: customs barriers or regulations, non-perishability, no smell, non-breakable and size, all rating at least 5 on a 7-point Likert scale of importance (Altintzoglou *et al.*, 2016). Transport considerations would obviously be a lesser consideration for foods consumed at the destination.

Lin (2017) identified pastries, cookies and confections as the most common type of food souvenir among Chinese visitors to Taiwan (up to 83% on recent trips), and she proposed this was because these foods are relatively cheap, lightweight, portable, suitable for young and old, and easy to share (Lin, 2017). The next most popular category was beverages (e.g. wine, tea, coffee). In a very small sample, Buczkowska (2014) found that the most common food souvenir purchases were sweets or alcohol, each purchased by at least half of respondents. At least 30% purchased spices, tea, fruit, coffee, cheeses and oil. In terms of gifts for others, alcohol and sweets were most popular.

Cultural distance and uniqueness/familiarity may also impact souvenir buying. Travelers may select food souvenirs that are not available, or that are different from those available at home (Altintzoglou *et al.*, 2016; Buczkowska, 2014). Food souvenir purchases may also be affected by familiarity with local foods, and this may depend on the number of previous visits (Altintzoglou *et al.*, 2016). Basically, familiarity leads to a higher likelihood of buying. Soukhathammavong and Park (2019) suggest that local product souvenirs are especially meaningful when they can only be purchased at the destination.

Where are Food Souvenirs and Retail Foods Purchased?

Food souvenir (and food retail) purchases while traveling have tended to focus on *what* was purchased and *why* it was purchased, rather than *where* it is purchased. However, purchase locations may provide additional insight. Buczkowska (2014) suggests that food souvenir purchases can categorized in two ways. A spur-of-the-moment purchase is influenced by a particular place where the tourist tasted the food or was attracted to its packaging. The second category is an intentional purchase influenced by previous knowledge, a prior visit, a guidebook, or a request to bring an item home for an acquaintance. This behavior, therefore, can be categorized as either serendipitous or planned.

Lin (2017) asked Chinese tourists to Taiwan where they purchased a food souvenir. Results (some categories overlapping) showed tourist destination (27.5%), souvenir store (14.6%), food specialty store (19.2%), convenience store/supermarket (11.3%), airport/train station (15.3%) and food vendor/market (11.3%). Travelers may also purchase food souvenirs at more than one location. Most of Lin's categories are retail locations, but tourists' food purchases have a direct connection with an onsite consumption or production experience. For example, travelers may purchase packaged food or drink from wineries, breweries and restaurants. The meaning that a traveler associates with a food product may differ based upon where the item was purchased, and in what context.

Tourists purchase retail food and drink as souvenirs or otherwise at food specialty stores, gourmet shops, gift shops, supermarkets, ethnic grocers, factories and production sites, farms and orchards, and farmers markets or food halls. Some (such as grocery stores) may be used more for local consumption, rather than souvenirs. Others, such as international branded food retail stores, may be used primarily for souvenir purchases.

Food specialty stores can be of general interest (a variety of food and drinks), specific to a product line (such as a wine store), or specific to a particular brand. This can be divided into smaller categories, including general food and drink, product specialty, branded stores (local, regional, national or international). General food and drink specialty stores are retailers that may not be considered large enough to be considered as a grocery or supermarket. Local branded retail stores are operated by a specific brand or producer, and primarily offer their brand (with perhaps a few related products). Examples include the Original Ghirardelli Ice Cream and Chocolate Shop (San Francisco, California, USA) and Björn's Colorado Honey (USA). Although they also may offer some consumer goods (e.g. skincare products and jewelry), their primary focus is food. Regional branded retail stores are also operated by the brand or the producer. However, they are regional, rather than local in scope. Fannie May Chocolates was founded in Chicago in 1920 and now has nearly 50 retail stores across the Midwestern US. National and international branded food retail stores include shops like the M&Ms Store, which can be found at many tourist destinations. These stores are operated by multinational brands, so their souvenirs may lack authenticity, originality, or a local flair, but they still attract tourists. These stores may seem more like souvenir shops, rather than food shops, because they offer many non-edible consumer goods such as t-shirts, along with food items.

Gourmet stores, gift shops and specialty retailers sell a variety of products, often with a focus on local goods including foods. Along with local crafts and art, Made in Chico (Chico, California, USA) and Made in Colorado (Denver, USA) offer food and drink specialties that are produced locally. These range from agricultural products (olive oil, almonds and walnuts in Chico) to honey, chocolate, teas, sausages and spices in both stores. Other stores may be more generally focused on souvenirs, but also include food. Airport retailers increasingly offer local food and drink in their stores, from the stroopwafels in Amsterdam's Schiphol Airport to the ice wine in Toronto's Pearson Airport. Hotel gift shops increasingly feature local food and drinks as souvenirs and for consumption while staying at the hotel.

The markets for grocery stores and supermarkets are not primarily tourists, but tourists often find food souvenirs there, as well as foods for local consumption. With the growth in peer-to-peer accommodations with kitchens (e.g. Airbnb), purchasing food at stores, rather than solely at restaurants, is a growing practice. In addition, tourists have increasingly added grocery stores to their travel agendas and consumer spaces. Nearly half (47%) of the leisure travelers in Stone *et al.*'s (2016) study shopped at a grocery or gourmet store on a recent trip. Travelers may seek products that locals consume, even if they are not manufactured locally. For example, an American traveler to London may find British sauces, candies, or digestive cookies that British people would enjoy, even if the products are not made in metropolitan London. Likewise, a visitor to the USA may pick up boxed macaroni and cheese or Pop Tarts as gifts, or to enjoy at home. These foods, while they are mass produced, are part of the lived American experience and difficult to find overseas.

Ethnic stores and groceries often share characteristics with grocery stores or specialty food shops, but they specialize in international or ethnic products. Cities have ethnic enclaves that attract visitors; popular ones in the US include San Francisco's Chinatown or New York's Little Italy. Although the main purpose of visiting these areas may be a cultural experience or dining, there are many stores that specialized in unique world foods. While these foods may be imported, travelers purchase them as food souvenirs or as gifts. Others may visit cultural areas with the primary purpose of visiting ethnic stores or groceries to buy items to take home. This type of travel is popular for expatriates, such as Asian Americans living in College Station, Texas, who drive to Houston to shop in international grocery stores.

Production sites and factories including food and drink producers – from wineries to breweries to candy makers – often have gift shops on their premises. This may be a mass-production facility, such as the Jelly Belly Factory Store in Fairfield, California, USA, which offers a variety of candy products. Or, it may be a craft producer, such as a local fudge shop in Mackinac Island, Michigan, USA, where tourists can see the sweets being made by hand. In beverages, wineries make money from cellar door retail sales (Everett, 2016a). Sierra Nevada Brewing (Chico, California and Mills River, North Carolina, USA) sell beers in their gift shop – some of which are available nationwide, with others only available in the store.

Farms or orchards are places where a food product is grown or raised. There may also be production functions, such as processing or packaging, at the site, but the primary purpose of this location is growing. In addition to selling fresh produce, many locations also sell a variety of processed products. The country store at Curtis Orchard in Champaign, Illinois, USA, sells apple butter, jams and jellies, sauces, dressings and popcorn products. Some consumables may be made onsite, while others are made by outside producers. The California State University, Chico Farm sells raw meats for local consumption, as well as packaged products like beef jerky. According to a study by Stone *et al.* (2020), 22% of leisure travelers visited a farm or orchard on a recent trip.

While there are distinctions between a farmers market and a food hall, they both comprise individual vendors. These vendors may be growers (farmers selling fruit and vegetables), producers (e.g. jam makers), and retail vendors (e.g. wine shops). Even traditional farmers markets have often expanded to include producers other than just growers. Travelers may find these locations to be especially authentic because they can often meet the producer or grower, they can shop along with locals (rather than just tourists), and it is often a learning experience about local agriculture. Halas Market in Vilnius, Lithuania, has a traditional vegetable, fruit and meat market but also stalls for wine, imported food and drink, baked goods and a few small restaurants. Acknowledging the market's tourist appeal, the city of Vilnius prepared a tourist brochure about the market in multiple languages. Over one-quarter of leisure travelers (29%) in Stone *et al.*'s (2020) study visited a farmers market or agricultural fair on a recent trip.

Conclusion

This chapter reviews food and drink retail purchases while traveling, with a focus on souvenirs. It places food retail purchases within the study of food tourism (culinary tourism), as well as the literature on souvenirs. Food souvenirs can be viewed through the lenses of authenticity and perceived authenticity, which may be placed on a continuum, although there are many different ways to view how authentic a product is. Souvenirs may also be perceived as authentic because they are part of a traveler's authentic lived experience.

Food souvenirs (as well as other food purchases while traveling) may be distinct from other souvenir purchases. Food souvenirs have a functional use, and consumption, including the sensory aspects of food, is a major element of experiencing a food souvenir. Therefore, food retail purchases in a tourism setting may be distinct from other retail purchases.

Finally, the chapter reviewed previous research on souvenirs and updated and expanded several areas of inquiry, including classifying food souvenirs, why travelers buy food souvenirs, and where food souvenirs are purchased. It is hoped this research will lead to a better understanding of tourists' food and drink retail purchases, which are important to travelers, local businesses and tourism destinations.

This chapter presents new ways of classifying and understanding tourists' food retail purchases. Researchers should use many of these characteristics and classifications when conducting future research, especially as it relates to the characteristics that separate food retail from other tangible retail and souvenir consumption. There are also many other ways to extend research on food retail.

More research could be undertaken on perceptions of authenticity. When considering consumer behavior, perceptions of authenticity may be more important than other measures of authenticity. What is important to consumers regarding authenticity in food souvenirs and retail food purchases? In a study in Norway, taste, quality and local production were found to be the most important factors to influence food souvenir purchases (Altintzoglou et al., 2016), which suggests that authenticity is not a particularly important factor for some consumers. Consumers may state that they prefer authenticity, but purchase decisions can be more complex. An experiment held in an actual retail setting, varying packaging and statements of authenticity and origin may yield more insight.

Researchers might consider investigating how the meaning of a food purchase during travel is different in different situations. For example, does the food hold more personal meaning or importance if it is experienced first on a vacation, rather than just purchased at a destination to bring home? Does it have more meaning or importance if the consumer visited the place where the food was grown or prepared? When given as a gift, do any of these meanings pass along to the recipient? For example, does 'I brought you this blackberry jam from a farm we visited' hold more meaning than 'I brought you this blackberry jam'? When a food is repurchased at home, what emotions and senses are heightened (if any)? All types of travelers may purchase food souvenirs, but Swanson and Timothy (2012) proposed that food (and other functional) souvenirs may be more prioritized as people travel more. This should be empirically investigated.

Focusing mainly on food experiences, such as restaurant dining, Kim et al. (2009) found many motivational factors for experiencing local food at a destination (e.g. escape from routine, authentic experience, prestige), and these findings can also be applied to tourists' retail food purchases. Similar motivational factors could be applied to retail food purchases. In addition, the selection of food experiences by tourists could be compared with food retail purchases by tourists to identify similarities and differences in consumer behavior.

This chapter identifies ways in which food souvenirs are different from non-food souvenirs – from the senses involved to the connection with local life. Further research could investigate the importance of these factors. Do these factors give food souvenirs more personal value or importance than other sorts of souvenirs, as suggested by Hazman-Wong and Sumarjan (2016). What attributes entice travelers to choose food souvenirs (for themselves or as gifts), rather than other consumer goods?

Relatedly, more research on gift recipients could help shed more light on not just the act of souvenir purchase and gift giving, but also the results of travelers' actions. Are food gifts preferred by recipients to other types of gifts from travel? It has been established that food and drink are purchased as gifts for social reasons. Researchers could uncover more about the perceptions of the gift receivers. Do food and drink gifts make recipients more knowledgeable about the culture of the place the items came from? Do they created interest in the culture of the destination (source of the food product) or increase a desire to visit a destination?

References

Altintzoglou, T., Heide, M. and Borch, T. (2016) Food souvenirs: Buying behaviour of tourists in Norway. *British Food Journal* 118 (1), 199–131.

Bernardo, E. and Kastenholz, E. (2023) Souvenirs in tourism studies: A thematic analytical framework. *Tourism Culture and Communication* 23 (4), 333–346.

Bernardo, E. and Rodrigues, V. (2020) Buying sweet memories: The heritagization of food souvenirs in northern Portugal. *Journal of Gastronomy and Tourism* 4 (3), 129–140.

Björk, P. and Kauppinen-Räisänen, H. (2016) Local food: A source for destination attraction. *International Journal of Contemporary Hospitality Management* 28 (1), 177–194.

Buczkowska, K. (2014) Local food and beverage products as important tourist souvenirs. *Turystyka Kulturowa* 1, 47–58.

Cacho, K.O. (2022, November 11) DOT brings gastronomy fest to Cebu. *SunStar.* Accessed from https://ph.news.yahoo.com/dot-brings-gastronomy-fest-cebu-113600149.html on December 10, 2022.

Chen, C., Chen, H.B., Yeh, S.S., Tseng, L.Y. and Huan, T.C. (2022) Exploring tourists' purchase intention of food-related souvenirs. *Tourism Management Perspectives* 44, 101035.

Choe, J.Y.J. and Kim, S.S. (2018) Effects of tourists' local food consumption value on attitude, food destination image, and behavioral intention. *International Journal of Hospitality Management* 71, 1–10.

Choi, T.M., Liu, S.C., Pang, K.M. and Chow, P.S. (2008) Shopping behaviors of individual tourists from the Chinese Mainland to Hong Kong. *Tourism Management* 29 (4), 811–820.

Cleave, P. (2013) Sugar in tourism: 'Wrapped in Devonshire sunshine'. In L. Jolliffe (ed.) *Sugar Heritage and Tourism in Transition* (pp. 159–174). Channel View Publications.

Cohen, E. (1988) Authenticity and commoditization in tourism. *Annals of Tourism Research* 15 (3), 371–386.

Dixit, S.K. (2019) Gastronomic tourism: A theoretical construct. In S.K. Dixit (ed.) *The Routledge Handbook of Gastronomic Tourism* (pp. 13–23). Routledge.

Eastham, J. (2019) Sustainable supply chains in gastronomic tourism. In S.K. Dixit (ed.) *The Routledge Handbook of Gastronomic Tourism* (pp. 225–233). Routledge.

Everett, S. (2016a) *Food and Drink Tourism: Principles and Practices*. Sage.
Everett, S. (2016b) Iconic cuisines, marketing and place promotion. In D.J. Timothy (ed.) *Heritage Cuisines: Traditions, Identities and Tourism* (pp. 119–131). Routledge.
Gordon, B. (1986) The souvenir: Messenger of the extraordinary. *Journal of Popular Culture* 20 (3), 135–146.
Hazman-Wong, N.F.S. and Sumarjan, N. (2016) The potentiality of food as tourism souvenir product. In M.F.S.B. Salamiah, A. Jamal, S.M. Radzi, N. Sumarjan and C.T. Chik (eds) *Innovation and Best Practices in Hospitality and Tourism Research* (pp. 305–308). Routledge.
Horodyski, G.S., Manosso, F.C., Bizinelli, C. and Gândara, J.M. (2014) Souvenirs Gastronômicos como Lembranças de Viagem: Um estudo de ca2so em Curitiba – Brasil. *Via@ Revista Internacional E Interdisciplinar de Turismo* 2, 1–16.
Kim, Y.G., Eves, A. and Scarles, C. (2009) Building a model of local food consumption on trips and holidays: A grounded theory approach. *International Journal of Hospitality Management* 28 (3), 423–431.
Kivela, J. and Crotts, J.C. (2006) Tourism and gastronomy: Gastronomy's influence on how tourists experience a destination. *Journal of Hospitality and Tourism Research* 30 (3), 354–377.
Lasusa, D.M. (2007) Eiffel Tower key chains and other pieces of reality: The philosophy of souvenirs. *The Philosophical Forum* 38 (3), 271–287.
Likoudis, Z., Sdrali, D., Costarelli, V. and Apostolopoulos, C. (2016) Consumers' intention to buy protected designation of origin and protected geographical indication foodstuffs: The case of Greece. *International Journal of Consumer Studies* 40, 283–289.
Lin, L. (2017) Food souvenirs as gifts: Tourist perspectives and their motivational basis in Chinese culture. *Journal of Tourism and Cultural Change* 15 (5), 439–454.
Lin, L. and Mao, P.C. (2015) Food for memories and culture – A content analysis study of food specialties and souvenirs. *Journal of Hospitality and Tourism Management* 22, 19–29.
Littrell, M.A., Anderson, L.F. and Brown, P.J. (1993) What makes a craft souvenir authentic? *Annals of Tourism Research* 20 (1), 197–215.
Long, L.M. (1998) Culinary tourism: A folkloristic perspective on eating and otherness. *Southern Folklore* 55 (3), 181–204.
MacCannell, D. (with Lippard, L.L.) (1999) *The Tourist: A New Theory of the Leisure Class*. University of California Press. (Original work published 1976.)
Medeiros, M.D.L., Horodyski, G.S. and Passador, J.L. (2017) Food souvenirs in the perception of the tourist: The case of the artisanal minas Serro cheese. *Revista Brasileira de Pesquisa em Turismo* 11, 347–364.
Park, M.K. (2000) Social and cultural fac-tors influencing tourists' souvenir-purchasing behavior: A comparative study on Japanese 'Omiyage' and Korean 'Sunmul'. *Journal of Travel and Tourism Marketing* 9 (1–2), 81–91.
Pizzichini, L., Temperini, V. and Gregori, G.L. (2020) Place branding and local food souvenirs: The ethical attributes of national parks' brands. *Journal of Place Management and Development* 13 (2), 163–175.
Simone-Charteris, M.T. (2019) Craft drinks tourism worldwide and in Northern Ireland. In S.K. Dixit (ed.) *The Routledge Handbook of Gastronomic Tourism* (pp. 420–430). Routledge.
Sims, R. (2009) Food, place and authenticity: Local food and the sustainable tourism experience. *Journal of Sustainable Tourism* 17 (3), 321–336.
Smith, S.L. and Xiao, H. (2008) Culinary tourism supply chains: A preliminary examination. *Journal of Travel Research* 46 (3), 289–299.
Sosianika, A., Suhaeni, T. and Wibisono, N. and Suhartanto, D. (2018) The dimension of food souvenir: An exploratory-confirmatory factor analysis. In *MATEC Web of Conferences* (Vol. 218, p. 04002). EDP Sciences. https://doi.org/10.1051/matecconf/201821804002

Soukhathammavong, B. and Park, E. (2019) The authentic souvenir: What does it mean to souvenir suppliers in the heritage destination? *Tourism Management* 72, 105–116.

Sthapit, E. (2017) Exploring tourists' memorable food experiences: A study of visitors to Santa's official hometown. *Anatolia* 28 (3), 404–421.

Stone, M.J. (2022) Culinary tourism. In D. Buhalis (ed.) *Encyclopedia of Tourism Management and Marketing* (pp. 696–698). Edward Elgar Publishing.

Stone, M.J. (2024) Not just another trinket: Defining unique attributes of food souvenirs. *Tourism Recreation Research*. https://doi.org/10.1080/02508281.2023.2296813

Stone, M.J., Migacz, S., Garibaldi, R. and Wolf, E. (2020) *2020 Food Travel Monitor*. Portland, OR: World Food Travel Association.

Stone, M.J., Migacz, S. and Wolf, E. (2016) *2016 Food Travel Monitor*. Portland, OR: World Food Travel Association.

Stone, M.J., Migacz, S. and Wolf, E. (2019) Beyond the journey: The lasting impact of culinary tourism activities. *Current Issues in Tourism* 22 (2), 147–152.

Stone, M.J. and Zou, S. (2023) Consumption value in food tourism: The effects on purchase involvement and post-travel behaviours. *Tourism Recreation Research*. https://doi.org/10.1080/02508281.2023.2246737

Supeková, S., Honza, M. and Kacenová, D. (2008) Perception of Slovak foodstuffs designated by the protected geographical indication by Slovak consumers. *Journal of Food and Nutrition Research* 47, 205–208.

Swanson, K.K. and Horridge, P.E. (2006) Travel motivations as souvenir purchase indicators. *Tourism Management* 27 (4), 671–683.

Swanson, K.K. and Timothy, D.J. (2012) Souvenirs: Icons of meaning, commercialization and commoditization. *Tourism Management* 33 (3), 489–499.

Timothy, D.J. (ed.) (2016) *Heritage Cuisines: Traditions, Identities and Tourism*. Routledge.

UN Tourism (2012) *Global Report on Food Tourism*. UN Tourism.

Wilkins, H. (2011) Souvenirs: What and why we buy. *Journal of Travel Research* 50 (3), 239–247.

Wolf, E. (2014) *Have Fork Will Travel: A Practical Handbook for Food and Drink Tourism Professionals*. World Food Travel Association.

World Food Travel Association (WFTA) (2022) What is food tourism? Online: https://www.worldfoodtravel.org/what-is-food-tourism.

Xu, Y. and McGehee, N.G. (2012) Shopping behavior of Chinese tourists visiting the United States: Letting the shoppers do the talking. *Tourism Management* 33 (2), 427–430.

Part 4

Conclusions: Past, Present and Future Perspectives

15 Shopping Tourism and Tourist Shopping: Looking Backward, Looking Forward

Dallen J. Timothy

Introduction

The contents of this book have addressed many key, disruptive concepts related to the multifarious relationships between tourism and shopping, including key management challenges. Among others, these themes and concepts include heritage and heritage-making, place and retail, landscapes of shopping, the transformation of traditional food from culinary experience to take-home souvenir, the importance of location to retail success, marginal places and shopping successes, the value of experience sometimes being more important than price, the changing demand for shopping tourism and emerging markets, the influence of cultural background and nationality in retail decision-making and behavior, interculturalism, urban renewal and gentrification, the parameters of luxury shopping, place-making, place-branding, the specifics of tourism shopping in different tourism contexts, governance and taxation, tax-exempt retail, customer loyalty, shopping festivals and events, and the crucial element of creating satisfied customers from both merchants' and destinations' perspectives.

Many of these concepts have been thoroughly studied, so much so that many new publications seem to offer little new knowledge or conceptual depth as they repeat much of the work that has been done already on client satisfaction, loyalty and trust, branding and general retail management successes. The chapters in this volume go beyond those normative perspectives to shed light on the opportunities for, and challenges to, retail and tourism.

Not every chapter intended for this volume came to fruition for various reasons. Thus, several key concepts were not included in detail, but they were not forgotten. Although the contents of this tome cover a wide range of issues and management challenges, additional discussion is warranted. The remainder of this chapter examines several notions that we know relatively little about and whose knowledge could be fruitfully brought to bear with additional research. These topics include spaces of shopping, souvenirs, tourism types and their connections to shopping, non-traditional shopper tourists, non-destination retail spaces, technology and shopping, sustainability and green retail, and shopping tourism in times of crisis.

Spaces of Shopping

In an effort to understand how merchants and destinations can best extract additional expenditures from shopping tourists and create positive destination experiences, much research has been done over the past 50 years on formal retail spaces and the characteristics of shops, malls and retail districts that educe eudemonic experiences in the shopping encounter. Window shopping, using one's imagination, feeling and handling merchandise, and acquiring new consumer products all contribute to the enjoyment of shopping. We now know a great deal about how physical design, merchandising, scents and aromas, visual stimuli and even sounds influence people's leisure and tourism shopping behaviors (Doucé & Janssens, 2013; El-Adly & Eid, 2016; Esfandiar *et al.*, 2024). We also know a lot about the environmental qualities of urban shopping districts (e.g. comfortable walkways, resting benches, landscaping and signage) that can create a sense of place, raise consumer comfort, and escalate people's interest in browsing, often inducing increased expenditures and overall destination satisfaction (McKercher, 2020; Murphy *et al.*, 2011). Much of this shopping appeal has come about since the 1980s with the increased Disneyfication of the retail experience – shoppertainment and retailtainment, in which the theatrical, ludic and entertainment factors associated with the world's mega-malls (e.g. cinemas, bowling alleys, skating rinks, gaming centers, and much more) are equal to, or possibly even exceed, the role of actual shopping in the largest shopping centers (Timothy, 2005; Zaidan, 2016, 2019). Yet, there are many traditional commercial venues that have not received adequate attention by shopping scholars from a retail and/or place perspective. Plenty has been written about these places from a sociospatial viewpoint, but we could learn more about their role in creating a holistic shopping environment that provides both a heritage narrative and a memorable retail experience for destination visitors.

One of these alternative retail spaces is traditional markets. These take many forms, from the souks and bazaars of the Middle East, to the farmers markets of the US Midwest. Likewise, the fish markets of coastal communities, and the flea markets of small towns and urban centers not only serve the needs of the local community, but also attract the gaze of out-of-town visitors. The souks and bazaars of Southwest Asia and North

Africa provide unique combinations of sounds, smells and flavors not found in any other place, and they are a salient element of local heritage that continue to play an important social role in those societies.

Although there is a growing literature on farmers markets, we know little about their potential as tourism assets from both supply and demand perspectives (Beer *et al.*, 2012; Garner & Ayala, 2019; Hall, 2016; Thompson, 2020). As Ripoll González *et al.* (2022) rightfully acknowledge, farmers markets may have an important role to play in fostering sustainable agriculture practices, sustainable tourism at the local level, and more direct economic earnings by cutting out the intermediaries in the distribution process.

Garner and Ayala (2019) study how farmers markets can be instrumental in creating regional brands and destination identities related to food production and culinary heritage. Consumers are generally willing to drive considerable distances to mingle with locals and buy locally produced organic food. Fishmongers and farmers markets provide many self-catering tourists with the sustenance they need while also providing a heritage experience.

The floating markets of Southeast Asia have long been included in cultural tours of Bangkok and other cities of the region (Nguyen & Huynh, 2024), but have they ever been studied from a retail perspective? While the original merchandise sold in these markets was generally utilitarian in nature, including fruits, vegetables and household products, the touristification of these spaces of commerce has seen an increase in the number of traditional vendors who now sell mass-produced souvenirs or locally made handicrafts. Researching the evolution of such retail, and often informal economy-oriented, spaces and the variables that effect such changes would provide deeper knowledge about the touristification of traditional retail spaces, how communities cope with tourism and the effects of tourism on the local retail economy.

Many tourist destinations are also well populated by street vendors hawking a wide range of consumer goods. Some studies have shed light on the potential role that street sellers play in tourism as retailers of small-ticket items such as food, clothing, toys, books, cigarettes and souvenirs (Correia & Kozak, 2016; Cukier & Wall, 1994; Steel, 2009, 2012; Truong, 2018), as well as their connections to formal and informal economies (Timothy & Wall, 1997), and the spaces they occupy as semi-private goods, common pool resources or collective goods (Damayanti *et al.*, 2018). Questions about the informal economy in tourist destinations need to be raised and answered, especially those pertaining to resident employment, government control and the rights of residents to earn a livelihood outside the formalized economic framework of government control. It is often difficult to study people involved in the informal economy, as such investigations could lead to increased government intervention and eventually formalization, which might defeat the purposes of the informal or shadow economy which so many destination residents rely on for their livelihoods. Thus, recruiting participants in a vendor study might face its own unique set of challenges.

Merchants in the informal economy are particularly vulnerable to the effects of crises. For them, economic, social, ecological and political disruptions put their livelihoods in jeopardy as tourists stop coming. Likewise, because of their work in the underground economy, they are frequently ineligible for social benefits, such as social security, insurance, unemployment benefits or sick leave, the way workers in the formal economy are. The socioeconomic implications (and many others) are relatively unknown in the shopping tourism domain of the shadow economy.

Souvenirs

There is a strong academic record of interrogating souvenirs, their meanings and their values. Several studies have developed souvenir typologies based on merchandise characteristics, locations, utilitarian value versus hedonic value and other related criteria. In addition, scholars have examined the role of souvenirs in memory-making and memory-keeping, as well as the process of retail decision-making in souvenir-buying situations (Mawufemor *et al.*, 2019; Swanson & Timothy, 2012). The approaches to, and conceptual underpinnings of, studying souvenirs are enormous. Recent work has focused on the place-meaning connotations of souvenirs or their dis-placement (He & Timothy, 2024c, 2024d; Hashimoto & Telfer, 2007).

The term 'souvenir' has Latin and French origins related to 'memory' or 'remembering'. The meaning of souvenirs in the touristic sense began in the 18th century to refer to something tangible that evokes memories or reminds its possessor of a special time and place. Swanson and Timothy (2012), however, remind us that a souvenir does not necessarily have to be tangible, wanted, or intentionally received or purchased. An airline boarding pass, a concert ticket, a sunburn, memories of a summer romance, hair braids received on the beach, or even a hangover can be counted among the wide range of 'souvenirs' people receive or experience during their journeys. A souvenir, then, can be anything received, purchased or otherwise acquired during a journey that reminds travelers of their experience. They may be fleeting and ephemeral, seared into one's memory or they may be tangible, occupying a prominent place in one's home or office (Collins-Kreiner & Zins, 2011). Despite this broad and somewhat abstract notion, most research on souvenirs has focused on tangible mementos and their meanings and values.

Through decades of research, we have learned much about different kinds of souvenirs, their forms and functions, differing levels of perceived authenticity and the emotive responses they elicit during the purchasing process and afterward, at home or office proudly displayed for others to see. Yet, there are other elements of 'the material culture of tourism' (Hitchcock & Teague, 2000) that we hardly understand. For instance, little is known about cross-cultural perspectives in relation to authenticity, tastes and preferences, and the purchasing behaviors of spendthrifts and conservative spenders. These have not only theoretical implications but also important management consequences that can affect marketing, staffing, merchandising and general sales.

Among several new trends in the souvenirs marketplace are personalized mementos, food souvenirs, cultural and creative products, and online souvenir shopping. These have received considerable research attention recently. Personalized souvenirs entail visitors designing their own merchandise that often reflects elements of their destination or a particular attraction, and their personal experiences there. These designs may be sent away to be manufactured by crafters and then mailed to the tourists once they return home, or they can be created immediately on the spot through various means, such as 3D printing or in situ with crafters waiting to do the work (Anastasiadou & Vettese, 2019). Cultural and creative products are particularly popular in China and other parts of Asia. These are a higher quality combination of mass-produced and site-specific mementos that represent the locality where they are sold, particularly in museum shops and other attractions. They are often deemed more authentic than typical mass-produced items and depict museum artifacts or other cultural icons and also often combine utilitarian uses with the appeal of typical souvenirs. Included in this category might be mobile phone covers, notepad holders and tea sets (He & Timothy, 2024a).

Food souvenirs have long been an important consumer product. Bringing home a bottle of wine from a beloved winery; a bottle of whiskey from a famous distillery; cookies make from a unique local ingredient; a special regional cheese; or an herbal mix for grilling fish – these sorts of mementos have long been a part of the retail practices of tourists, but the idea of food as souvenir has gained traction in the past few years (e.g. Altintzoglou *et al.*, 2016; Lin, 2017). As tourists dine on traditional destination cuisine, they often desire to take with them some of the flavors of the destination. There is growing interest among the traveling public in the 'experiential souvenir' rather than simply something to adorn the mantel. Food and drink are an important part of this movement, as these can figuratively extend the holiday experience into one's homelife after a journey. The 1992 establishment of 'protected designation of origin' standards and labeling has caused an upsurge in this phenomenon. Likewise, the growth of culinary tourism or food tourism has stimulated this trend. In some countries, regional and very place-specific alimentary specialties draw consumers to taste and take home a piece of local culinary heritage. In China, for instance, many localities are famous for specialized and famous noodles – because of the type of starch used to make them, a unique color or physical quality, a special flavor, or an exclusive production process. Buying these noodles is a common part of the retail experience in this type of cultural tourism. The same takes place with regard to cheeses in Italy and Switzerland, and bread and wine in Portugal.

We now know a great deal about tourists' inclination to consume souvenirs and the values and meanings ascribed to the mementos people buy. However, we know relatively little about souvenir makers and other suppliers. Only very recently have the suppliers of souvenirs been studied to understand their role in place-making and place-making (He & Timothy, 2024a; Light et al., 2024; Soukhathammavong & Park, 2019) and their motives beyond pure economics. Light and his colleagues (2024) examine the non-economic motivations of souvenir vendors. While earning a living is an obvious motive for most sellers, Light et al. identified additional influential factors, including the vendors seeing themselves as cultural intermediaries, having a passion for what they do, satisfying their hobby interests, and the opportunity such an activity gives them to interact with their customers.

Some early work focused on the role of local souvenir production on the status of women and as an economic empowerment mechanism for female destination residents (Farver, 1984; Nason, 1984). Since these earlier studies, relatively little research has tried to shed light on the role of handicrafts in empowering women and children and improving their livelihood options. In many destinations, the creation of handicrafts is in the domain of women and children, yet we know little about how this manifestation of retail can affect their lives. There may be a degree of exploitation associated with the makers of souvenir arts and crafts (Dlaske, 2014; Youkhana, 2010), but additional research is needed to verify the types and degrees of exploitation if it indeed exists.

Tourism Types and Their Shopping Connections

Some contributors to this volume expounded upon the core relationships between heritage tourism and shopping. Research shows that serious (versus casual) heritage tourists are among the biggest spenders on shopping and prefer to buy mementos that represent elements of a visit that were important to them. This often ranges from customized souvenirs that depict artifacts in a beloved museum (He & Timothy, 2024c) or inexpensive, mass-produced merchandise that may be manufactured far from the places they are meant to depict. A lot has been written about cultural-heritage based tourism and its shopping implications, but what of other types of tourism? What sorts of retail manifestations do they possess? Although many different types of tourism exist, which are based on the experiences sought and the places visited, only a few will be discussed here to illustrate these points.

Sport tourists are heavy consumers of memorabilia that supports an event or a particular team. Pilgrims and other religious tourists are enormously enthusiastic shoppers for a wide range of products. The most devout pilgrims might only buy religious souvenirs, depending on their preferences. These include religious icons, pieces of holy places (e.g. rocks or water), religious attire, sacred food, books and postcards, prayer beads, flowers and similar items associated with devotion. Most pilgrims and other religious tourists, however, also buy secular, 'touristic' souvenirs that commemorate the places they visit (Ron & Timothy, 2019). Retail establishments in most pilgrimage destination sell a broad range of religious and secular merchandise to appeal to all visitor tastes.

Wine tourism and culinary tourism are expanding exponentially with the traveling public's growing interest in food sources, food production methods, sustainable rural livelihoods and similar other movements. Visits to wineries or distilleries almost always result in alcohol purchases (Lin & Mao, 2015; Telfer & Hashimoto, 2000). Food tours frequently result in tourists buying food items, herbs and spices, cookbooks, and cooking and eating utensils. People often eat something in the destination that they want to try again later at home, which can cause them to buy food items for use at home to re-live a positive memory of the flavors of the destination (Stone *et al.*, 2022).

Hobbies and personal interests also stimulate shopping behavior in the destination. Numismatists frequently seek coin and banknote sets of the countries they visit, or ancient coin aficionados may buy from dealers abroad. Car enthusiasts enjoy touring car manufacturing facilities, buy scale models, and even purchase cars as a result of their visits (Coles, 2004). Train and railway enthusiasts are known to buy model trains during their travels, and the list could be endless.

All types of tourism have an associated retail element, which might include equipment and gear (e.g. mountaineering, hiking or canoeing); items to donate to a worthy cause (e.g. volunteer tourism or solidarity tourism); fresh fruits and vegetables (e.g. agritourism); books, figurines and costumes (e.g. film-induced tourism); and bathing suits, suntan lotion and sunglasses (e.g. sun, sea and sand tourism). Additional research is needed to understand the retail nuances of different types of tourism and different experiences tourists seek. Such knowledge may be critical for destination management organizations and individual retailers as they plan for tourism and their marketing budgets, as well as for the academy as we continue to expand our knowledge about tourists' behaviors and the variables that influence their retail choices.

Non-Traditional Shopper Tourists

Although not always numbered as tourists by data analysts, government agencies, or even tourism scholars, petty traders and street vendors do in fact play an important role in shopping and tourism, as both shopping tourists and merchandise providers. Timothy and Teye (2005) examined the role of petty traders who cross borders in West Africa to re-sell items they purchase or grocery items they produce themselves. Timothy and Teye conceptualized these people as one of the Global South's versions of business tourists, as they cross international boundaries for trade and livelihoods, just as formal businesspeople do everywhere. They spend money buying supplies abroad, they often stay overnight and they sometimes engage in social activities in a neighboring country. Nevertheless, they remain outside the scope of tourism research. Perhaps it is time to expand our vision of tourism to look more broadly at human mobilities, as the mobilities paradigm has tried to do during the past decade (Bianchi *et al.*, 2020; Hannam *et al.*, 2006; Sheller & Urry, 2006). Tourism is not, as many people still erroneously believe, exclusively about leisure and pleasure. There are many types of tourism that may in fact be antithetical to leisure and pleasure but which nonetheless are an important part of the tourism system (e.g. medical tourism and petty trade). What are the shopping implications of non-traditional 'tourisms'? How does the role of retail manifest in other mobilities beyond pleasure and leisure and how can destinations capitalize on these often-marginalized manifestations of mobility?

Non-Destination Spaces of Retail

A holistic view of tourism includes pre-travel activities in tourists' hometowns, during transit time and space and in the destination. Retail activity takes place in all three spatial contexts, yet nearly all shopping studies focus exclusively on the destination. Nonetheless, there is an increasing literature about shopping in transit, with most attention being given to duty-free shopping and airport shopping generally (Creed *et al.*, 2021; Liang & Yu, 2024). Most of the work on in-transit retail focuses on duty-free sales of high-end items, gifts for family and friends, and last-minute items forgotten at home, such as electrical adapters, hats or backpacks. Although a significant amount of tourism-related consumption takes place before the journey begins, such as buying clothes, hats, shoes, extra batteries, haircuts and even weight-loss supplements to prepare for the beach, pre-trip, at-home shopping has largely been ignored by retail and tourism specialists. Understanding pre-travel retail behavior would reveal a great deal about people's expectations and provide a more holistic examination of the economic impacts of tourism in places outside the final destination.

Artificial Intelligence and Other Technologies

Missing from this volume is a dedicated chapter on technology and shopping, although it was alluded to in a few works herein. This is not a reflection of a lack of importance of artificial intelligence and other technologies in relation to shopping and tourism. Thus, this section discusses some of the technological advancements of the past decade and their implications for shopping tourism and tourist shopping.

Perhaps more impactful than any other influence on shopping tourism today is technology. For example, live chat features enable customers to 'chat' virtually with merchants and retail staffers, facilitating consumers asking questions about products and digital browsing. Virtual shopping allows consumers to find the products they need at the best prices with only a few clicks of a mouse. This has the potential to save time in the destination if products are known and tried ahead of time, and such chat features might also act as a conversion tool that encourages people to browse more and buy more while in the destination.

Social media likewise plays an extremely significant role in retail tourism today (Egresi, 2017). Extensive research on the influences of social media on tourism has shown its ability to influence travel decision-making, consumer choices, behaviors and quality of experience. In the area of shopping, social media has also been influential (Atsız & Seyitoğlu, 2023; Hyun *et al.*, 2022). Ratings and reviews can make or break a retailer's reputation. Therefore, social media content has the potential to increase merchants' accountability and quality control to avoid receiving negative online reviews and detrimental exposure through various social channels. Likewise, positive reviews and social media commentary can be among the most effective promotional tools in existence today (Dolega *et al.*, 2021). Many travelers seek retail advice (user-generated content) from various social media with regard to where they should shop, the quality of service and the range of retail products available in the destination (Phucharoen *et al.*, 2022; Ting, 2022).

Social commerce is also trending now in relation to social media. Social commerce refers to live shopping on social media. It blends entertainment and e-commerce, with social media facilitating e-commerce wherein online 'window shopping', product selections and purchasing all take place on a single social platform (Shamim *et al.*, 2024). Erdly (2024) estimates that 98% of all social media users participated in social commerce in 2024. This adds depth and breadth to the ways in which recreational shopping takes place, including in tourism settings (Pham *et al.*, 2023). Will social commerce and other forms of virtual retail take the place of shopping tourism? It seems unlikely for now, but it has the potential to reduce people's need to travel physically to a destination to buy what they want to buy.

Online shopping is another manifestation of relatively recent technology during the past three decades. Television-based shopping continues but has largely been supplanted by online retail. Large online retailers, such as eBay, Amazon, Temu, Alibaba and Etsy, have come to monopolize the virtual shopping marketplace in recent years. However, even traditional brick-and-mortar stores have begun to provide simultaneous onsite and online shopping opportunities, which has proved to be extremely profitable in addition to the physical stores that still exist. Souvenir shops in popular destinations have done the same, extending their retail impact beyond the destination to include post-trip purchases (He & Timothy, 2024b).

From a tourism and place perspective, virtual/online shopping has somehow deterritorialized the shopping experience and created placeless retail spaces that have both diminished and enhanced leisure shopping (He & Timothy, 2024b). Clearly the social element and sense of place associated with shopping might have diminished, but the convenience and usability of shopping has improved for consumers. For sellers, there may be a cost savings associated with online retail, reducing inventory storage costs, employee expenses, and the physical maintenance of stores. From a tourism perspective, online options provide opportunities for people retroactively to buy the souvenir they missed in the destination, the gift they forgot to buy a friend, or higher-value and larger items they might not have dared carry home.

Augmented and virtual reality play a critical role in retail commerce in general but in particular in specific tourism contexts. Shopping tourism and tourist shopping have benefited significantly from the development of augmented and virtual reality. Some research shows that retail experiences that involve augmented reality (AR) are more highly favored among shopping tourists and add value to the retail experience (Dogra *et al.*, 2023; Punzon, 2021). Augmented reality enables shoppers to shop virtually in a more interactive and intimate way. AR enables people to virtually place products, including souvenir items, in their home to see how and where they might fit best. With regard to clothing, the same technology can enable people to see what fits best with their body shape and skin tones (Zackariya, 2023). With this technology, consumers can 'see the size and fit in real-time, comparing it with elements around them' (Threedium, 2024, n.p.). 'Virtual Try On' is a growing mobile phone technology that allows consumers to try on their potential purchases (e.g. jewelry, clothing, cosmetics, shoes and even cars) before actually making the purchase.

In some cases, augmented reality allows consumers to design their own products based on what looks good or feels good as derived from the virtual experience (Alimamy & Gnoth, 2022). With increasing demand for personalized and self-designed travel souvenirs or luxury products, the idea of VR enabling people to design their own souvenirs to buy online or to pick up in the destination once one's personal touches have been added, is gaining popularity (Do *et al.*, 2020). The 3D printing of personalized souvenirs relates to this and may be driven by AR technology together with 3D technology (Berjozkina & Karami, 2021).

Artificial intelligence is not new, but it has received increasingly widespread media and academic attention since 2020. Its main use in the shopping domain is generative text for small businesses in advertising, newsletter writing and press releases (Zackariya, 2023). In 2023 and 2024, ChatGPT grew remarkably fast and was in fact the fastest growing app to that point in history (Erdly, 2024). Retail industry experts argue that AI can have the potential to maximize customer service and profits, satisfy changing demands during times of increased competition, streamline operations, improve merchandising and supply chain efficiency, improve point-of-sale systems and embrace future changes all for better service and customer satisfaction (Cognizant Industries, 2024, n.p.).

Many questions have been raised regarding how retailers can utilize AI to become more efficient, while keeping the human side of operations management in balance. The rapid growth of technology raises many questions in the realm of tourism more generally, but how it affects shopping and retail in the tourism sector requires more research from both supply and demand perspectives to understand the product mix, customer satisfaction, tourists' behaviors and conversion processes before, during and after a shopping tourism experience.

Sustainability and Green Shopping

Consumers in general are becoming increasingly concerned about environmental sustainability and buying ethically-produced merchandise. A recent survey concluded that 75% of millennials claim to consider sustainability in their shopping behaviors, and general retail customers want manufacturers and retailers to demonstrate proof of their ecological commitment (Erdly, 2024). Likewise, another report indicates that 60% of consumers do not believe corporations' guarantees of eco-friendliness. The same report concludes that '…merely operating with less harm to the planet than your competitors is not enough – shoppers want to buy from stores which offer real climate solutions and are genuinely working to be regenerative' (Erdly, 2024, n.p.).

There is a growing sentiment towards fair-trade and otherwise ethically-sourced merchandise, particularly with regard to animal welfare and the exploitation of human workers. Coffee and tea products, gold and gems, as well as any sort of animal product are at the fore of these sorts of discussions (Pecoraro *et al.*, 2021; Wang & Chou, 2020). Although the black market for products (e.g. skins, skulls, bone and ivory carvings) made from exotic animal species continues to exist, global efforts to quash the trade in animal products have accelerated since the 1990s. Similarly, the trade in illicit archaeological artifacts is thriving as collectors travel the world in search of unique relics and pieces of cultural heritage that are looted from burial sites and archaeological remains (Timothy, 2020b). Most countries have enacted legislation to protect tangible and intangible heritage and to reduce the use of endangered animals for food and souvenirs, but a lot more work is needed to understand people's motives for buying these underground products and what can be done to stop it. Additional research is also needed to understand better the latent value tourists attach to green and sustainable retailing.

Crises and Shopping Tourism

Crises are significant disrupters of consumption. Natural disasters, economic downturns, political instability, rioting and strikes, and many other disruptions all affect people's ability to shop and therefore retailers' livelihoods. Economic challenges such as high inflation and unemployment rates will inevitably affect a region's retail-based economy, and in a tourism context, shopping may be one area tourists have to sacrifice to make a trip affordable. These sorts of limitations on tourists' retail activity can have devastating consequences for destination communities. By the same token, recessions can cause an increase in certain types of retail tourism (e.g. cross-border shopping), as people may turn to crossing a national boundary to buy less-expensive goods and services, even while luxury shopping declines simultaneously as a result of the same financial crisis.

The COVID-19 pandemic from 2020 to 2022 severely limited people's ability to travel, let alone travel for shopping purposes. Shops were closed for everyone – tourists and non-tourists alike – and most national borders were sealed to contain any spread of the Coronavirus. Border towns were especially hard hit, particularly those that rely overwhelmingly on cross-border consumers. Many Mexican border towns suffered tremendously when the US–Mexican frontier was closed to Mexicans crossing northward. Many towns and cities in California, Arizona, New Mexico and Texas are highly dependent on Mexican shoppers, many of whom cross daily to shop and participate in other activities. Likewise, although Americans were permitted to cross into Mexico and return to the US during most of the pandemic period, the idea of crossing lost its appeal for most American shoppers. Thus, owing to this rebordering process, the retail-dependent Mexican towns on the country's northern border were economically devastated (Timothy, 2020a). Along the US–Canada border, an interesting phenomenon took place. When that border closed, Canadians and Americans were not permitted to cross either direction for many months. This provided an economic windfall for communities north of the border, as Canadians were required to shop in their own country rather than cross into the United States for lower prices and a greater selection of merchandise (Baggs *et al.*, 2022). Many examples of the positive and negative effects of COVID-induced border closures can be found all over the world (Paül *et al.*, 2022; Primc & Slabe-Erker, 2020; Tömöri & Staniscia, 2023).

Additional research is needed to help shopping tourism-oriented communities prepare for future crises. Each type of crisis requires specific planning mechanisms and responses, but it is crucial for retailers and communities to plan for activities such as recovery marketing and other counter measures to ensure people's safety and ensure that customers return.

Final Remarks

This concluding chapter, and indeed the entire book, has raised many questions but answered only a few. There is deep scope for much more research on retail and tourism beyond the normative studies about satisfaction, experience and behavioral intentions. As the chapters in this book have shown, these and many other multifarious relationships between shopping and tourism have received considerable research attention, and the work in those areas continues to grow unabated (Choi *et al.*, 2016; Jin *et al.*, 2017; Michalkó, 2004; Timothy, 2005).

However, we have not yet reached far enough in understanding the destination's perspectives or the intentions and experiences of makers and sellers. Addressing questions of community development, sustainable tourism, the importance of place, the value of otherness and similar ideas in the context of retail and tourism is crucial as shopping-based tourism continues to grow and as retail experiences continue to play an increasingly fundamental part in every other type of tourism. To reach a more holistic level of understanding, we need to think differently about shopping tourism to include non-traditional retail spaces and places, non-normative products, including cars, real estate and vacation properties, as these too are manifestations of retail tourism with unique perspectives on production and consumption (Coles, 2004; Timothy, 2024), and atypical spaces and activities that are often overlooked in shopping studies. Only then will we have more answers to the sociocultural, psychological, spatial and development implications of shopping tourism and tourist shopping.

References

Alimamy, S. and Gnoth, J. (2022) I want it my way! The effect of perceptions of personalization through augmented reality and online shopping on customer intentions to co-create value. *Computers in Human Behavior* 128, 107105.

Altintzoglou, T., Heide, M. and Borch, T. (2016) Food souvenirs: Buying behaviour of tourists in Norway. *British Food Journal* 118 (1), 119–131.

Anastasiadou, C. and Vettese, S. (2019) 'From souvenirs to 3D printed souvenirs' Exploring the capabilities of additive manufacturing technologies in (re)-framing tourist souvenirs. *Tourism Management* 71, 428–442.

Atsız, O. and Seyitoğlu, F. (2023) A netnography approach on the daily local-guided shopping tour experiences of travellers: An unexplored facet of the sharing economy. *Journal of Vacation Marketing* 29 (1), 103–118.

Baggs, J., Fung, L. and Lapham, B. (2022) An empirical examination of the effect of COVID-19 travel restrictions on Canadians' cross-border travel and Canadian retailers. *Canadian Public Policy* 48 (1), 162–185.

Beer, S., Murphy, A. and Shepherd, R. (2012) Food and farmers' markets. In C. McIntyre (ed.) *Tourism and Retail: The Psychogeography of Liminal Consumption* (pp. 111–142). Routledge.

Berjozkina, G. and Karami, R. (2021) 3D printing in tourism: An answer to sustainability challenges? *Worldwide Hospitality and Tourism Themes* 13 (6), 773–788.

Bianchi, R.V., Stephenson, M.L. and Hannam, K. (2020) The contradictory politics of the right to travel: Mobilities, borders and tourism. *Mobilities* 15 (2), 290–306.

Choi, M.J., Heo, C.Y. and Law, R. (2016) Progress in shopping tourism. *Journal of Travel and Tourism Marketing* 33 (1), 1–24.

Cognizant Industries (2024) The new foundations of retail. Online: https://www.cognizant.com/us/en/industries/retail-technology-solutions?cid=pse17213875670001-CMP-003582andgad_source=1andgclid=CjwKCAjwl6-3BhBWEiwApN6_kqvkL9WxbenSjLM-5tXlhtb4dkxAstrUJxOjexUy_5dpghO3nbRAGRoCDsUQAvD_BwE

Coles, T. (2004) Tourism and retail transactions: Lessons from the Porsche experience. *Journal of Vacation Marketing* 10 (4), 378–389.

Collins-Kreiner, N. and Zins, Y. (2011) Tourists and souvenirs: Changes through time, space and meaning. *Journal of Heritage Tourism* 6 (1), 17–27.

Correia, A. and Kozak, M. (2016) Tourists' shopping experiences at street markets: Cross-country research. *Tourism Management* 56, 85–95.

Creed, B., Shen, K.N., Ashill, N. and Wu, T. (2021) Retail shopping at airports: Making travellers buy again. *Journal of Business Research* 137, 293–307.

Cukier, J. and Wall, G. (1994) Informal tourism employment: Vendors in Bali, Indonesia. *Tourism Management* 15 (6), 464–467.

Damayanti, M., Scott, N. and Ruhanen, L. (2018) Space for the informal tourism economy. *The Service Industries Journal* 38 (11–12), 772–788.

Dlaske, K. (2014) Semiotics of pride and profit: Interrogating commodification in indigenous handicraft production. *Social Semiotics* 24 (5), 582–598.

Do, H.N., Shih, W. and Ha, Q.A. (2020) Effects of mobile augmented reality apps on impulse buying behavior: An investigation in the tourism field. *Heliyon* 6 (8), e04667.

Dogra, P., Kaushik, A.K., Kalia, P. and Kaushal, A. (2023) Influence of augmented reality on shopping behavior. *Management Decision* 61 (7), 2073–2098.

Dolega, L., Rowe, F. and Branagan, E. (2021) Going digital? The impact of social media marketing on retail website traffic, orders and sales. *Journal of Retailing and Consumer Services* 60, 102501.

Doucé, L. and Janssens, W. (2013) The presence of a pleasant ambient scent in a fashion store: The moderating role of shopping motivation and affect intensity. *Environment and Behavior* 45 (2), 215–238.

Egresi, I. (2017) Tourists' satisfaction with shopping experience based on reviews on TripAdvisor. *Tourism* 65 (3), 330–345.

El-Adly, M.I. and Eid, R. (2016) An empirical study of the relationship between shopping environment, customer perceived value, satisfaction, and loyalty in the UAE malls context. *Journal of Retailing and Consumer Services* 31, 217–227.

Erdly, C. (2024) Four major trends that will shape retail in 2024. *Forbes*, online: https://www.forbes.com/sites/catherineerdly/2024/01/26/four-major-trends-that-will-shape-retail-in-2024/?sh=1c03a4aa64a5

Esfandiar, K., Rahmani Seryasat, M. and Kozak, M. (2024) To shop or not to shop while traveling? Exploring the influence of shopping mall attributes on overall tourist shopping satisfaction. *Tourism Recreation Research* 49 (6), 1411–1426.

Farver, J.A.M. (1984) Tourism and employment in the Gambia. *Annals of Tourism Research* 11 (2), 249–265.

Garner, B. and Ayala, C. (2019) Regional tourism at the farmers' market: Consumers' preferences for local food products. *International Journal of Culture, Tourism and Hospitality Research* 13 (1), 37–54.

Hall, C.M. (2016) Heirloom products in heritage places: Farmers' markets, local food and food diversity. In D.J. Timothy (ed.) *Heritage Cuisines: Traditions, Identities and Tourism* (pp. 88–103). Routledge.

Hannam, K., Sheller, M. and Urry, J. (2006) Mobilities, immobilities and moorings. *Mobilities* 1 (1), 1–22.

Hashimoto, A. and Telfer, D.J. (2007) Geographical representations embedded within souvenirs in Niagara: The case of geographically displaced authenticity. *Tourism Geographies* 9 (2), 191–217.

He, L. and Timothy, D.J. (2024a) Souvenirs and place: Suppliers' perspectives. *Current Issues in Tourism* 27 (24), 4463–4478.

He, L. and Timothy, D.J. (2024b) Tourists' perceptions of buying souvenirs in online museum shops. *International Journal of Tourism Research* 26 (5), e2761.

He, L. and Timothy, D.J. (2024c) Tourists' perceptions of 'cultural and creative souvenir' products and their relationship with place. *Journal of Tourism and Cultural Change* 22 (2), 143–163.

He, L. and Timothy, D.J. (2024d) Understanding souvenirs from a place–product perspective: Territorialization, deterritorialization, and reterritorialization. *Tourism Review International* 27 (3–4), 3–4.

Hitchcock, M. and Teague, K. (2000) *Souvenirs: The Material Culture of Tourism.* Ashgate.

Hyun, H., Thavisay, T. and Lee, S.H. (2022) Enhancing the role of flow experience in social media usage and its impact on shopping. *Journal of Retailing and Consumer Services* 65, 102492.

Jin, H., Moscardo, G. and Murphy, L. (2017) Making sense of tourist shopping research: A critical review. *Tourism Management* 62, 120–134.

Liang, C.C. and Yu, A. (2024) Customer impulse shopping in airports. *International Journal of Retail and Distribution Management* 52 (3), 372–385.

Light, D., Lupu, C., Crețan, R. and Chapman, A. (2024) Unconventional entrepreneurs: The non-economic motives of souvenir sellers. *Tourism Review* 79 (8), 1442–1456.

Lin, L. (2017) Food souvenirs as gifts: Tourist perspectives and their motivational basis in Chinese culture. *Journal of Tourism and Cultural Change* 15 (5), 439–454.

Lin, L. and Mao, P.C. (2015) Food for memories and culture – A content analysis study of food specialties and souvenirs. *Journal of Hospitality and Tourism Management* 22, 19–29.

Mawufemor, K., Eshun, G. and Tichaawa, T.M. (2019) Factors influencing choice of souvenirs by international tourists. *African Journal of Hospitality, Tourism and Leisure* 8 (5), 1–10.

McKercher, B. (2020) Anatomy of successful tourism shopping districts. *International Journal of Tourism Cities* 6 (4), 831–846.

Michalkó, G. (2004) *A Bevásárlóturizmus* (Shopping Tourism). Székesfehérvár: Kodolányi János Főiskola.

Murphy, L., Moscardo, G., Benckendorff, P. and Pearce, P. (2011) Evaluating tourist satisfaction with the retail experience in a typical tourist shopping village. *Journal of Retailing and Consumer Services* 18 (4), 302–310.

Nason, J.D. (1984) Tourism, handicrafts, and ethnic identity in Micronesia. *Annals of Tourism Research* 11 (3), 421–449.

Nguyen, T.N. and Huynh, V.D. (2024) Floating market history, status, and changes: Insights from Vietnam. *Tourism, Culture and Communication* 24 (1), 71–88.

Paül, V., Trillo-Santamaría, J.M., Martínez-Cobas, X. and Fernández-Jardón, C. (2022) The economic impact of closing the boundaries: The lower Minho Valley cross-border region in times of Covid-19. *Journal of Borderlands Studies* 37 (4), 761–779.

Pecoraro, M., Uusitalo, O. and Valtonen, A. (2021) Experiencing ethical retail ideology in the servicescape. *Journal of Marketing Management* 37 (5–6), 520–547.

Pham, L.H., Woyo, E., Pham, T.H. and Truong, D.T.X. (2023) Value co-creation and destination brand equity: Understanding the role of social commerce information sharing. *Journal of Hospitality and Tourism Insights* 6 (5), 1796–1817.

Phucharoen, C., Jarumaneerat, T. and Sangkaew, N. (2022) Comparing shopping experiences in department stores and street markets: A big data analysis of TripAdvisor reviews. *International Journal of Culture, Tourism and Hospitality Research* 16 (1), 259–275.

Primc, K. and Slabe-Erker, R. (2020) The success of public health measures in Europe during the COVID-19 pandemic. *Sustainability* 12 (10), 4321.

Punzon, J.G. (2021) Augmented reality in shopping tourism: Boosting tourism development through innovation in Barcelona. *European Journal of Tourism, Hospitality and Recreation* 11 (1), 1–10.

Ripoll González, L., Belén Yanotti, M. and Lehman, K. (2022) Local focus: Farmers' markets as an approach to sustainable tourism. In A. Farmaki, L. Altinay and X. Font (eds) *Planning and Managing Sustainability in Tourism: Empirical Studies, Best-practice Cases and Theoretical Insights* (pp. 95–113). Springer.

Ron, A.S. and Timothy, D.J. (2019) *Contemporary Christian Travel: Pilgrimage, Practice and Place*. Channel View Publications.

Shamim, K., Azam, M. and Islam, T. (2024) How do social media influencers induce the urge to buy impulsively? Social commerce context. *Journal of Retailing and Consumer Services* 77, 103621.

Sheller, M. and Urry, J. (2006) The new mobilities paradigm. *Environment and Planning A* 38 (2), 207–226.

Soukhathammavong, B. and Park, E. (2019) The authentic souvenir: What does it mean to souvenir suppliers in the heritage destination? *Tourism Management* 72, 105–116.

Steel, G. (2009) Dishing up the city: Tourism and street vendors in Cuzco. In M. Baud and A. Ypeij (eds) *Cultural Tourism in Latin America: The Politics of Space and Imagery* (pp. 161–176). Brill.

Steel, G. (2012) Local encounters with globetrotters: Tourism's potential for street vendors in Cusco, Peru. *Annals of Tourism Research* 39 (2), 601–619.

Stone, M.J., Migacz, S. and Sthapit, E. (2022) Connections between culinary tourism experiences and memory. *Journal of Hospitality and Tourism Research* 46 (4), 797–807.

Swanson, K.K. and Timothy, D.J. (2012) Souvenirs: Icons of meaning, commercialization and commoditization. *Tourism Management* 33 (3), 489–499.

Telfer, D.J. and Hashimoto, A. (2000) Niagara icewine tourism: Japanese souvenir purchases at Inniskillin Winery. *Tourism and Hospitality Research* 2 (4), 343–356.

Thompson, M. (2020) Farmers' markets and tourism: Identifying tensions that arise from balancing dual roles as community events and tourist attractions. *Journal of Hospitality and Tourism Management* 45, 1–9.

Threedium (2024) See it, try it on, love it! Online: https://threedium.io/en-us/enhance-brand-experiences-with-ar-virtual-try-ons?utm_source=Googleandutm_medium=searchandutm_campaign=Septemberandutm_id=acquisitionandutm_term=usaandutm_term=augmented%20reality%20ecommerceandutm_campaign=andutm_source=adwordsandutm_medium=ppcandhsa_acc=8141413262andhsa_cam=21661025436andhsa_grp=167637390392andhsa_ad=712636944608andhsa_src=gandhsa_tgt=kwd-467335027423andhsa_kw=augmented%20reality%20ecommerceandhsa_mt=eandhsa_net=adwordsandhsa_ver=3andgad_source=1andgclid=CjwKCAjwl6-3BhBWEiwApN6_kgAqjf6IN_5BmUSMVw-x_H-R8qIWZcmyzNeTD-1_R0O6siT-9ZmAOBoCdgAQAvD_BwE

Timothy, D.J. (2005) *Shopping Tourism, Retailing and Leisure*. Channel View Publications.

Timothy, D.J. (2020a) La pandemia ha devastado al turismo médico, necesitamos soluciones creativas. *El Universal*, October 23, 2020. https://www.eluniversal.com.mx/opinion/dallen-j-timothy/la-pandemia-ha-devastado-al-turismo-medico-necesitamos-soluciones-creativa

Timothy, D.J. (2020b) Plundering the past: Tourism and the illicit trade in archaeological remains. In D.J. Timothy and L.G. Tahan (eds) *Archaeology and Tourism: Touring the Past* (pp. 134–151). Channel View Publications.

Timothy, D.J. (2024) Tourism, shopping and connotations of place. In C.M. Hall (ed.) *The Wiley Blackwell Companion to Tourism* (2nd edn, pp. 501–514). Wiley.

Timothy, D.J. and Teye, V.B. (2005) Informal sector business travelers in the developing world: A borderlands perspective. *Journal of Tourism Studies* 16 (1), 82–92.

Timothy, D.J. and Wall, G. (1997) Selling to tourists: Indonesian street vendors. *Annals of Tourism Research* 24 (2), 322–340.

Ting, T.Y. (2022) Mundane citizenship on the move: A counter-public response to inbound shopping tourism via mobile social media applications use. *Mobile Media and Communication* 10 (3), 531–551.

Tömöri, M. and Staniscia, B. (2023) The impact of the COVID-19 pandemic on cross-border shopping tourism: The case of Hungary. *Hungarian Geographical Bulletin* 72 (2), 147–161.

Truong, V.D. (2018) Tourism, poverty alleviation, and the informal economy: The street vendors of Hanoi, Vietnam. *Tourism Recreation Research* 43 (1), 52–67.

Wang, E.S.T. and Chou, C.F. (2020) Norms, consumer social responsibility and fair trade product purchase intention. *International Journal of Retail and Distribution Management* 49 (1), 23–39.

Youkhana, E. (2010) *Gender and the Development of Handicraft Production in Rural Yucatán/Mexico*. ZEF Working Paper Series.

Zackariya, S. (2023) *Leading Travel and Tourism Retail: How Businesses Can Sustainably Capture New Profits in Shopping Tourism*. Kogan Page.

Zaidan, E. (2016) Tourism shopping and new urban entertainment: A case study of Dubai. *Journal of Vacation Marketing* 22 (1), 29–41.

Zaidan, E. (2019) Shopping, tourism and hyper-development in the Middle East and North Africa. In D.J. Timothy (ed.) *Routledge Handbook on Tourism in the Middle East and North Africa* (pp. 365–377). Routledge.

Index

Abu Dhabi 6, 127
accommodations 3, 4, 80, 82, 85, 94, 99, 102, 103, 117, 125, 128, 147, 180, 187, 203, 206, 218, 223, 238, 241
affluence 3, 9, 20, 67, 78, 79–81, 107, 147
Africa 41, 112, 218, 256
African consumers 41
agriculture 168, 200, 230, 236, 241, 242, 251
agritourism 209, 255
airlines 57, 216, 221
airports 5, 61, 65, 96, 98–99, 102, 147, 182, 200, 217, 235, 240–241, 256
Alibaba 116, 257
alliances *see* collaboration
Amazing Thailand Grand Sale 215–217
Amazon 116, 257
American consumers 38, 45, 47, 49
Amsterdam 241
Andorra 96–97, 113, 216–218
Andorra Shopping Festival 216–217
apps (mobile phone) 62–63, 65, 70–71, 102, 146, 219, 258
artificial intelligence 5, 28, 186, 256–259
Asia 11, 100, 170, 196, 220, 221
augmented reality 25, 28, 30, 71, 258
Australia 40, 45, 61–62, 64, 96, 97, 99, 101, 196–197, 200, 205, 207–208
Australian consumers 45
Austria 39, 112–113
authenticity 5, 25, 27, 28, 30, 44, 62, 67, 69, 71, 79, 82–83, 199–200, 202, 205–206, 207, 219, 228, 232–235, 239, 243, 252–253

baby boomers 4
Bangkok 251
Bangladesh 64
Barcelona 169
bargain-hunting 42, 111

bazaars *see* marketplaces
behavior, consumer 2, 8–9, 19, 21–23, 37–38, 47, 146–147, 178–179, 220
Belarus 116
Belgian consumers 47, 112
Belgium 112
Belize 114
Berlin 161, 177, 182
biosecurity 200
borders 10, 95, 98–99, 108–110, 113, 114–115, 118, 147–148, 200, 256, 260
See also cross-border shopping
brand awareness 6, 8, 20–21, 26, 40–41, 65, 67, 78–80, 82, 107, 153–154, 161, 165, 214, 222, 230, 240
branding 6–8, 9, 11, 78, 116, 119, 126, 131, 160, 217, 249
brand value *see* brand awareness
breweries 96, 196, 236, 240, 241
British consumers 45–46, 49
browsing 20, 23, 47, 57, 100, 103, 187, 220, 229, 230, 250, 257
Bulgaria 110
business tourism 125, 256

Canada 64, 97, 101, 109, 118, 260
Canadian consumers 38, 49, 118, 260
ChatGPT *see* artificial intelligence
Chile 39, 46, 99
Chilean consumers 46
China 3, 38, 40, 60, 64, 65, 67–68, 86, 99, 101, 114, 115, 253–254
Chinese consumers 3, 9, 40, 42, 45, 57, 61–62, 63–67, 100, 114, 181, 201, 202, 237, 239–240
climate change 259
clustering, retail 114, 115, 119, 127, 137, 144, 145, 148, 218, 224

co-creative experiences 4–5, 26, 30, 51, 71
collaboration, public-private 11, 28–29, 30–31, 79, 83, 95, 101, 119, 138, 188, 215, 220, 223
commercialization 2, 205
commodity chain 159
consumer capitalism 161, 170
corporate social responsibility 50
COVID-19 pandemic 3, 9, 57, 58–60, 64–65, 66, 69, 98, 136, 146, 150, 152, 185–186, 200, 215, 260
crafts *see* handicrafts
creative industries 10, 65, 66, 129, 160, 167–168
credit cards 5, 8, 21–22
crises 116, 159, 250, 260–261
cross-border shopping 9, 107–119, 144, 147, 260
cross-cultural perspectives *see* intercultural perspectives
cruises 7, 11, 95, 98, 195, 200–203
culinary tourism *see* food tourism
cultural appropriation 206
cultural capital 165, 199, 204, 209
cultural creative city 161, 162–163, 169
cultural distance 40, 239
cultural influences 3, 8, 27, 37, 118, 199
cultural experiences 23, 24–25, 79, 83–84, 128–129
See also heritage
cultural tourism *see* heritage
currencies 5, 109, 118, 128
customer relationship management (CRM) 21
customer service *see* service quality

day-trips 10, 108–109, 112, 115, 150
Denmark 39, 45, 96, 150
department stores 10, 65, 85, 160, 164–165, 177, 178, 182
destination management organizations (DMOs) 2, 9, 20, 25, 30, 81, 83, 198, 217, 221–222, 223
destination marketing organizations *see* destination management organizations
digital shopping *see* online shopping
discounts *see* special offers
Disneyfication 250
distance-decay 110–111, 145
distilleries 196, 233, 253, 255

Doha 6, 84, 127
domestic tourism 57, 60, 63–64, 71–72, 94, 97, 99, 101, 182, 214
Dubai 6, 8, 84–85, 99, 127, 177, 214, 217, 218, 220
Dutch consumers 47, 112
duty-free 5, 9, 40, 41, 44, 65, 93–103, 113, 147, 182, 201, 256
dynamic creative optimization (DCO) 29

eBay 116, 257
e-commerce *see* online shopping
economy of scale 108, 113, 150
El Salvador 64
emotional connections 20–32
employment generation 2, 80, 94, 101, 117, 124, 131, 155, 213, 230
empowerment 1, 19, 27, 29, 32, 49, 66, 254
entrepreneurship 1, 137, 138, 146, 203, 230, 234
environment, retail 4, 5, 23, 40, 42, 43–44, 48, 79, 128, 120, 179, 181, 202, 218, 219, 250
environmental taxes 95
ethical consumption 60, 70–71, 81, 159, 199, 259–260, 299
Etsy 257
Europe 3, 9, 10, 64, 95–96, 111, 112, 116, 145, 160–171, 199
European consumers 38, 41, 48
European Union *see* Europe
events, shopping *see* festivals
exchange rates 10, 109, 112, 118, 128
experience quality 6, 62, 81–82, 250

factory outlets 7, 65, 126, 182, 183
fair-trade *see* ethical consumption
fantasy city 10, 161, 162
farmers markets 7, 127, 198, 217, 218, 236, 240, 242, 250, 251
fashion 7, 69, 79, 85–86, 125, 126, 152–153, 164, 165, 215
festivals, shopping 1, 8, 11, 28, 83, 128, 161, 198, 213–224, 249
Fiji 201, 203
Finland 96, 109, 148, 238
flagship stores 62, 69, 78–79, 82, 85, 161
flea markets 7, 217, 250
food as souvenir 5, 7, 11, 65, 107, 198–199, 228–244

food consumption 3
food halls 10, 160, 168–169, 240, 242
food tourism 11, 168, 195, 198–200, 228, 229–230, 253, 255
France 86, 97, 113, 164, 236
free-trade zone 114, 119
French consumers 41, 47, 112
frequent shopper programs 8, 21–22
    See also loyalty

Gen Y see millennials
Gen Z 4, 69–71
gentrification see urban regeneration
geopolitics 2, 60, 103, 119
German consumers 113, 115
Germany 38, 64, 112–113
gift-giving 38, 40, 42, 183, 199, 202, 229–230, 232, 237, 256
globalization 3, 9, 37, 57, 159, 161–162, 219, 230
Global North 1, 125
Global South 1, 125, 219, 256
government role 220–222
Great Singapore Sale 215
green shopping see sustainability
grocery stores see supermarkets
GST see taxes

haggling 42, 185, 195, 219
handicrafts 1, 7, 63, 69, 83–84, 107, 126, 128, 166, 196, 206, 213, 231–232, 235, 251, 254
hedonic shopping see leisure shopping
heritage 7, 8, 10–11, 43, 44, 80, 84, 94, 128–139, 160–171, 182, 195, 203, 205, 218, 219, 230, 251, 254, 259
heritagization 10, 160–171, 249
high-end retail see luxury shopping
historic cities 8, 10, 124–139, 161, 169
Hollywood 85
home delivery 187
Hong Kong 8, 40, 86, 97–98, 115, 179, 180, 181, 214
Hong Kong Shopping Festival 216–217
Hungary 64, 96, 112

Iceland 96
identity 62, 71, 81–82, 84, 87, 127, 131, 132, 135, 161, 163, 165, 170, 203, 209, 230, 233, 235
IKEA 149–150

illicit trade 259–260
    See also ethical consumption
image see brand awareness
impulse buying 5, 43, 202
incentives 94–95, 96, 98, 100, 102–103, 118, 137, 186, 221
India 3, 38, 40, 60, 64
Indian consumers 9, 45, 60, 63, 67–69
Indonesia 3, 115
innovation 188
intercultural perspectives 8–9, 37–51, 249
international tourism 57, 60, 78, 99–101, 182
Iran 10, 115, 126, 129–139, 205
Ireland 97
Istanbul 44, 177, 217
Italian consumers 41
Italy 8, 20, 113–114, 161, 166, 236, 254

Jamaica 202
Japan 3, 38, 40, 96, 202, 237–238
Japanese consumers 24, 38, 40–41, 44, 96

Korea Grand Sale 216–217

language, staff ability 5, 179–180
Liechtenstein 96, 113
leisure shopping 4, 7, 46, 49, 57, 58–60, 63–64, 68–69, 70, 108, 110–111, 113, 116, 118, 147, 159, 161, 177–178, 181, 183, 219–220, 222, 250
Lisbon 182
Lithuania 242
lodging see accommodations
London 6, 148, 165, 169, 177, 182, 203, 204
loyalty 6–7, 8, 19–32, 41, 94, 154, 179, 219, 222
Luxembourg 96, 112
luxury shopping 2, 3, 6, 9, 20, 38, 43, 78–87, 100, 107, 113, 127, 161, 164, 177, 206, 215, 218, 258

Macau 8, 115, 181
Malaysia 3, 99, 101, 115
Malaysian consumers 42, 45
Maldives 3
malls see shopping centers
marketing 7–8, 19, 23, 27, 29–30, 37, 40, 44–47, 48, 78, 80–83, 109, 114,

115–116, 119, 147, 150, 161, 178, 196, 203, 257, 261
marketing mix 9, 37, 40, 42, 44, 50, 178, 259
marketplaces 7, 10, 23–24, 43, 113, 113, 115, 125, 127–139, 136, 161, 168, 170, 177, 182, 183, 203, 217, 250–251
mass tourism 57, 95, 127, 202
McDonaldization 219
McDonald's 46
medical tourism 115–116, 256
mementos *see* souvenirs
memorable experiences 4, 26, 62, 78, 94, 110, 184, 218, 250
merchandising 4, 40, 43, 109, 152, 179, 222, 250, 253, 259
metaverse 186
Mexican consumers 114
Mexico 64, 103, 114, 115, 260
Michelin-star restaurants 86
microstates 113–114
Middle East 3, 11, 125, 218, 220, 250
Milan 84, 161, 177
millennials 4, 69–71
mobility 3, 9, 107, 109, 112, 119, 144, 203, 224, 256
Monaco 84–85
Mongolia 24
Moscow 46
multipliers 101, 102, 201
museums 62, 69, 169, 203–206, 208, 254
museum shops 10, 43, 165–167, 204, 205

nationality, influence of 3, 8, 38, 128
Netherlands 40, 42
New York 6, 84–86, 177, 182, 203, 234, 241
New Zealand 39, 200
Nigeria 39, 64
Norway 45, 96, 115, 148, 150, 201, 239
nostalgia 10, 184

Oman 203
online shopping 6, 65, 116, 118, 119, 124, 136, 143, 146, 152, 154, 185–187, 221, 223–224, 257–258
outshopping *see* cross-border shopping
overtourism 95, 201

package tours 7, 41, 60, 68, 101
packaging 5, 200, 206, 229, 231, 235, 238, 249, 242, 243
Pakistan 39, 64
pandemic *see* COVID-19
Papua New Guinea 115
Paris 6, 84, 86, 125, 148, 164–165, 170–171, 181, 182, 203
periodic markets 213
personalized messaging 29
petty trade 109, 217, 251, 255–256
pharmacies 64, 66, 113, 115
Philippines 3, 64, 238
pilgrimage *see* religious tourism
Poland 64, 111, 112–113, 114–115, 116
political instability *see* geopolitics
Portugal 24, 199, 254
price 4, 24, 42–43, 63, 65, 79, 109, 126, 128, 144, 150, 151, 155, 179, 220–221, 249
product quality 67, 109, 117
promotion *see* marketing
protected designation of origin 199, 236, 253
protected geographical indication 199, 236–237

quality of life 131, 135–136

recreational shopping *see* leisure shopping
religious tourism 209, 255
repeat visitation 23, 27, 28, 94, 202, 214, 219–220
restaurants 80, 86, 94, 102, 113, 115, 125, 146, 147, 150, 196, 228, 240
retail landscapes/retailscapes 2, 11, 21, 93, 101, 126–127, 136, 160, 215, 249
retail management 4, 49, 164, 167, 249
retail location 145, 152–153, 240
retail spaces 1, 2, 43, 44, 108, 113–115, 136, 182, 184, 219, 250–252, 258, 261
retailtainment *see* shoppertainment
revitalization of historic cities *see* urban regeneration
risk aversion 42, 43, 47, 68, 202
 See also uncertainty avoidance
Romania 64, 112
Russia 38, 40, 116
Russian consumers 41, 48, 111

safety 11
sales *see* special offers
San Marino 113
satisfaction 4, 5, 10–11, 20–21, 23–24, 27, 48, 50, 154, 177–190, 202, 218, 221, 249, 259
Saudi Arabia 3, 46
Saudi consumers 46
second homes 116, 127
sense of place 7–8, 129, 219, 250
sensory marketing 26
service failures 50
service quality 4, 23–24, 26–27, 30, 48–49, 50, 79, 109, 117, 152, 180–181, 257, 259
shop design *see* environment, retail
shoppertainment 7, 126–127, 250
shopping centers 7, 10, 43, 65, 69, 81, 83, 86, 115, 126, 128, 138, 145–155, 161, 162, 177, 182, 217, 219, 221, 222, 250
shopping districts 10, 82, 111, 114, 128, 135–139, 183, 221
sightseeing 2, 99, 126–127, 147
Singapore 3, 6, 8, 42, 86, 127, 179, 181, 214–215, 220, 221
Singaporean consumers 3
Slovakia 112
smart cities 180
smartphones 65, 66, 97, 146
social commerce 257
social media 6, 7, 9, 26, 28–29, 31–32, 59, 66, 71, 78, 109, 146, 257
social networks 64, 72, 130, 204
social status 40, 42, 45, 79
souqs *see* marketplaces
South Africa 64
South Korea 3, 198, 202, 238
South Korean consumers 43, 47
souvenirs 1, 2, 5, 6, 7, 38–39, 42, 58, 63, 65–67, 99, 107, 115, 128, 165, 183, 195, 196, 198, 200–201, 202, 203–206, 217, 219, 225–244, 249, 250, 251, 252–255, 258, 259
Spain 97, 112, 113, 188
Spanish consumers 47
special offers 4, 5, 22, 28, 43, 45, 111, 185, 215–216, 220–221
sport tourism 11, 195, 208–209, 255
staff, retail 4, 179
staged events 41

stakeholders 4, 19, 20, 26, 28, 30, 31, 119, 164, 222, 223
staycations 64
storytelling 26, 28
street markets *see* marketplaces
street vendors *see* petty trade
supermarkets 64, 65, 112–113, 136, 238, 240–241
sustainability 9, 59–60, 62–63, 69, 71–72, 81, 159, 169, 201, 202, 205, 209, 230, 250, 251, 259–260
Sweden 10, 40, 42, 96, 109, 115, 144–155
Swiss consumers 112
Switzerland 3, 42, 96, 98, 113, 254

Taiwan 3, 237, 239–240
target marketing *see* marketing
tariffs 108, 110, 117
taxes 10, 93–103, 110, 112–114, 116, 118, 128, 147, 155, 188, 221, 249
tax-free *see* duty-free
tax refunds 96, 97, 99–100, 101–102
technology 1–2, 9, 21, 25, 28, 62, 70–71, 102, 146, 250, 256–259
   See also virtual reality; augmented reality; artificial intelligence
Temu 116, 257
Thailand 3, 46
Thai consumers 24, 46
Timor Leste 115
Tokyo 64, 84
Tourist Refund Scheme (Australia) 97
tourist shopping villages 84, 188
tour operators 94, 101, 180, 181, 202
tracking 28, 29, 30
trails 168, 198, 219
transit spaces 5, 93, 97–98, 103, 108, 256
transportation 1, 4, 94, 99, 101, 103, 125, 128, 137, 180, 214, 218, 221
trust, consumer 6, 7, 21, 25–26, 27, 190
Turkey 24, 44, 99, 110
Turkish consumers 41, 49
Turkmenistan 115

Ukraine 116
Ukrainian consumers 111
uncertainty avoidance 39, 41, 68
   See also risk aversion
United Arab Emirates 3, 85, 99
   See also Dubai; Abu Dhabi

United Kingdom 3, 38, 40, 42, 46, 49, 64, 101, 127, 167, 185, 200, 208
United States of America 38, 40, 42, 45, 64, 95, 97–98, 100, 109, 118, 147–148, 182, 196, 208, 237, 240–242, 260
urban regeneration 7, 8, 10, 124–139, 161, 169–170, 249
 See also waterfront development
user-generated content 29, 257
utilitarian shopping 4, 24, 40, 46, 64, 108–109, 111, 116, 151, 219–220, 222, 251

value for money 42, 67, 117
 See also price
variety-seeking 23, 38, 41–42, 97
VAT *see* taxes
Vatican City 113–114
Venice 166–168, 171

Vietnam 114
virtual reality 2, 25, 28, 30, 71, 146, 162, 258
virtual retail *see* online shopping
virtual space 2, 146, 180
visiting friends and relatives (VFR) 209, 220

waterfront development 7, 127, 138
 See also urban regeneration
wellness tourism 11, 137, 206–208
window shopping *see* browsing
wineries 64, 195–197, 228, 235, 240, 242, 253
wine routes 168
wine tourism 11, 64, 195, 255
word-of-mouth 23, 27, 82, 109, 150, 219
World Food Travel Association 229

zones, retail *see* shopping districts

For Product Safety Concerns and Information please contact our EU Authorised Representative:

Easy Access System Europe

Mustamäe tee 50

10621 Tallinn

Estonia

gpsr.requests@easproject.com